A Guide
to the Implementation
of the ISO 14000 Series
on Environmental Management

PRENTICE HALL
PTR ENVIRONMENTAL
MANAGEMENT AND
ENGINEERING SERIES

RITCHIE AND HAYES: *A GUIDE TO THE IMPLEMENTATION OF THE ISO 14000 SERIES ON ENVIRONMENTAL MANAGEMENT*

LOUVAR AND LOUVAR: *HEALTH AND ENVIRONMENTAL RISK ANALYSIS VOLUME 2: FUNDAMENTALS WITH APPLICATIONS*

MCBEAN AND ROVERS: *STATISTICAL PROCEDURES FOR ANALYSIS OF ENVIRONMENTAL MONITORING DATA AND RISK ASSESSMENT*

A GUIDE
TO THE IMPLEMENTATION
OF THE ISO 14000 SERIES
ON ENVIRONMENTAL MANAGEMENT

Ingrid Ritchie, Ph.D.
School of Public and Environmental Affairs
Indiana University

William Hayes, C.I.H.
Indiana Department of Environmental Management

To join a Prentice Hall PTR Internet mailing list, point to:
http//www.prenhall.com/mail_lists/

Prentice Hall, Upper Saddle River, New Jersey 07458
http://www.prenhall.com

Library of Congress Cataloging-in-Publication Data

Ritchie, Ingrid.
    A guide to the implementation of the ISO 14000 series on
environmental management / Ingrid Ritchie, William Hayes.
        p.   cm.
    Includes bibliographical references and index.
    1. ISO 14000 Series Standards.   2. Production management—
Environmental aspects.   I. Hayes, William, C.I.H.   II. Title.
TS155.7.R58   1997
658.4'08—dc21   97-10725
                                                          CIP

Editorial/Production Supervision: Nicholas Radhuber
Acquisitions Editor: Bernard Goodwin
Manufacturing Buyer: Alexis Heydt
Marketing Manager: Miles Williams
Cover Design: Bruce Kenselaar
Cover Design Direction: Jerry Votta
Art Director: Gail Cocker-Bogusz

© 1998 by Ingrid Ritchie and William Hayes
Published by Prentice Hall PTR
Prentice-Hall, Inc.
A Simon & Schuster Company
Upper Saddle River, NJ 07458

The creator, author, and copyright owner of the following materials in this book is Ingrid Ritchie:
Chapters 1, 7, 8, 10, 12, 13, 14, 15, 16, and 17. The creators, authors, and copyright owners of the following materials in this book are Ingrid Ritchie and William Hayes: Chapters 2, 3, 4, 5, 6, 9, 11, and appendixes.

Prentice Hall books are widely used by corporations and government agencies
for training, marketing, and resale.

The publisher offers discounts on this book when ordered in bulk quantities.
For more information, contact: phone: 800-382-3419; fax: 201-236-7141; e-mail: corpsales@prenhall.com
or write: Prentice Hall PTR, Corporate Sales Department, One Lake Street, Upper Saddle River, NJ
07458

Printed in the United States of America
10   9   8   7   6   5   4   3   2   1

ISBN 0-13-541097-5

Prentice-Hall International (UK) Limited, London
Prentice-Hall of Australia Pty. Limited, Sydney
Prentice-Hall Canada Inc., Toronto
Prentice-Hall Hispanoamericana, S.A., Mexico
Prentice-Hall of India Private Limited, New Delhi
Prentice-Hall of Japan, Inc., Tokyo
Simon & Schuster Asia Pte. Ltd., Singapore
Editora Prentice-Hall do Brasil, Ltda., Rio de Janeiro

Dedicated to
my sister, Barbara

—I.R.

TABLE OF CONTENTS

PART 3

EVALUATING AND IMPROVING PERFORMANCE   131

CHAPTER 8
CHECKING AND CORRECTIVE ACTION   132

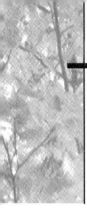

LIST OF FIGURES

PREFACE

The International Organization for Standardization (ISO) Environmental Management standards were developed by dedicated men and women from around the world who recognized the need for a new model of environmental management—a shift from the view that environmental issues are a drain on profits to a realization that profitability and environmental protection are intertwined. The holistic approach to environmental management embodied in ISO 14001 has the potential for moving the global community toward economic development that is sustainable and restorative.

The ISO 14001 standard is a common-sense approach to managing environmental issues that integrates management controls into media-specific problems. Its principles can be adopted by large, small, and medium-sized organizations throughout the world to improve their effectiveness and efficiency, not just in the management of environmental issues, but throughout their operations. The goal of this book is to build on the work of the ISO committees and to provide useful information to anyone who wants to consider developing an Environmental Management System (EMS) under the guidelines of the ISO 14000 standards. The intention is not to teach environmental issues but to provide a practical guide around which the systems that make up the standard can be developed. This book is not a substitute for the standards, and those wanting to develop and implement an EMS are strongly encouraged to obtain a copy from the ISO or from sponsors of the standard for the country in which they do business. Addresses for the ISO and the United States co-sponsors of the standard follow.

International Organization for Standardization
1, rue de Varembe
Case postale 50
CH-1211
Geneva 20,
Switzerland
Telephone: 41-22-749-0111
Fax: 41-22-733-3430
Internet: http://www.iso.ch
E-mail: central@isocs.iso.ch

American National Standards Institute
11 West 42nd St.
New York, NY 10036-8002
Telephone: 212-642-4900
Fax: 212-642-4969
Internet: http://www.ansi.org

American Society for Quality Control
611 E. Wisconsin Avenue
Milwaukee, WI 53201-3005
Telephone: 800-248-1946
Fax: 414-272-1734
Internet: http://www.asqc.org

American Society for Testing and Materials
100 Barr Harbor Drive
West Conshohocken, PA 19428-2959
Telephone: 610-832-9585
Fax: 610-832-9555
Internet: http://www.astm.org
E-mail: service@astm.org

NSF International
P.O. Box 130140
Ann Arbor, MI 48113-0140
Telephone: 800-673-6275 ext. 5744
Fax: 313-769-0109
Internet: http://www.NSF.org
E-mail: info@NSF.org

The book is divided into four parts that explain the sections of the standard and take the reader from writing a policy statement through continual improvement. The appendixes contain examples of an environmental management policy manual, procedures related to the standard, and checklists to help evaluate the status of planning and implementation activities. Additional examples of documentation—flowcharts, procedures, and records—are contained in the body of the book. Part 1 provides an overview of the standards. Part 2 gives instructions on how to write a policy statement, analyze and chart processes, and write and organize documentation, implementation issues, and operational controls. Part 3 focuses on the elements needed to continually improve the EMS—quality assurance/quality control, corrective and preventive actions, management review and continual improvement, tools for problem solving, auditing, and the selection of auditors. The last chapter in this section provides an overview of the American Industrial Hygiene Association's draft standard for an occupational health and safety management system. Part 4 provides an overview of the regulatory framework in the U.S. and highlights factors to consider when preparing for regulatory compliance audits of selected topics. Although this part of the book is targeted to audiences in the U.S., it contains information useful to others.

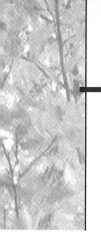

ACKNOWLEDGMENTS

The authors gratefully acknowledge the American Society for Quality Control (611 E. Wisconsin Avenue, Milwaukee, WI 53201-3005) for allowing the use of excerpts from the ISO 14000 Series of Standards.

The authors gratefully acknowledge the American Industrial Hygiene Association (2700 Prosperity Avenue, #250, Fairfax, VA 22031) for allowing the *Occupational Health and Safety Management System: An AIHA Guidance Document* to be summarized.

The authors gratefully acknowledge the assistance of the following individuals in reviewing parts of the manuscript:

Michael Anderson
Senior Environmental Engineer
Office of Environmental Response
Indiana Department of Environmental Management

Lonnie Brumfield
Chief, Permits Section
Office of Water Management
Indiana Department of Environmental Management

Pamela O'Rourke, CHMM
Senior Environmental Manager
Office of Enforcement
Indiana Department of Environmental Management

ABOUT THE ICONS

When you see this icon, refer to the disk that accompanies this book. The referenced material has been included on the disk. The disk contains the questionnaires in Appendix A to help assess the status of planning and implementation activities for an environmental management system. The disk also includes Appendix B which is an example policy manual. The policy manual is not intended to be used as is, but it can provide guidance for those who wish to develop similar documentation. Additional forms, worksheets, and checklists from the book that can be used for specific programs are also included on the disk. For the full contents of the disk and instructions for installation and use, see the section entitled *About the Software.*

When you see this icon, you will be directed to other volumes in the *Prentice Hall PTR Environmental Management and Engineering Series.* The concept behind the Environmental Management series is to provide professionals and those in training with comprehensive information linking the scientific principles of environmental science and engineering, governmental regulations, and international standards with their practical applications in order to manage the diverse and complex problems facing today's environmental and engineering professionals. For a list of the volumes in this series and their corresponding volume icons, see the series page opposite the title page.

When you see this icon, you will be alerted to the wealth of information that is available through the Internet. The World Wide Web (WWW) icon will draw your attention to specific addresses in the text or to a resource section elsewhere in the book. Appendix F provides an introduction to using the Internet and includes useful addresses that span many topics and fields.

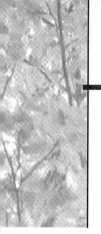

ABOUT THE SOFTWARE

The disk that is included with this book provides access to examples of questionnaires, forms, and checklists that can be used to plan, implement, and audit an environmental management system. When you see the disk icon in the text, you will be able to locate the material that is referenced in the book in files on the accompanying disk. Once you open a file, you will be able to read, edit, and print it. To use the files, you will need an IBM-compatible PC and Microsoft Word (the files were created in Word for Windows Version 6.0, but earlier versions can be used). After you have opened Word for Windows, place the enclosed disk into the A-drive. Select *Open* from the *File* menu. You will see a list of all the files on the disk. Scroll down to the file you wish to open, then double click or click on OK to open the file. Now you are ready to use the file.

The following files are included on the disk:

File Name	Figure/Appendix Title	Page Number in the Book
APPXA-1.DOC	Environmental Management System Planning Questionnaire	344
APPXA-2.DOC	Environmental Management System Implementation Questionnaire	353
APPXB.DOC	Environmental Management Policy Manual	362
APPXE.DOC	Air Pollution Lists	425
FIG5–7.DOC	Example of a Training Record	63
FIG5–11.DOC	Example of Documentation for EMS Responsibilities by Job Title or Position	68

ABOUT THE AUTHORS

 **Ingrid Ritchie, Ph.D.,** is an Associate Professor in the School of Public and Environmental Affairs, Indiana University, where she teaches courses in environmental and hazardous materials management. She received her doctoral and master's degrees in environmental health from the School of Public Health, University of Minnesota. With over twenty years of experience in the environmental field, Dr. Ritchie's areas of expertise include the management of air, land, and water pollutants; environmental site assessment; quality assurance/quality control; environmental and health-risk assessment; worker health and safety; and policy analysis.

Dr. Ritchie is the author of technical reports and publications in scholarly journals, two U.S. Environmental Protection Agency publications on the investigation and management of indoor air quality, and a book on environmental hazards related to properties. Dr. Ritchie consults and lectures on the management of environmental hazards for the National Association of Environmental Risk Assessors, Indiana University's Hazardous Materials Managers Executive Education Program, various real estate professional organizations, and other clients.

William Hayes, C.I.H., is currently employed by the Indiana Department of Environmental Management where he is developing a risk management program to incorporate risk assessment/risk analysis into the decision-making processes for the department, and he consults in the areas of management and occupational safety and health. Mr. Hayes received his Bachelor of Science degree in Public Health and a Master's in  Public Affairs degree in Public Management from Indiana University. He is a practicing Certified Industrial Hygienist and a registered Environmental Health Professional. He has worked for the U.S. Air Force, health and environmental agencies, and industry in the areas of management, industrial hygiene, and environmental protection. Mr. Hayes is also an adjunct faculty member at Indiana University in the School of Public and Environmental Affairs where he teaches courses in industrial hygiene and management.

GETTING STARTED

Part 1 sets the stage for the planning and implementation activities discussed in the later sections of the book. Chapter 1 begins with an overview of the International Organization for Standardization (ISO) and the factors leading up to the development of ISO 14001, the standard setting process, and the relationship between ISO 14001 and other standards. It closes with a discussion of the pros and cons of the standard and an overview of the ISO 14000 series of standards. Chapter 2 presents the elements of ISO 14001 and introduces the requirements of each section of the standard.

CHAPTER 1

PROACTIVE MANAGEMENT

21ST-CENTURY THINKING

As we push closer to the 21st century, it is clear that new ways of thinking are needed to tackle the enormous environmental and societal issues that mark our global community. Futurists, such as Paul Hawken, believe that an enduring society must be based on a system of commerce and production that is sustainable and restorative.[1] This is not a new viewpoint. In 1983, the General Assembly of the United Nations (UN) asked the UN World Commission on Environment and Development to formulate a global agenda for change in development practices. The final report, *Our Common Future*, published in 1987, calls on the global community to move toward sustainable development—an approach that uses the earth's resources in such a way that future generations' needs will not be compromised.[2] In other words, sustainable development seeks a balance between economic growth and environmental protection. This means that countries (and business) need to integrate economic, biologic, and human systems to create a sustainable system of commerce, and that governments need to incorporate flexibility that rewards proactive environmental management. Hawken believes that we must move from systems of penalizing or taxing pollution and consumption to rewarding conservation of resources and waste reduction, and he recognizes that the burden cannot be shouldered by the private sector alone.

Making these changes is a daunting task that must overcome strongly held beliefs. Many environmentalists believe that improving the bottom line is the primary concern of business and that business must be "encouraged" to preserve the environment through a strong system of command and control regulation. Environmentalists often view government as not doing enough, hiding behind the uncertainty inherent in science, and granting favors to politically connected businesses. Government bureaucrats must contend with shifting political agendas that constantly seek to reinvent environmental programs and their administration with the net result that accomplishing goals becomes increasingly difficult. At the other end of the spectrum, many business managers view strong environmental management programs as deal killers rather than deal makers, and they believe that advocates of strong environmental programs do not fully appreciate the significant barriers that must be overcome to keep a business viable.

Today's manager must grapple with an endless array of obstacles in the quest for position in the marketplace. National and international competition, identifying and satisfying consumer and customer demands, integrating technological advances, and coping with workforce capability and availability issues are major concerns. Negotiating the path of increasingly complex regulations, particularly environmental regulations, can be overwhelming. Uncertainty about the scope of regulations and costs of compliance leads many managers to believe that environmental management is an undercurrent that sucks up resources and drains productivity. These managers are struggling for survival, and thinking about ways to become a "sustainable" company is not a top priority.

Robert Repetto, a World Resources Institute economist, challenged some of this prevailing thinking in a thought-provoking 1995 report, *Jobs, Competitiveness, and Environmental Regulation.*[3] Repetto compared environmental performance to profitability using thousands of U.S. industrial operations and found that there was no overall tendency for plants with superior environmental performance to be less profitable than were plants with poor environmental performance, even when age, size, and technology were controlled.[4] Repetto argues that the real issue is getting good value for dollars spent on environmental management. He suggests that government could help make environmental management programs less burdensome by: 1) providing increased flexibility to industry by setting goals and allowing industry to decide how to meet those goals, 2) promoting emissions trading in permits or other kinds of entitlements to the use of natural resources, 3) instituting government rewards for proactive environmental management practices in all sectors of the economy, and 4) building the cost of environmental degradation into the price structure of markets.[5]

Managing the "known and familiar" and reacting, often in crisis mode, to the "unknown and unfamiliar" have been traditional responses to environmental issues. Forward-thinking companies, however, are realizing that proactive environmental management can result in a more effective organization and an improved bottom line. New approaches to environmental management have shifted the business community's view of what constitutes an effective and efficient program from short-term expediency to planning over the long term. The best-managed companies have demonstrated that implementing a systematic environmental management program can produce significant increases in productivity and profitability. For example, the 3M Company's Pollution Prevention Pays (3P) program reduced the company's air, land, and water emissions by about 1 billion pounds and saved $500 million dollars over a period of about 15 years![6] A more modest example is given by a savings of $123,000 in energy costs achieved by the Upjohn company at its Portage, Michigan facility simply by changing its lighting system.[7]

Although these companies may not be "sustainable," they are moving in a direction many believe to be essential for global stability. To some, futurists like Hawken are ivory-tower liberals who are out of touch with the pressures facing the global economy, but representatives of the international business and governmental communities have taken a significant step in the direction of sustainable and restorative commerce by developing the ISO 14000 series of environmental management standards.

WHAT ARE THE ISO 14000 STANDARDS?

The ISO 14000 standards, which are listed in Figure 1–1, are international voluntary, consensus standards. This means that the standards are developed by agreement among participants and that adoption of or compliance with the standards occurs on a voluntary basis. A standard has regulatory, legal standing only if it is formally adopted by a country. These standards, which could revolutionize industrial environmental management, give managers a proactive generic structure for initiating, improving, or sustaining environmental protection programs.

The standards were developed by the ISO with input from industry, government, and other interested parties. Located in Geneva, Switzerland, the ISO was founded in 1947 to promote the development of international manufacturing, trade, and communication standards. The ISO accomplishes this goal by developing standards for these industries. Industries related to electrical and electronic engineering have their own standards organization, the Geneva-based International Electrotechnical Commission (IEC). The IEC is made up of representatives from more than 40 countries, including the U.S.

The ISO is not affiliated with the UN or wih the European Union (EU). It is privately funded and currently has member standards organizations from more than 100 countries. The U.S. representative to the ISO and the IEC is the American National Standards Institute (ANSI); other countries are represented by similar organizations.

All standards developed by the ISO are *voluntary consensus standards*. Because the process for developing the standards is iterative and represents input from a diverse clientele, many countries and trade groups elect to adopt ISO standards, and they quickly become accepted as a condition of doing business.

WHAT LED TO THE DEVELOPMENT OF THE ISO 14000 STANDARDS?

Since the start of the environmental era in the 1970s, industry, governments, and the public have sought solutions to environmental problems. Initially, these ef-

FIGURE 1–1.   The ISO 14000 Series of Standards:

ISO 14004[a]	Environmental management systems—general guidelines on principles, systems and supporting techniques
ISO 14001[a]	Environmental management systems—specification with guidance for use
ISO 14010[a]	Guidelines for environmental auditing—general principles
ISO 14011[a]	Guidelines for environmental auditing—audit procedures—Part 1: Auditing of environmental management systems
ISO 14012[a]	Guidelines for environmental auditing—qualification criteria for environmental auditors
ISO 14024	Environmental labeling—guidance principles, practices and criteria for multiple criteria-based practitioner programmes (Type I)—guide for certification procedures
ISO 14040	Life cycle assessment—principles and guidelines
ISO 14060	Guide for the inclusion of environmental aspects in product standards

[a]*Issued as final standards in 1996; remaining standards not issued in final form as of 12/96.*

forts resulted in a national model of regulation that attempted to stop pollution at the discharge pipe. These command and control efforts generally focused on a single medium, such as air pollution. Beginning in the mid-1980s, there was a growing recognition that environmental problems were interrelated and required interdisciplinary solutions. Regulators responded with multimedia approaches to reducing and preventing pollution. At the same time, the global nature of some types of environmental problems such as ozone depletion and global warming led to the emergence of international efforts to control pollution.

Events such as the chemical disaster at Bhopal, India; radiation releases at Chernobyl, USSR; and oil spilled by the *Exxon Valdez* off the coast of Alaska spurred global concern about industry's impact on the environment and motivated international interest in new ways of preventing pollution. A series of international meetings that addressed environmental issues culminated in the demonstration of significant international interest in protecting the environment at the 1992 Rio Conference on the Environment. This conference produced the *Rio Declaration On Environment and Development*,[8] which enumerates guiding principles that emphasize sustainable development.

The General Agreement on Tariffs and Trade (GATT) talks also spurred interest in international *voluntary* environmental standards. Beginning with the 1986 talks, there was concern that differing national environmental regulations posed nontariff barriers to trade. National standards that impose conflicting and confusing requirements can increase business costs and create hardships, particularly for developing economies.

An important factor was the success of ISO 9000, the worldwide quality standard. This standard provides organizations with a process for producing quality products through a systems approach that involves all phases of production. The foundation of the standard is the quality theory set forth by W. Edwards Deming and others: Consistent products and improved quality will result from developing and maintaining the process used to produce the product. In other words, good business practices throughout an organization can improve quality.

From the time the standard was finalized in 1987 until 1994, more than 30,000 ISO 9000 certificates were issued worldwide.[9] By the end of 1995, that number climbed to over 95,000 certificates; 12,100 of these certificates were issued in North America.[10] Many countries have adopted ISO 9000, and the widespread acceptance of ISO 9000 sent a positive message to all types and sizes of businesses that profitability and quality systems were interdependent.

All of these factors contributed to a favorable climate for ISO 14000. In 1991, the Strategic Advisory Group on the Environment (SAGE) was established by the ISO to make recommendations on the need for an international environmental standard. In 1993, SAGE recommended the formation of an ISO technical committee to develop an international environmental management standard and other related standards. In June 1993, the ISO began to draft the ISO 14000 series of standards, which would help companies throughout the world to improve environmental performance by showing them how to take a systems approach to their management of environmental issues.

HOW DOES ISO DEVELOP STANDARDS?

The ISO develops standards through a system of technical committees and working groups. Any member can propose a new work item, which is assigned to a technical committee for review. If the work item is accepted for further study, it progresses through an orderly system of development and review. After the technical subcommittees reach consensus on a standard, it is elevated to the status of draft international standard. It then proceeds to the ISO and technical committee vote to finalize the draft to the status of international standard.

The ISO 14000 series was developed by Technical Committee (TC) 207. The Committee's official scope is "standardization in the field of environmental management tools and systems."[11] The following areas are excluded from its scope:

- test methods for pollutants,
- setting limit values regarding pollutants or effluents,
- setting environmental performance levels, and
- standardization of products.

Leadership for TC 207 is provided by Canada and subcommittees are led by Australia (environmental labeling), France (life cycle assessment), Germany (environmental aspects in product standards), the Netherlands (environmental audits), Norway (terms and definitions), the UK (environmental management systems), and the U.S. (environmental performance evaluation).

WHAT DOES THE ISO 14001 CERTIFICATION MEAN?

Certification is a procedure by which a third party gives written assurance that a product, process, or service conforms to specific requirements. A company that gains ISO 14001 certification can claim that it has a documented environmental management system (EMS) that is fully implemented and consistently followed. The company cannot claim that its products or services are more environmentally friendly or more environmentally sound than those of its competitors. The ISO 14001 certification is based on an audit of the company's EMS, not the product or service provided by the company.

Certification, registration, and *accreditation* are related, but different, terms that are sometimes used interchangeably. *Registration* is a procedure by which a body indicates relevant characteristics of a product, process, or service, or particulars of a body or person, then includes or registers the product, process, or service in an appropriate publicly available list. *Accreditation* is a procedure by which an authoritative body gives formal recognition that a body or person is competent to carry out specific tasks.

In Europe, *certification* is the preferred term for gaining recognition for compliance with voluntary standards. Registration is favored in the U.S. because certification typically carries legal liability. In the context of the ISO 14000 standards, an accreditation authority (for example, ANSI in the U.S.) accredits an organization or person to grant certification (or registration) that a company has met the requirements of ISO 14001. A registrar is a person or organization authorized to certify (or register) companies that want to qualify for ISO 14001 recognition.

WHAT IS THE RELATIONSHIP BETWEEN ISO 14001 AND OTHER STANDARDS?

Because ISO 14001 is process oriented and a guide for managing environmental programs, it has many elements in common with ISO 9001. Both require a policy statement, commitment from top management, training, corrective action, document control, management review, and continual improvement. ISO 14001, however, imposes additional requirements on the policy statement and requires setting objectives and targets, identifying environmental impacts and benefits (aspects), and a commitment to conforming to applicable regulations.

ISO 14001 has its roots in BS 7750: *Environmental Management Systems,* which is the national environmental management system standard for the UK. BS 7750 was developed by the British Standards Institution (BSI) in 1992 (and revised in 1994) as a companion to BS 5750, the quality standard that served as the prototype for ISO 9000. BS 7750 was seen as a more prescriptive and restrictive standard than ISO 14001. In the US, there were significant objections to the public disclosure requirements in BS 7750.

Another related standard, the Eco-Management and Audit Regulation (EMAR), was developed by the European Commission (EC), which is the executive branch of the EU, a confederation of fifteen European member nations. EMAR contains the European Eco-Management Audit Scheme (EMAS) which establishes specifications for environmental management systems for companies with sites in the EU. The EMAS is site-specific and voluntary, but it is anticipated that companies wishing to do business with EU members may need to be registered under EMAS. It is anticipated that ISO 14001, combined with a bridge document, will be recognized as meeting EMAS requirements. Figure 1–2 shows how ISO 14001, EMAS, and BS 7750 are related.

WHAT ARE THE PROS AND CONS OF IMPLEMENTING THE ISO 14000 STANDARDS?

PROS

A business that implements the standards can expect a number of benefits that will ultimately improve the bottom line. First, a carefully designed and fully implemented environmental management program provides evidence to stakeholders that the company is serious about environmental management. Tangible benefits include maintaining good relations with the public and government organizations, enhancing the company's image, and increasing its market share. Market pressure is one of the strongest reasons for becoming certified. Companies that become ISO 14001 certified will have a competitive advantage over non-

FIGURE 1–2.   Differences among ISO 14001, EMAS, and BS 7750:

ISO 14001	EMAS	BS 7750
An international standard	An EU legislative instrument, i.e., a regulation	A national standard
Applies to the international arena	Applies across the whole of the EU	Applies in the UK, but can be applied anywhere developed
Can apply to the whole organization or part of an organization	Applies to sites only	Can apply to the whole organization
Applicable to an organization's activities, products, and services in any sector	Restricted to site-specific industrial activities	Open to any sector or activity
Applicable to nonindustrial activities, e.g., transport and local government	Nonindustrial activities can only be included on an experimental basis	Open to nonindustrial activities, e.g., transport and local government
Focuses on organizations implementing environmental management systems; indirect link to environmental improvements emerging from the system	Direct focuses on environmental performance improvements at a site and the provision of information to the public	Focuses on organizations implementing systems; environmental improvements emerge from the system
Identification of aspects is required in the specification; an initial review is suggested in the informative annex	Initial environmental review essential	Preparatory environmental review advisable but not a specification of the standard
Environmental policy commitment to continuous improvement of environmental management systems and compliance with relevant environmental legislation	Environmental policy commitment to continuous improvement of environmental performance and compliance with relevant environmental legislation	Environmental policy commitment to continual improvement of environmental performance; no reference to legislation in the standard's specification
Environmental management audits concerned with the assessment of environmental management systems only	Environmental audit assesses management systems, processes, factual data, and environmental performance	Environmental management audits concerned with the assessment of environmental management systems, its operation and results only
Frequency of audits not specified	Maximum audit frequency-specific at 3 years	Frequency of audits not specified
Only environmental policy must be publicly available	A description of the environmental policy, program, and management system made publicly available in the statement	Only environmental policy must be publicly available
Public statement not required, consideration must be given to external communication (subclause 4.3.3) but left up to management as to how much information to disclose	Public environmental statement and annual simplified statement, including factual data essential	Not required, left up to management as to how much information to disclose
Document is more clearly structured	Confusing arrangement (a lot of cross-references)	

Sources: Lloyds Register Quality Assurance and *United Nations Industrial Development Organization.* Reprinted with permission from *What is ISO 14000? Questions and Answers,* 3rd edition, published by CEEM Information Services, Fairfax, VA.

certified companies, which, in turn, creates pressure for the noncertified companies to become certified. Paralleling the ISO 9000 experience, ISO 14001 certification may be needed to fulfill contractual requirements or as a condition of placing a purchase order. Some companies may use only those subcontractors who are certified.

Because the EMS emphasizes prevention, savings can be realized through waste minimization and prevention of pollution activities that result in a reduction in the use of raw materials, energy, and hazardous materials. Companies that implement an EMS often find new opportunities to increase efficiency, to reduce paperwork, and to lower costs in other ways.

The EMS should also result in fewer uncontrolled releases of hazardous materials and should demonstrate "due care" in environmental matters, which will reduce potential liability. In the U.S., responsible party transfer laws and the Securities and Exchange Commission's (SEC) environmental disclosure requirements can make property sales and mergers difficult unless a comprehensive environmental management program is in place. Other related benefits include a more favorable relationship with lending institutions, improved access to capital, and reasonably priced insurance.

Companies that participate in the global marketplace can expect a level international playing field. ISO 14001-certified companies will be able to minimize multiple inspections and registrations, differing labeling requirements, environmental trade barriers, and the costs associated with these problems. The acceptance of a single standard by the global community will allow companies to implement an EMS without fear of being placed at a competitive disadvantage.

Companies can take pride in their contribution to a cleaner, healthier environment for their communities and in improving the global environmental health. Closer to home, implementation of ISO 14000 allows each participating company to contribute in a real and substantial way to cleaning up the environment in their communities and to improving the health of their workers and others in the community.

One example of success on a local level in preventing pollution is a medium-sized Borden facility that produces formaldehyde, resins, and adhesives. This facility, which is located in California, made operational changes to reduce the amount of phenol in its waste stream. These changes, which were phased in over a period of time, included segregating resin wastes and wastewaters, modifying cleaning procedures for filter housings, using two-stage rather than one-stage rinsing procedures, retraining plant personnel to increase their awareness for waste-reduction opportunities, tracking waste generation, and changing formulations. The company's pollution prevention efforts allowed it to slash by 93% the major phenol-laden waste stream and saved more than $150,000 per year in waste disposal and potential legal costs![12]

Another example of individual success in implementing prevention of pollution strategies is given by the Hyde Manufacturing Company, which is located in Southbridge, Massachusetts. The company produces putty knives, surface preparation tools, and machine blades using 30 different processes; in 1993, it had sales in excess of $30 million. This company embraced the prevention of pollution in 1989 with an environmental goal of zero discharge of hazardous materials to all media—air, land, and water—and production of the smallest amount of waste possible for its operations. Some of its strategies include conserving water, increasing recycling of wastes and materials, using design for environment principles, using less hazardous materials and biodegradable absorbents, and instituting operational changes for filtration and fluid handling. Over a 3-year period, environmental expenses exceeded $100,000 but savings or cost avoidance from environmental programs exceeded $200,000. Benefits to the community included reducing water resources from 27 million to 5 million gallons per year; eliminating the use of ozone-depleting chemicals; eliminating the use of toxic chemicals, including kerosene and 1,1,1-trichloroethane; reducing waste coolant discharges by 80%; and reducing the material sent to the town landfill by about 135 tons per year.[13]

The Borden and Hyde examples show what modestly sized individual facilities can accomplish on a voluntary basis. Compliance with regulations also produces important environmental benefits. For example, successful implementation of air pollution regulations in the U.S. reduced lead emissions by a total of 88% during the period 1984–1993; the decrease in ambient concentrations of lead was 89% during this same period of time.[14] Elevated levels of lead in the blood can contribute to a variety of health problems in children, including mental retardation, seizures, decrements in intelligence, behavioral problems, and organ system damage. During a somewhat broader period beginning in 1976 and ending in 1991, the National Health and Nutrition Examination Survey in the U.S. reported a 78% decrease in blood lead levels [from 12.8 micrograms per deciliter (μg/dL) in 1976–1980 testing period to 2.8 μg/dL in the 1988–1991 testing period].[15] Children in the U.S. still have significant exposure to lead-based paint and a number of stationary lead smelters. The dramatic progress toward reducing the average blood lead levels in children is attributed primarily to the gradual reduction of lead in gasoline and to the removal of lead from soldered cans.

Another example of successful pollution control in the U.S. is exposure to ozone, an air pollutant that damages the respiratory tract and causes signficant crop losses. In 1988, about 112 million people in the U.S. lived in counties that did not meet the ozone air pollution standards.[16] By 1993, this estimate dropped to 51.3 million people exposed.[17] Efforts to reduce ozone levels were attributed primarily to reductions in emissions of nitrogen oxides and volatile organic compounds. Although the U.S. still has signficant problems with lead and ozone pollution, these examples dramatically show the improvements that can be realized through pollution control efforts. Decreases in the levels of particulates, sulfur dioxide; mercury, lead, arsenic, and other metals; pesticides; PCBs and other organic chemicals; and other hazardous materials in local and regional environments will have substantial and signficant impact on health and quality-of-life indicators such as clean air and water.

Cons

There are a number of concerns surrounding implementation of the ISO 14000 standards. In the U.S., companies are concerned about the confidentiality of data collected through the certification audit. Even though the U.S. Environmental

Protection Agency (U.S. EPA) has a long-standing practice of not requesting voluntary audit reports to trigger enforcement actions, the agency makes it clear that environmental audits do not shield companies from federal regulatory action.[18] A related concern is that ISO 14000 could become the legal standard of due care, and the failure to conform to it could result in a legal ruling of environmental negligence.

Implementing a comprehensive EMS increases costs initially, and these costs may be critical for small- to medium-sized companies. Market-driven pressure for ISO certification may pose a trade barrier for these companies if they cannot commit the time and resources to the process.

An important goal of the ISO 14000 movement is to create a level playing field by facilitating trade and minimizing trade barriers. Although ISO 14000 is a voluntary standard, it could create barriers to trade if a country requires all companies doing businesses within its borders to become certified. Foreign companies located in countries that do not mandate certification may be excluded from trade if they are unable to meet the requirements of the standard. Market-driven or mandated ISO certification could pose significant barriers for companies from developing countries. The companies may not have the resources to achieve certification, and the infrastructure for certification, and registration may be absent.

Concerns also exist about the registration process and the overall system for assessing and assuring conformity to the standards. How will the registrars (certification bodies) be accredited? Will ISO 14000 certificates be recognized throughout the world? What is the role of self-declaration versus third-party registration? Will ISO 14000 auditors be uniformly competent and will they consistently interpret the standards?

Finally, there is a concern that ISO certification will not lead to improved environmental performance. Companies that make a genuine and serious commitment to the ISO process will improve performance. But there are no foolproof certifications. Some companies may slip through, and others may slack off between auditor visits. Overall, the system should work. If it does not, stakeholders will lose confidence in the process.

A CAPSULE DESCRIPTION
OF THE ISO 14000 STANDARDS

The ISO series includes standards for: 1) environmental management systems, 2) environmental auditing, 3) environmental performance evaluation, 4) life cycle assessment (LCA), and 5) environmental aspects in product standards (Figure 1–1). The cornerstones of the series are *ISO 14004: 1996(E) Environmental Management Systems—General Guidelines on Principles, Systems and Supporting Techniques* and *ISO 14001: 1996(E) Environmental Management Systems—Specification with Guidance for Use*. Together, these standards provide a framework for building or enhancing an EMS for any size and type of organization. A glossary of some of the basic terms contained in the standard is given in Figure 1–3 at the end of this chapter. More information about the ISO standards can be found through the World Wide Web sites which are given in the Preface for the sponsoring organizations.

ENVIRONMENTAL MANAGEMENT SYSTEM STANDARDS, ISO 14001, AND ISO 14004

ISO 14001 establishes a model environmental management system based on a set of guiding principles contained in ISO 14004 that should be adopted by an organization. These principles require the organization to:

- define and adopt an environmental policy;
- ensure organizational commitment to environmental improvement;
- formulate a plan with objectives and targets to fulfill the environmental policy;
- implement the plan by developing the capabilities and resources to achieve the environmental policy, objectives, and targets;
- measure, monitor, and evaluate environmental performance; and
- review and continually improve the EMS.

ISO 14004 provides general guidance for establishing the EMS whereas ISO 14001 specifies the core elements for an EMS that can be objectively audited for certification, registration, or self-declaration purposes. ISO 14001 is the only standard in the series against which a company's environmental management system is audited for ISO certification/registration; all other standards are guidance or informative documents. The standard provides additional flexibility by allowing an organization to utilize an internal auditor for self-auditing purposes. However, if the organization chooses to have its EMS certified/registered, it must use a third-party auditor.

An important feature of ISO 14001 is that it does not establish absolute requirements for environmental performance beyond a commitment, in the policy statement, to continual improvement and prevention of pollution and compliance with relevant environmental legislation and regulations. It should also be noted that the standard encourages but does not require integration of occupational safety and health management into the EMS and implementation of best available technology (BAT), where appropriate and economically viable. This approach of commitment and implementation parallels the quality movement and the ISO 9000 quality standards. Although adoption of the standard does not guarantee environmental compliance or a nonpolluting company, it is believed that companies that implement these commitments will improve environmental performance. In other words, a company says what it will do, then does what it has said.

ENVIRONMENTAL AUDITING STANDARDS, ISO 14010, ISO 14011, AND ISO 14012

ISO 14010 provides guidance on the general principles for conducting environmental audits. An *environmental management system audit* is defined in the auditing standard (ISO 14011, Section 3.2) as the "systematic, documented verification process of objectively obtaining and evaluating audit evidence to determine whether specified environmental activities, events, conditions, management systems, or information about these matters conform with audit criteria, and communicating the results of this process to the client." The guiding principles for environmental audits include:

- basing the audit on objectives defined by the client;
- utilizing an audit team that is independent of the activities they audit and utilizing an auditor who meets the qualification criteria given in ISO 14012;
- use of due professional care by the auditor to maintain confidentiality and adequate quality assurance;
- use of systematic procedures for conducting the audit;
- developing audit criteria, evidence, and findings;
- ensuring the process provides the desired level of confidence in the reliability of the audit findings and conclusions; and
- providing an adequate report of findings.

ISO 14011 elaborates on the general principles of ISO 14010 by providing a generic framework for conducting the EMS audit. The audit process determines whether or not the management system has the requisite elements and is functioning properly. It answers the questions: Does the company do what it says it will do? and, Does the EMS conform to the ISO 14001 standard?

ISO 14012 provides guidance on qualification criteria for the individual environmental auditors. These criteria apply to both internal and external auditors and include education, on-the-job training, work experience, personal attributes and skills, maintenance of competency, and due diligence.

ENVIRONMENTAL LABELING, ISO 14024

ISO 14024 is a multiple-criteria-based voluntary environmental labeling program that requires third-party verification. It was developed to reduce the burdens that companies and consumers experience because of diverse labeling programs. There are currently many separate national labeling programs. Because these programs lack uniform criteria, it is difficult for companies to comply with requirements, and the significance of any particular label is hard to understand. Also, independent labeling programs have the potential to create trade barriers throughout the world.

The coverage of this standard (committee draft as of April 1997) is quite broad. It includes goods or services for consumer, commercial, and industrial purposes. The ISO environmental labeling standard requires the applicant, product, or product category to meet certain criteria for the award of an environmental label. These criteria are likely to go beyond national or international standards, but a demonstration of compliance with applicable national standards is a general requirement.

For a program with this scope to be effective, it must have credibility. To this end, adequate scientific information must be supplied for the evaluation, and a third party must license the manufacturer to use the label. The third party that issues the label may be a governmental entity or a private firm. These practitioners must be independent of the sale of the product.

Guiding principles and practices for third-party environmental labeling programs include:

- Standards and criteria applicable to environmental labels must be developed through a consensus process, and the program must be voluntary.
- A product can be considered for an environmental label if it complies with relevant environmental regulations of the country in which it is manufactured and marketed.
- Environmental labeling criteria should, whenever appropriate, incorporate a life cycle approach to assessment. Elements to be considered are extraction of resources, manufacturing, distribution, use, recoverability, and disposal.
- Environmental labeling programs should be selective and should distinguish leading products from alternatives.
- The product criteria developed by the practitioners should be periodically reviewed to account for new developments or technologies.
- There should be consultation with stakeholders in the selection of criteria and product categories, and the results should be made public.
- Excepting confidential information, the process should be transparent, i.e., open for examination. This implies availability of information on criteria, certification and award procedures, periodic revision of criteria, and demonstration that the funding sources for the program do not create conflicts of interest or undue influence.
- Environmental labeling programs should not create artificial trade barriers or discriminate in the treatment of domestic and foreign products and services.
- The labeling programs should include a conformity assessment using ISO or other appropriate standards; the elements of the assessment must be verifiable by the practitioner.
- Labeling programs should be accessible, objective, affordable, and confidential.

The requirements for awarding a label are divided into general rules that apply to all products and applicants, and to product criteria that set requirements for each product category. These are the only criteria that may be considered as a basis for awarding the label. The practitioner awards the label when satisfied that the applicant is in compliance with the general rules of the program and that the product is in compliance with the specific product criteria for the product category. The practitioner must maintain a publicly available list of products currently licensed to carry the label.

When the label has been awarded, it is the responsibility of the practitioner to take any necessary steps to ensure ongoing compliance with the product criteria. The practitioner will require the licensee to take corrective action if compliance monitoring indicates that compliance is not being maintained.

LIFE CYCLE ASSESSMENT, ISO 14040

ISO 14040 (final draft standard as of April 1997) provides guidelines for incorporating LCA into environmental management programs. *Life cycle assessment* is defined as a "systematic set of procedures for compiling and examining the inputs and outputs of materials and energy and the associated environmental impacts directly attributable to the functioning of a product or service system throughout its life cycle, from the acquisition of raw materials through final disposal. Life cycle assessment is done so that a complete picture of the environmental impacts throughout the lifetime of products and services can be developed. This provides significantly more useful information than does evaluating the impact from the manufacturing process alone; it also provides a systematic way to evaluate the costs and benefits associated with product or service changes at various points in their life cycle.

The phases of an LCA are: 1) establishing goals and the scope of the assessment, 2) conducting the inventory analysis, 3) conducting the impact assessment, and 4) improvement assessment. Impact assessment has three elements: classification, characterization, and valuation. The goals of the LCA study should include the reasons for carrying out the study, the intended applications, the intended audience, the initial data quality objectives, and the type of critical review that will be conducted for the LCA. The scope should include background information for the product or service being evaluated, boundaries of the study, method of impact assessment, data requirements, assumptions, and limitations of the study.

The standard recognizes that there is no single method to conduct LCAs, and it allows organizations flexibility in implementing the LCA. The assessment can be suited to the particular product or services, but it is important for the assessment to be systematic, understandable, and transparent. The methodology should be based on current scientific findings and improvements in LCA. The results should not be reduced to a simple overall conclusion, but they should reflect the complexities and trade-offs inherent in the process. The assessment must respect confidentiality and proprietary matters. The results of the LCA must be reported to the intended audience and to interested third parties. The report should cover the goals and scope of the study, the methods employed, the results (and critical elements), and conclusions and recommendations.

Although LCAs can provide useful information for decision making, the assessment process does have limitations, and it may not be appropriate in all situations. Parties using LCAs should recognize that the assessment methodology is still developing and that the outputs are only as good as the inputs and method of assessment. In some instances, the data may not be available to conduct a complete assessment, economic and social issues may not be addressed, or the data may not be appropriate for the circumstances (for example, a focus on global impacts may not be appropriate for a localized project).

Some of the potential problems with LCAs can be minimized by the critical review recommended by the standard. The intention of the critical review is to ensure that the study was conducted according to the requirements of the LCA standard and to ensure the validity of the results. The review can be conducted in several ways, and it might include self-review, expert or practitioner review, or peer and stakeholder review.

PRODUCT STANDARD DEVELOPMENT, ISO 14060

ISO 14060 (committee draft standard as of April 1997) provides general guidelines that should be taken into account when developing standards to reduce environmental effects and to achieve the intended performance of the product or service. It is intended to be used by ISO in developing ISO product standards, but any entity that writes standards can use it.

The guide recognizes the complexity of identifying and establishing environmental effects from products and services throughout their life cycles. It recognizes that environmental effects should be balanced against factors that include product function, performance, safety and health, cost, marketability, and quality. The guide also recognizes that environmental provisions in product standards must be reviewed and changed to reflect innovation and technology, but not so frequently that innovation, productivity, and environmental improvements are jeopardized. The provisions of a product standard should be no more stringent than necessary to avoid excessive or inefficient material or energy use.

The environmental provisions in product standards can incorporate many considerations including material and energy inputs; wastes and other emissions; impacts from transportation, packaging, distribution, and use; reuse and recycling potential; and product disposal and associated wastes. Impact can be assessed using LCA, risk assessment, or other appropriate methods, providing that the method is appropriate to the product or service and that the data are available for conducting the assessment.

However, product standards should not specify the materials to be used in products, because doing so may hamper innovation and actually prevent environmental improvements. A standard might include a recommendation to use secondary or recycled materials, but consideration must be given to total life-cycle effects.

Resource conservation, pollution prevention, and design for environment are three strategies that can be used in product standards to protect the environment. Each has unique contributions to make to environmental protection. Resource conservation focuses on preventing depletion of resources, which ensures the resource for future use. Renewable resources can be replenished if managed carefully. Examples are timber resources, soil fertility, and crops with industrial value. Nonrenewable resources present different challenges. Recycling and waste minimization are two strategies that can be used to preserve these resources and extend them as much as possible. Pollution prevention utilizes strategies such as source reduction and material substitution to reduce emissions to the environment. Pollution prevention can result in numerous financial benefits, including lower cleanup and disposal costs, reduced cost for pollution control equipment and its operation, insurance savings, and many others. Design for environment (DFE) is a developing technique being applied in some product sectors. Design for environment considers potential environmental impact during the early stages when the concept, need, and design of a product are considered. Design for environment methods include the consideration of material substitution, reuse, maintainability, and design for disassembly and recycling.

SUMMARY

The ISO 14000 series of standards is a strong beginning to global environmental management and has the potential of moving the global community toward a sustainable and restorative system of commerce. The benefits of implementing the standards are more immediate to individual companies and include savings from better management of raw materials, reduced waste production, increased operational efficiencies, reduced costs of controlling releases, reduced costs of mitigating impacts, and other potential savings. All of these factors can combine to improve corporate image and to strengthen the company's position in the marketplace. Equally important, but sometimes overlooked, are the two tangible benefits of reduction in worker and community illness, injury, and death, and improved quality of life.

FIGURE 1–3.   Selected Definitions Used in the ISO Standards:

Continual improvement—process of enhancing the environmental management system to achieve improvements in overall environmental performance in line with the organization's environmental policy (Note—The process need not take place in all areas of activity simultaneously.)

Environment—surroundings in which an organization operates, including air, water, land, natural resources, flora, fauna, humans, and their interrelation (Note—Surroundings in this context extend from within an organization to the global system.)

Environmental aspect—element of an organization's activities, products, or services that can interact with the environment (Note—A significant environmental aspect is an environmental aspect that has or can have a significant environmental impact.)

Environmental impact—any change to the environment, whether adverse or beneficial, wholly or partially resulting from an organization's activities, products, or services

Environmental management system—that part of the overall management system that includes organizational structure, planning activities, responsibilities, practices, procedures, processes, and resources for developing, implementing, achieving, reviewing, and maintaining the environmental policy

Environmental management system audit—a systematic and documented verification process to objectively obtain and evaluate evidence to determine whether an organization's environmental management system conforms to the environmental management system audit criteria set by the organization, and communication of the results of this process to management

Environmental objective—overall environmental goal, arising from the environmental policy, that an organization sets itself to achieve and which is quantified where practicable

Environmental performance—measurable results of the environmental management system, related to an organization's control of its environmental aspects, based on its environmental policy, objectives, and targets

Environmental policy—statement by the organization of its intentions and principles in relation to its overall environmental performance which provides a framework for action and for the setting of its environmental objectives and targets

Environmental target—detailed performance requirement, quantified where practicable, applicable to the organization or parts thereof, that arises from the environmental objectives and that needs to be set and met to achieve those objectives

Interested party—individual or group concerned with or affected by the environmental performance of an organization

Organization—company, corporation, firm, enterprise, authority, or institution, or part or combination thereof, whether incorporated or not, public or private, that has its own functions and administration (Note—For organizations with more than one operating unit, a single operating unit may be defined as an organization.)

Prevention of pollution—use of processes, practices, materials, or products that avoid, reduce, or control pollution, which may include recycling, treatment, process changes, control mechanisms, efficient use of resources, and material substitution (Note—The potential benefits of prevention of pollution include the reduction of adverse environmental impacts, improved efficiency, and reduced costs.)

Source: Reprinted from ISO 14001 Environmental Management Systems—Specification with Guidance for Use, *with permission.*

ENDNOTES

[1] Paul Hawken, *The Ecology of Commerce*. New York, NY: HarperBusiness, 1993.

[2] World Commission on Environment and Development, *Our Common Future* (Oxford, Great Britain: Oxford University Press, 1987), pp. 1, 8.

[3] Robert Repetto, *Jobs, Competitiveness, and Environmental Regulation: What Are the Real Issues?* (Washington, D.C.: World Resources Institute, 1995).

[4] Repetto, *Jobs, Competitiveness, and Environmental Regulation*, p. 18.

[5] Repetto, *Jobs, Competitiveness, and Environmental Regulation*, pp. 29–34.

[6] D. Keith Denton, *Enviro-Management* (Englewood Cliffs, NJ: Prentice-Hall, 1994), p. 138. This book provides many additional examples of turning environmental costs into profits.

[7] Denton, *Enviro-Management*, p. 89.

[8] United Nations Conference on Environment and Development, *Rio Declaration on Environment and Development* (New York, NY: United Nations Department of Public Information, 1993).

[9] James G. Patterson, *ISO 9000 Worldwide Quality Standard* (Menlo Park, CA: Crisp Publications, Inc.), p. 4.

[10] Vince Harper, Irwin Special Publications (*personal communication*), May 22, 1996.

[11] International Standards Organization, ISO/TC 207, *Section 1 Strategic Plan ISO/TC 207 on Environmental Management N47*, ISO/TC 207 N45 (International Standards Organization, January 24, 1994), p. 3.

[12] U.S. Environmental Protection Agency, "Pollution Prevention—It's a Whole New Way of Doing Business . . ." *EPA Journal:* 19(3): pp. 11, 18–19, 1993.

[13] U.S. Environmental Protection Agency, "Pollution Prevention—It's a Whole New Way of Doing Business . . ." *EPA Journal:* 19(3): pp. 17–18, 1993.

[14] Office of Air Quality Planning and Standards, U.S. Environmental Protection Agency, *National Air Quality and Emissions Trends Report, 1993* (Research Triangle Park, NC: U.S. Environmental Protection Agency, 1994) p. 33.

[15] Office of Air Quality Planning and Standards, U.S. Environmental Protection Agency, *National Air Quality and Emissions Trends Report, 1993*, p. 32.

[16] Office of Air Quality Planning and Standards, U.S. Environmental Protection Agency, *National Air Quality and Emissions Trends Report, 1993*, p. 96.

[17] Office of Air Quality Planning and Standards, U.S. Environmental Protection Agency, *National Air Quality and Emissions Trends Report, 1993*, p. 96.

[18] U.S. Environmental Protection Agency, "Incentives for Self-Policing: Discovery, Disclosure, Correction, and Prevention of Violations; Notice," *Federal Register*: 60(246): 66705–66712, December 22, 1995.

CHAPTER 2

A SNAPSHOT OF THE PROCESS

INTRODUCTION

One of the strengths of ISO 14001 is that it is not a performance standard. It does not specify how the requirements of any section should be satisfied, nor does it specify levels of environmental performance that a company must achieve. Because the standard establishes a *process* for achieving a quality program, it can be applied successfully to any type or size of business. To gain certification, an organization must demonstrate that it conforms to the requirements of the standard, which are embodied in the Environmental Management System Model (see Figure 2–1).

The complexity of the task and the starting point for developing the EMS will depend on previous efforts and existing programs. It is important to emphasize that the EMS must be tailored to each organization. Attempting to superimpose someone else's program on your organization will not work, but the path will be easier and, perhaps, smoother if an ISO 9000 quality program is already in place or if the company has already gone through a complete or partial environmental audit. Many of the documents already prepared for ISO 9000 can be used with minor modifications to meet the specifications of ISO 14001 (see Figures 2–2 and 2–3) for a comparison of the two standards), and previous experience with the quality process will save time and money.

INITIAL ASSESSMENT

It is important to establish a structured process for developing or improving the environmental management program as the organization moves through the ISO 14001 process. Remember that the standard is based on the principle that *a company says what it will do, then does what it has said.* The initial assessment is part of the overall process and it provides a benchmark for evaluating the organization's current environmental efforts against the requirements of the standard. It can be performed by an outside consultant or by in-house staff, but it should be absolutely neutral and objective. The initial assessment can consider a number of areas, including:

FIGURE 2–1.   Environmental Management System Model:

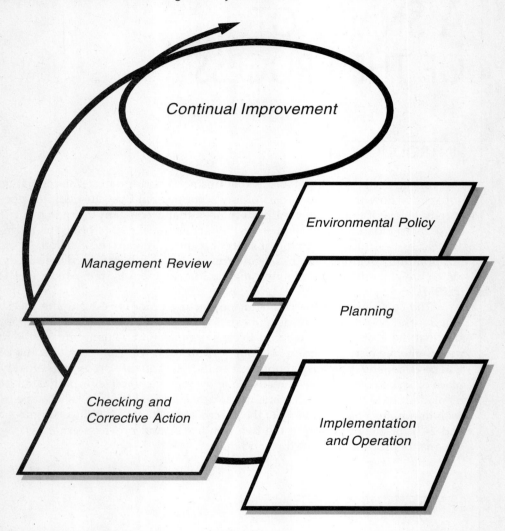

Reprinted from *ISO/DIS 14001 Environmental Management Systems—Specification with guidance for use,* with permission.

Note: 1) Environmental Policy corresponds to Commitment and Policy in 14004

2) Planning corresponds to planning in 14004; includes environmental aspects, legal and other requirements, objectives and targets, environmental management program(s)

3) Implementation and Operation corresponds to Implementation in 14004; includes structure and responsibility; training, awareness, and competence; communication; EMS documentation; document control; operational control; emergency preparedness and response

4) Checking and Corrective Action corresponds to Measurement and Evaluation in 14004; includes monitoring and measurement, nonconformance and corrective and preventive action, records, EMS audit

5) Management and Review corresponds to Review and Improvement in 14004

FIGURE 2–2.   Correspondence between ISO 14001 and ISO 9001:

ISO 14001: 1996		ISO 9001:1994	
General Requirements	4.1	4.2.1 ᵃ	General
Environmental Policy	4.2	4.1.1	Quality policy
Planning			
Environmental aspects	4.3.1	–	
Legal and other requirements	4.3.2	– 1)	
Objective and targets	4.3.3	– 2)	
Environmental management programme(s)	4.3.4	4.2.3	Quality planning
Implementation and Operation			
Structure and responsibility	4.4.1	4.1.2	Organization
Training awareness and competence	4.4.2	4.18	Training
Communication	4.4.3	–	
Environmental management system documentation	4.4.4	4.2.1 ᵇ	General
Document control	4.4.5	4.5	Document and data control
Operational control	4.4.6	4.2.2	Quality system procedures
	4.4.6	4.3 3)	Contract review
	4.4.6	4.4	Design control
	4.4.6	4.6	Purchasing
	4.4.6	4.7	Control of customer-supplied product
	4.4.6	4.9	Process control
	4.4.6	4.15	Handling, storage, packaging, preservation, and delivery
	4.4.6	4.19	Servicing
	4.4.6	4.8	Product identification and traceability
Emergency preparedness and response	4.4.7	–	
Checking and Corrective Action			
Monitoring and measurement	4.5.1ᶜ	4.10	Inspection and testing
	–	4.12	Inspection and test status
	–	4.20	Statistical techniques
Monitoring and measurement	4.5.1 ᵈ	4.11	Control of inspections, measurement, and test equipment
Nonconformance and corrective and preventive action	4.5.2 ᵉ	4.13	Control of nonconforming product
Nonconformance and corrective and preventive action	4.5.2 ᶠ	4.14	Corrective and preventive action
Records	4.5.3	4.16	Control of quality records
Environmental management system audit	4.5.4	4.17	Internal quality audits
Management Review	5.6	4.1.3	Management review

1) Legal requirements addressed in ISO 9001, 4.4.4
2) Objectives addressed in ISO 9001, 4.1.1
3) Communication with the quality stakeholders (customers)

ᵃ1st sentence
ᵇwithout 1st sentence
ᶜ1st & 3rd paragraph
ᵈ2nd paragraph
ᵉ1st part of 1st sentence
ᶠwithout 1st part of 1st sentence

Reprinted from ISO 14001 Environmental Management Systems—Specifications with guidance for use, with permission.

21

FIGURE 2–3. Correspondence Between ISO 9001 and ISO 14001:

ISO 9001:1994			ISO 14001:1996
Management Responsibility			
Quality policy	4.1.1	4.2	Environmental policy
	-	4.3.1	Environmental aspects
	- 1)	4.3.2	Legal and other requirements
	- 2)	4.3.3	Objectives and targets
	-	4.3.4	Environmental management programme(s)
Organization	4.1.2	4.4.1	Structure and responsibility
Management review	4.1.3	4.6	Management review
Quality System			
General	4.2.1 a	4.1	General
	4.2.1 b	4.4.4	Environmental management system documentation
Quality system procedures	4.2.2	4.4.6	Operational control
Quality planning	4.2.3	-	
Contract review	4.3 3)	4.4.6	Operational control
Design control	4.4	4.4.6	Operational control
Document and data control	4.5	4.4.5	Document control
Purchasing	4.6	4.4.6	Operational control
Control of customer-supplied product	4.7	4.4.6	Operational control
Product identification and traceability	4.8	-	
Process control	4.9	4.4.6	Operational control
Inspection and testing	4.10	4.5.1 c	Monitoring and measurement
Control of inspection, measuring, and test equipment	4.11	4.5.2 d	Monitoring and measurement
Inspection and test status	4.12	-	
Control of nonconforming product	4.13	4.5.2 e	Nonconformance and corrective and preventive action
Corrective and preventive action	4.14	4.5.2 f	Nonconformance and corrective and preventive action
	-	4.4.7	Emergency preparedness and response
Handling, storage, packing, preservation, and delivery	4.15	4.4.6	Operational control
Control of quality records	4.16	4.5.3	Records
Internal quality audits	4.17	4.5.4	Environmental management system audit
Training	4.18	4.4.2	Training, awareness, and competence
Servicing	4.19	4.4.6	Operational control
Statistical techniques	4.20	-	
	-	4.4.3	Communication

1) Legal requirements addressed in ISO 9001, 4.4.4
2) Objectives addressed in ISO 9001, 4.1.1
3) Communication with the quality stakeholders (customers)

a 1st sentence
b without 1st sentence
c 1st & 3rd paragraph

d 2nd paragraph
e 1st part of 1st sentence
f without 1st part of 1st sentence

Reprinted from ISO 14001 Environmental Management Systems—Specifications with Guidance for use, with permission.

- regulatory requirements that apply to the facility;
- organizational support, including environmental and related personnel, funding, management practices and procedures, and top management commitment;
- sources of contaminants and environmental impacts resulting from all aspects of the operation, including use of raw materials, processing, waste generation and disposal, suppliers, contractors, and so forth;
- environmental controls and their effectiveness;
- evaluation of performance compared with applicable regulations and internal policies;
- current gaps and needed program areas, equipment, staff, training, documentation, and other needs; and
- estimated costs and benefits of certification.

The planning and implementation questionnaires in Appendix A can help develop the initial environmental management profile for the organization. Interviews, checklists such as those in Chapters 13–17, and audit results are additional tools that can be used to develop an initial assessment. The assessment should produce a written report that thoroughly reviews all of the organization's environmental management options. Done with care, it will reveal strengths, weaknesses, and any gaps in the environmental program. It should identify specific requirements that are needed or areas that must be upgraded (for example, a waste minimization program is needed). If the assessment is not complete or if it does not produce a frank appraisal of the organization's current environmental programs, the time and effort devoted to gaining certification may not be successful. A good initial assessment will maximize resources and position the organization for success as it pursues improvement of environmental management or the ISO certification.

REQUIREMENTS FOR THE ENVIRONMENTAL MANAGEMENT SYSTEM

Information gained from the initial assessment allows the development team to focus on a plan for implementing specific sections of the standard. Haphazard approaches to developing or improving the EMS will not be successful. It is important to formulate a systematic process for the overall implementation and follow it, recognizing that adjustments may be needed along the way. The plan should identify the players and their responsibilities, tasks to be accomplished, time frame for implementation, costs, and other resources. If additional resources are needed, these should be obtained early in the process.

The ISO EMS model shown in Figure 2–1 contains only those elements that can be objectively audited for the purposes of certification. Putting these elements together is a step-by-step process that involves the following activities (the applicable sections in ISO 14001 and ISO 14004 are included for easy cross-referencing):

1. Development of a corporate vision and environmental policy statement (ISO 14001, Section 4.2; ISO 14004, Section 4.1.4, Environmental Policy)
2. Process analysis and planning (ISO 14001, Section 4.3, Planning; ISO 14004, Section 4.1.3, Initial Environmental Review)

3. Activity-specific policies (ISO 14001, Section 4.3, Planning; ISO 14004, Section 4.1.4, Environmental Policy)
4. Procedures (ISO 14001, Section 4.3, Planning)
5. Work instructions (ISO 14001, Section 4.3, Planning)
6. Implementation and operation (ISO 14001, Section 4.4, Implementation and Operation)
7. Record keeping (ISO 14001, Section 4.4, Implementation and Operation)
8. Program review (ISO 14001, Section 4.5, Checking and Corrective Action; ISO 14004, Section 4.5, Review and Improvement)
9. Continual improvement (ISO 14001, Section 4.6, Management Review; ISO 14004, Section 4.5.3, Continual Improvement)

LAYING THE FOUNDATION: VISION AND POLICY STATEMENT

Among the elements that are essential to implementing the standard, none is more important than management's commitment to a quality EMS program. Management must clearly and enthusiastically communicate its commitment to the program, and it must be visible in its support of the EMS. This commitment must be supported by making available the needed resources and holding everyone in the organization accountable for the successful implementation of the program.

Establishing and communicating a management vision and policy statement for the EMS is a key element of the program's success. All else flows from this vision. The environmental policy statement should be specific, but not detailed or overly long. The environmental policy statement is the overarching policy for the organization. It allows the organization to develop meaningful, activity-specific operating policies such as procurement, training, or spill response.

PROCESS ANALYSIS, PLANNING, AND DOCUMENTATION

Process analysis is essential to good planning, and it is an integral part of the continual improvement program. Process analysis means that the activities in the organization are broken down into discrete steps, and each is examined to identify opportunities to eliminate or minimize environmental impacts. Processes that should be considered include, but are not limited to, purchasing, contracting, manufacturing, servicing products after sale, energy management, waste management, and training. If a quality program already exists, some of this work has been done. Once the processes are charted, the other components of the EMS can be developed. The specification for planning in ISO 14001 requires the organization to identify environmental aspects, evaluate their impact, and use the gathered information to set objectives and targets. An environmental aspect is any activity, product, or service that can have an impact, either beneficial or adverse, on the environment. Additional requirements include identifying legal and other requirements and establishing an implementation program.

Documentation is required to demonstrate conformance with ISO 14001. The type of documentation that the organization needs will be revealed as the processes are charted. Carefully planned documentation will support the EMS, not overwhelm it. Although the standard does not prescribe a specific documentation system, the documentation procedures that are accepted for ISO 9000 programs provide a logical model for ISO 14001. This system includes a policy manual that provides a blueprint or plan of the organization's practices. This manual includes the environmental policy, provides information about the organization, and shows how the EMS conforms to the different elements of the standards.

Procedures that carry out the environmental policy should also be documented. These operating procedures cover the who, what, when, and where of operations. They should be complete but not lengthy. Operating procedures are needed for training, calibration, emergency response, maintenance, purchasing, contracting, documentation, and any other element of the organization's processes that can have a signficant environmental impact.

Work instructions detail exactly how individual tasks, such as the calibration of a monitor, cleanup of asbestos waste, and other activities, will be done. Work instructions can take many different forms, including step-by-step instructions in narrative form, invoices, flowcharts, engineering documents, standardized forms, or other types of directions. The key is to provide specific directions for performing a task. A work instruction is needed any time a process might cause an environmental impact if specific instructions are not followed. An auditor will not say that the organization needs more or fewer work instructions. Instead, the auditor will determine whether the work instructions control the system. If they do not, more effort is needed to identify the problem and correct it.

IMPLEMENTATION

Implementation of the EMS is an ongoing process that begins with the commitment to develop the program. Implementation is more than simply having written policies, procedures, and work instructions. A fully implemented program means that the systems are operated as documented or, in simpler terms, the organization does what it has said it will do. Successful implementation means that management and staff understand the overall program and their specific functions and responsibilities; staff members are using work instructions, generating records, controlling documents, and actively following all of the agreed-upon procedures. Everyone has the resources to perform the work, and they are accountable for their performance.

Good record keeping is one way of demonstrating that the EMS is implemented. Examples of records include training records, incident reports, audit results, copies of regulations, and other documents. Each procedure will identify records to be maintained, by whom, and for how long. Records that are easy to use, well managed, and available are the living proof of a successful EMS. They make it easy for an audit team to verify that the organization is doing what it said it would! Incomplete, sloppy, or poorly maintained records will jeopardize certification.

PERIODIC REVIEW AND CONTINUAL IMPROVEMENT

Periodic reviews of the environmental program are part of the continual improvement required by the standard. The measurement and evaluation of the

EMS determines whether the organization actually does what it says it will do. A formal process that includes audits should be in place to evaluate the EMS on a periodic and ongoing basis. Examples of components that would be evaluated include, but are not limited to, operation and maintenance activities, compliance with regulations, record keeping, information management, and training. When problems are identified, corrective and preventive actions should be planned and performed promptly.

Continual improvement means the organization reviews the overall EMS using available data to determine whether the EMS is meeting established objectives and targets and, if not, whether the organization makes the necessary changes. Continual improvement is not a one-time event, but part of an integrated effort to achieve excellence. The reference list at the end of this book contains some helpful resources that discuss continual improvement, quality programs, and other management issues.

CERTIFICATION

The length of time needed to operationalize the EMS depends on the size of the organization and the maturity of the program at the outset. It could range from as little as 6 months to about 2 years from the start of the process.

After the EMS appears to be operational and ready, it is best to perform a precertification audit prior to seeking certification. The precertification audit can be performed by in-house staff or by an external auditor. The precertification audit identifies any outstanding issues and allows these to be corrected.

Certification is usually done by a third party to ensure objectivity, but self-declaration is an option. A third-party certification assessment could involve one to four auditors, depending on the complexity of the operation, and they might spend 2 to 4 days at the facility. The cost of third-party certification also varies with the size and complexity of the operation, but estimates could range from $10,000 to $30,000, based on experience with ISO 9000.[1] These estimates do not include additional costs that might be incurred in establishing the EMS or in conducting the precertification audit. The auditing process is discussed further in Part 3 of this book. For more information about the certification process or to obtain copies of the ISO 14000 series of standards, contact the ISO or a member organization, such as the ANSI in the US or other sponsors such as the American Society for Quality Control (ASQC), the American Society for Testing and Materials (ASTM), and the NSF (National Sanitation Foundation) International in the U.S.[2] (addresses are given in the preface).

SUMMARY

The implementation of ISO 14001 requires an organization to take a long-term, systematic management approach to improving its environmental performance. The process begins with a preassessment to evaluate the status of its technical programs and management controls. Other key elements include the development of a corporate environmental management policy, planning activites, implementation and operation, checking and corrective action, management review, and continual improvement. None of these steps is particularly difficult, but they take time and depend on a strong management commitment to environmental excellence. An organization does not have to seek a third-party certification—it may elect to self-certify or to simply implement the ISO management model without certification. Whether or not certification is the end goal, implementation of an effective EMS has its own rewards in improved environmental performance and potentially significant cost savings.

ENDNOTES

[1] James G. Patterson, *ISO 9000 Worldwide Quality Standard* (Menlo Park, CA: Crisp Publications, Inc., 1995), p. 48.

[2] ANSI is a nonprofit organization that develops consensus standards and coordinates the US voluntary standards system. ASQC is a nonprofit organization dedicated to developing, promoting, and applying quality concepts and methods. ASTM develops and publishes voluntary industry standards, tests, practices, guides, and definitions for a variety of industries. The NSF International is a nonprofit organization dedicated to research, education, and service in the areas of health and environment. The NSF evaluates equipment, products, and services for compliance with NSF standards and criteria. The ASQC, ASTM, and the NSF are co-sponsors of the ISO 14000 standards in the United States.

PLANNING AND IMPLEMENTATION

Part 2 covers the nuts and bolts of planning and implementing an EMS that meets the requirements of ISO 14001. Chapter 3 discusses how to develop an environmental policy statement and ensures that it reflects the organization's philosophy on environmental management. Chapter 4 shows how to perform a process analysis which, in turn, allows appropriate documentation to be written. Appendix A at the end of the book contains a related questionnaire that focuses attention on whether or not planning elements are in place for the EMS. How to write and organize documentation is covered in Chapter 5. Additional examples of documentation are contained in Appendix B, which shows an example of an Environmental Policy Quality Manual, and Appendix C, which contains examples of procedures; more examples of documentation are shown throughout the book. Chapter 6 covers implementation issues relating to personnel, communications, and documentation, and Chapter 7 discusses operational controls that are needed for implementing the EMS. Appendix A also contains a questionnaire that helps to evaluate the status of implementation activities.

CHAPTER 3

VISION, MISSION, VALUES, AND ENVIRONMENTAL POLICY

INTRODUCTION

In Stephen Covey's book, *The Seven Habits of Highly Effective People*, habit number one is "be proactive," and habit number two is "begin with the end in mind."[1] These practices are applicable to all successful management programs, and they provide a solid foundation for building the management system specified by the standard.

Crafting and implementing an environmental policy statement is easier when it flows from the company's vision, mission, and core values. In fact, if the policy is to be effective, it must derive from these core statements. A brief explanation of the relationship among these concepts follows.

A useful resource for developing or modifying an organization's foundation policies is the *Business Charter for Sustainable Development*, which is a set of guiding principles for environmental management programs. This charter, which was developed by the International Chamber of Commerce, is reprinted in Figure 3–1. These guiding principles, or others, can be incorporated into the organization's vision, mission, and value statements; the environmental management policy; and process-specific policies.

VISION, MISSION, AND VALUES

Vision is the future to which an organization aspires. It is a long-term view that provides a platform upon which the organization is built. Without it, an organization cannot effectively conduct its business. The mission is the organization's approach to achieving this vision: its purpose, design, objective or plan. Values are what the organization stands for, such as integrity, honesty, fairness, innovation, and providing timely service. An organization's vision, mission, and values are the keys to understanding its future direction.

A common belief is that management's vision should be communicated throughout the company so that everyone understands what is to be done. In reality, this type of approach is likely to result in limited success. A vision that is owned only by management is likely to meet with resistance. Try to develop a

FIGURE 3–1. International Chamber of Commerce (ICC) Business Charter for Sustainable Development:

1. Corporate Priority
To recognize environmental management as among the highest priorities and as a key determinant to sustainable development; to establish policies, programs, and practices for conducting operations in an environmentally sound manner.

2. Integrated Management
To integrate these policies, programs, and practices fully into each business as an element of management in all its functions.

3. Process of Improvement
To continue to improve policies, programs, and environmental performance, taking into account technical developments, scientific understanding, consumer needs and community expectations, with legal regulations as a starting point; and to apply the same environmental criteria internationally.

4. Employee Education
To educate, train, and motivate employees to conduct their activities in an environmentally responsible manner.

5. Prior Assessment
To assess environmental impacts before starting a new activity or project and before decommissioning a facility or leaving a site.

6. Products or Services
To develop and provide products or services that have no undue environmental impact and are safe in their intended use, that are efficient in their consumption of energy and natural resources, and that can be recycled, reused, or disposed of safely.

7. Customer Advice
To advise and, where relevant, educate customers, distributors, and the public in safe use, transportation, storage, and disposal of products provided; and to apply similar considerations to the provision of services.

8. Facilities and Operations
To develop, design, and operate facilities and conduct activities taking into consideration the efficient use of energy and materials, the sustainable use of renewable resources, the minimization of adverse environmental impact and waste generation, and the safe and responsible disposal of residual wastes.

9. Research
To conduct or support research on the environmental impacts of raw materials, products, processes, emissions, and wastes associated with the enterprise and on the means of minimizing such adverse impacts.

10. Precautionary Approach
To modify the manufacture, marketing, or use of products or services or the conduct of activities, consistent with scientific and technical understanding, to prevent serious or irreversible environmental degradation.

11. Contractors and Suppliers
To promote the adoption of these principles by contractors acting on behalf of the enterprise, encouraging and, where appropriate, requiring improvements in their practices to make them consistent with those of the enterprise; and to encourage the wider adoption of these principles by suppliers.

12. Emergency Preparedness
To develop and maintain, where significant hazards exist, emergency preparedness plans in conjunction with the emergency services, relevant authorities and the local community, recognizing potential transboundary impacts.

13. Transfer of Technology
To contribute to the transfer of environmentally sound technology and management methods throughout the industrial and public sectors.

14. Contributing to the Common Effect
To contribute to the development of public policy and to business, governmental and intergovernmental programs and education initiatives that will enhance environmental awareness and protection.

15. Openness to Concerns
To foster openness and dialogue with employees and the public, anticipating and responding to their concerns about potential hazards and impacts of operations, products, wastes, or services, including those of transboundary or global significance.

16. Compliance and Reporting
To measure environmental performance; to conduct regular environmental audits and assessments of compliance with company requirements, legal requirements and these principles; and periodically to provide appropriate information to the Board of Directors, shareholders, employees, the authorities, and the public.

Source: Reprinted with permission of the International Chamber of Commerce, Paris, France.

shared vision that permeates all parts of the organization. Success requires commitment at all levels. When people do not feel ownership of a vision, they may grudgingly comply with an organization's policies, pretend to believe, or even express hostility. These and other negative behaviors are detrimental to the long-term health and viability of the organization.

The best way to develop successful vision, mission, and value statements is to get input from all levels of the organization. These statements do not need to be long. In fact, simply formulated, concise statements are most effective. A team of managers can draft a statement and seek comments from staff in the organization, the draft can be initiated by a team of staff-level employees, or a combination of staff and management may work together to establish the statement. The key to success is soliciting input from everyone in the organization and giving serious consideration to all ideas. Involving the entire company in this first step helps build trust and enthusiasm. A supplementary benefit is that people may also share their views about the shortcomings of the organization if they feel that they can do it without retribution. This information can be more valuable than the advice of a dozen consultants.

It is sometimes difficult to maintain momentum once the organization's vision has been implemented. Keeping a diverse group of people working toward the same goals can be difficult. Continual improvement is the best tool for maintaining momentum. The group must establish feedback loops, evaluate the vision, determine whether established policies are effective, review management performance, and take whatever steps are necessary to stay on track. A good vision is one that has commitment from people at all levels and creates an energy that will lead to significant accomplishments.

WRITING THE STATEMENTS

The vision statement is the future, to which the organization aspires (even if it may not be attainable), whereas the mission statement explains the business in which the organization is engaged, why the organization exists, and how the vision will be accomplished. Neither of these statements should be burdened with detail. Detail comes through company policies such as the environmental policy, activity-based policies, procedures, and work instructions.

Environmental management is not typically mentioned in traditional vision statements. When developing or revising a vision statement with the intention of incorporating environmental management, do not limit the organization's vision to simply complying with regulations. For example, a company's vision may be "to make quality products while contributing to a healthier environment." To contribute to a healthier environment, regulatory compliance is needed, but this company will strive to do more.

The vision and mission statements should have a positive tone. The organization should state what it wishes to do, not what it wishes to avoid. Statements that work for one company may not work for another. Resist the temptation to use canned statements and boilerplate language. Take the time to develop and craft statements that fit the organization. Examples of vision, mission, and value statements follow.

Example Vision Statements:

It is the vision of XYZ Company to produce the highest quality of products without harming the environment.

ABC Incorporated is proactive in protecting the environment while producing products of a quality even higher than our customers demand.

ZZZ Products work for continual improvement in the quality of our products while helping to create a cleaner, healthier environment for the people in our community.

Our vision is a world where sustainable development and a healthy economy are created without destruction of our natural environment.

We envision a world where the highest-quality products are produced in a way that preserves the quality of the air and water for use by people and wildlife.

Example Mission Statements:

Our mission is commitment to produce quality products for our customers through a program of innovative development, customer service, and continual improvement.

Our mission is to provide cost-effective electrical power service to our customers while minimizing the environmental impacts caused by providing this service.

By holding true to our vision, we create a healthy work environment in which our people can utilize their abilities to the fullest and provide engineering services to our customers in a cost-effective way.

Smith, Inc. continually works toward the creation and implementation of manufacturing processes that produce semiconductors with zero defects and that produce zero emissions to the environment.

Our mission is to provide our customers with the highest quality of toasters while minimizing environmental impact through a program of continual improvement.

Example Value Statements:

The XYZ Company values innovation in product design, creativity in problem solving, and honesty and integrity in communicating with customers.

The New Corporation values productivity, performance, innovation, and customer service. We value our employees, our community, and our environment. We strive to help our employees grow and develop, participate in strengthening our community, and work to improve our environment.

The core values of ABC Company are:

- honesty
- customer focus
- continual improvement.

ENVIRONMENTAL POLICY STATEMENT

The foundation of the EMS is the organization's environmental policy and management's support for it. The environmental policy statement is a formal declaration of the company's commitment to environmental protection. Careful work on this part of the EMS will simplify all that follows (Figure 3–2 shows a flowchart of the policy development process). Top management should be involved in developing the policy instead of simply giving it a "rubber stamp." Managers must believe in the policy and demonstrate its commitment in an active and visible way. Managers who believe in the policy will view the EMS as an integral, contributing member of the company's operations, not as a drain on resources.

The environmental policy statement should be signed by top management, ideally, the chief executive officer (CEO) and by managers, such as vice presidents, who have executive responsibility for the organization. Everyone in the company should understand the meaning and intent of the policy. A formal training session is an ideal way to communicate the contents of the policy and to discuss its meaning. A common mistake is to introduce the policy with great fanfare and then let it gather dust. Make sure that the importance of the policy is reinforced by making it visible. Post it in prominent locations, include it in newsletters, and make it available to the public. Consider using the policy in marketing efforts. All of these actions help integrate the policy into everyday activities.

ISO 14001 (Section 4.1) requires top management to define the organization's environmental policy and to ensure that it:

- is appropriate to the nature, scale, and environmental impact of its activities, products, or services;
- includes a commitment to continual improvement and prevention of pollution;
- contains a commitment to comply with relevant environmental legislation and regulations, and with other requirements to which the organization subscribes;
- provides the framework for setting and reviewing environmental objectives and targets;
- is documented, implemented, maintained, and communicated to all employees;
- is available to the public.

These are the benchmarks against which an organization will be audited if it seeks certification. There are other issues, which are enumerated in the guideline standard (ISO 14004, Subsection 4.1.4), that should be considered as the policy is developed. For example, the policy should reflect the organization's vision, mission, and values, and it must have approval and visible support of top management. The policy should tell who is responsible for implementing it. It should provide a framework from which to develop activity-specific policies and procedures and should show how the overarching environmental policy is related to other corporate policies, such as quality, health, and safety.

An example of a policy statement is:

FIGURE 3–2.   The Policy Development Process:

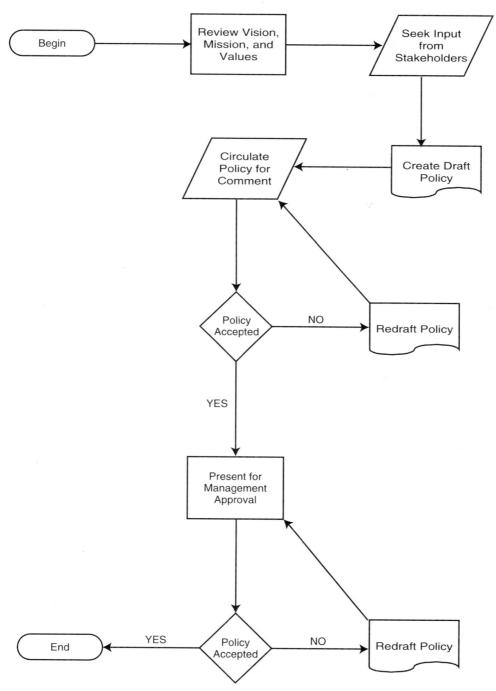

ENVIRONMENTAL POLICY OF ZZZ CORPORATION

ZZZ Corporation is dedicated to the inclusion of environmental factors in all decision making. As we work to improve the quality of our products and services, we also strive to preserve and improve the environment of our community. We view this as our responsibility as corporate citizens.

Our policy is to attain and, where possible, exceed the standards that have been established for environmental performance through the implementation and continual improvement of a comprehensive Environmental Management System (EMS) that encompasses all aspects of our operation.

The Board of Directors hereby empowers and instructs the corporate Director of Environmental Health and Safety to evaluate the environmental aspects of our operation and to make recommendations to senior management on policies to minimize environmental impact, prevent pollution, minimize waste, and protect the health and welfare of our workers, the community, and the environment.

All employees, including management staff and the Chief Executive Officer (CEO), are hereby assigned the task of working with the Director of Environmental Health and Safety to develop and implement a system of management that achieves the goals and targets established in the EMS. These goals and targets include protection from occupational injury and illness, and their achievement creates positive economic and environmental impact.

The CEO resolves any and all conflicts that may arise during the development and implementation of the EMS. The resolution of these conflicts is documented in the appropriate fashion so that there is a clear policy to guide future actions.

Signature

Chairman, Board of Directors
January 1996

IMPLEMENTATION ISSUES

THE TEAM APPROACH

It is important to organize a team very early in the process of planning and implementing the EMS. The team can be organized in many ways, but team members must have the right expertise, and the process must be overseen by a senior manager who has the authority and resources to ensure that the planning and implementation activities are carried out properly. The team can be composed entirely of employees from within the company, or it may be appropriate to use consultants for some activities. Of course, the corporate team should include representatives from the environmental health and safety departments, but other departments, such as engineering, purchasing, legal, and shipping and receiving, should be included. Some team members may come and go as they are needed; others may serve for longer periods.

COMMITMENT IS EVERYTHING

There are numerous examples of companies that have paid large sums of money to have consultants help them develop vision and mission statements, strategic plans, total quality management programs, and other management tools, only to have those plans shelved and never used. All of these tools can help the bottom line, but only when they are used on a day-to-day basis.

To be successful, the EMS should not be an add-on program; it should be completely integrated into the organization's way of doing business. Occasionally, gaining acceptance for the program and its auditing activities can be difficult for a company that has not been proactive in the past or whose program is not perceived to be a high priority for senior managers. For these reasons, it is often easier to implement the environmental management program as part of a comprehensive effort to improve quality. Everyone needs to know that sound environmental management is a top priority for the company, now and in the future. An important test of commitment is whether the policy is backed by the resources to do the job. Faith is broken quickly when management fails to provide the personnel and money to plan and implement the EMS. Providing adequate resources sends a strong signal to workers (and, ultimately, to outside stakeholders) that the statement is more than rhetoric.

SUMMARY

An organization's vision, mission, and values form the basis for developing an environmental policy statement. The policy statement, in turn, establishes the basis for the organization's ongoing programs to manage its operations in an environmentally sound manner. ISO 14001 requires the active involvement and support of top management in the development of the environmental policy statement and the EMS because management support is essential to the success of the program. Although the structure of the policy statement is flexible, it must include a commitment to continual improvement, pollution prevention, and compliance with relevant environmental regulations, and it must provide a framework for setting and reviewing environmental objectives and targets.

ENDNOTE

[1] Stephen R. Covey, *The Seven Habits of Highly Effective People* (New York, NY: Fireside/Simon & Schuster, 1989), pp. 66–144.

CHAPTER 4

PROCESS ANALYSIS AND PLANNING

INTRODUCTION

At this stage in the development of the EMS, an organization would have developed vision, mission, and value statements and a general policy that covers the total environmental management program. The next step is to analyze the organization's processes and provide the information needed to develop activity-specific policies, procedures, and work instructions to carry out the EMS. Remember that the overall goal of the EMS is to achieve sound environmental performance. A successful EMS depends on how well the organization identifies opportunities for reducing environmental impact. This means that environmental aspects and impact must be characterized for each process, activity, product, or service. Once the process is described, objectives and targets—which may be linked to regulations or other standards—can be established. All of these activities are part of the planning process required to design the EMS.

The key to success at this stage is to be systematic and to thoroughly analyze each process. Take advantage of existing process analysis information from ISO 9000 or other planning efforts. Start with available information and view it from the perspective of an environmental management program. A planning questionnaire in Appendix A can help identify the status of an existing environmental management program and stimulate ideas for future consideration.

CHARTING THE PROCESS

This section describes two widely used process analysis tools, the flowchart and the cause-and-effect chart. During the planning phase of developing the EMS, these or other analytical tools form the basis for identifying impacts and establishing objectives and targets. Once the EMS is developed, they can be used to review the process and to diagnose problems. Chapter 9 provides a brief discussion of other techniques that can be used for planning, evaluation, and continual improvement activities.

FLOWCHARTS

A flowchart is a diagram that uses connecting lines and a set of symbols to show the steps from beginning to end of an activity or procedure. Standard symbols or custom symbols and figures or pictures can be used to make the flowchart. The key is consistency—once a symbol is used, it should always have the same meaning. Some commonly used symbols are shown in Figure 4–1. The relationship be-

FIGURE 4–1.   Flowchart Symbols:

Symbol	Name	Description
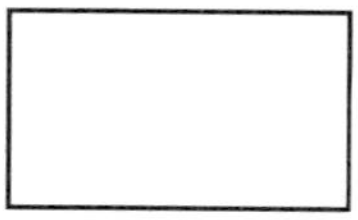	Process	Represents any type of process or activity, such as writing a memo, purchasing equipment, or interviewing a job candidate
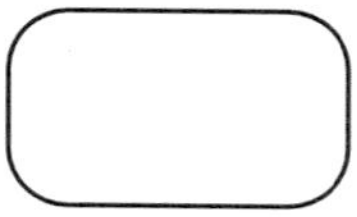	Alternate Process	Represents an alternate type of process, such as external (versus internal) or static (versus variable). Although popular, this shape is not an ISO-standard or ANSI-standard shape.
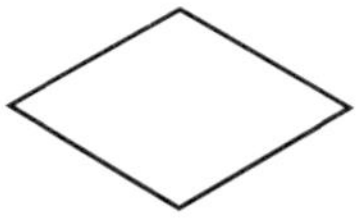	Decision	Indicates a point at which a decision must be made. Generally, two flow lines point out of the shape—one out of the bottom and one out of the side. Each line will be marked with a decision option, such as "Yes" and "No" or "True" and "False." These lines also can show branch options such as "Make" versus "Buy."
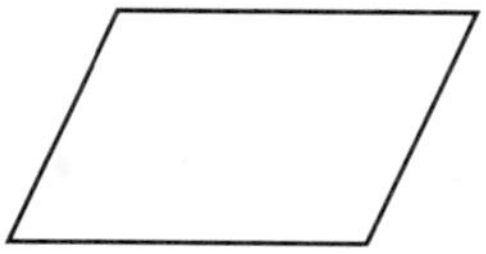	Input/Output	Represents the information that goes into or comes out of a process. Examples of input are orders and inquiries. Examples of outputs are reports and products.
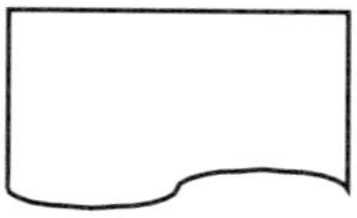	Document	Represents an activity recorded in a document, such as a computer file or printed report.
	Connector	Links a shape to another point in the flowchart without using a line. A letter or number in the circle links to the corresponding letter or number elsewhere in the chart. It is also used to connect multiple lines at one point.
	Terminal	Indicates the start or end of a process. The beginning terminal shape generally is labeled "Start" or "Begin." The ending terminal shape is labeled "Stop" or "End."

tween activities or procedures can also be diagrammed. Macroscale flowcharts describe the main steps for an overall process whereas microscale flowcharts dissect an individual step. For example, Figure 4–2 shows a macroscale flowchart that describes the manufacturing/distribution process from the receipt of raw material through distribution of the product whereas Figure 4–3 shows a microscale flowchart of the steps involved in one phase of the macroscale chart. Figure 4–3 describes the acquisition and distribution of raw materials. The microscale chart begins with the request for materials and, if the material is in stock, it is sent to the department issuing the request. If the part is not in stock, it is ordered, and its quality control is checked upon receipt. If the part is satisfactory, it is sent to the requesting department; if it is defective, it is returned to the supplier.

Flowcharts can be drawn by hand or prepared using readily available, moderately priced ($50 to $500) computer software such as *ABC FlowCharter*®.[1] Electronically prepared flowcharts are easily modified, and they can be linked to an electronic documentation system. Contact a computer/software center for recommendations on appropriate software for particular applications; before purchasing, ask current users about the performance of the software, ease of use, and support services.

Creating a flowchart is not difficult. Begin with a terminal shape labeled *start*, then ask a series of questions: What is the first step in this process? What happens next?

Typically, the first step is an activity, and it is represented by the process symbol (a rectangle). Each step builds on the previous one until the process has been completely documented. When a flowchart is complete, it should be distributed to all of the people who perform the process. They, in turn, review it for accuracy and make recommendations for revisions.

Figure 4–4 shows a sample flowchart that describes the consensus process, which can help teams come to agreement in an organized way on a variety of topics during the course of developing the EMS. The chart shows many of the common flowchart symbols. The beginning and ending symbols are the terminals, rectangles are activities or operations, and diamonds are the team's decision points. The document shape is shown for the draft proposal and for clarified proposals. Sometimes, connector circles are used to avoid cluttering the chart with too many lines. These circles link one point on a chart to another with the same number. For example, connector number 1 in this chart is from the decision point at which dissenters agree to stand aside. The *1* in the circle means that the next action goes back up to the other *1* in the diagram and continues through the flowchart.

The book by Louvar provides additional helpful information for developing and interpreting process flow sheets and diagrams.

CAUSE-AND-EFFECT CHART

Another popular tool for analyzing a process is the cause-and-effect chart, which is also known as an *Ishikawa diagram* (after its creator, Kaoru Ishikawa)[2] or a *fishbone diagram* (because its final form generally resembles a fish skeleton). The ef-

text continued on page 44

FIGURE 4–2.   Macroscale Flowchart Showing the Manufacturing/Distribution Process:

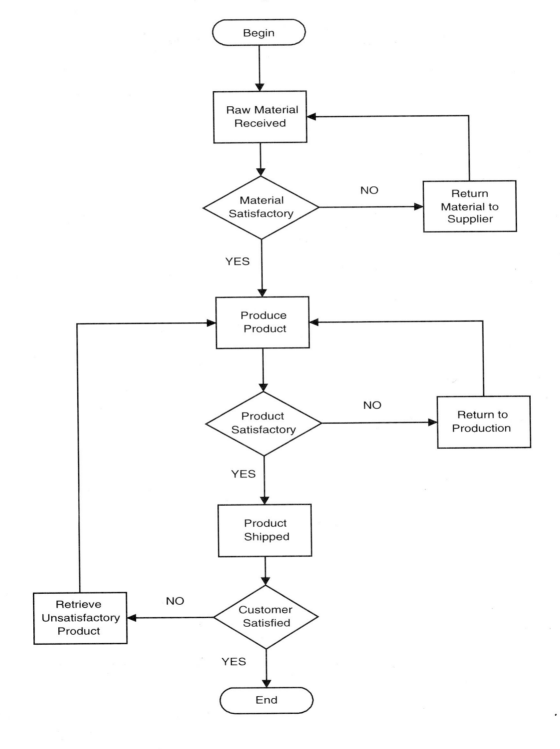

FIGURE 4–3. Microscale Flowchart Showing the Acquisition and Distribution of Raw Materials:

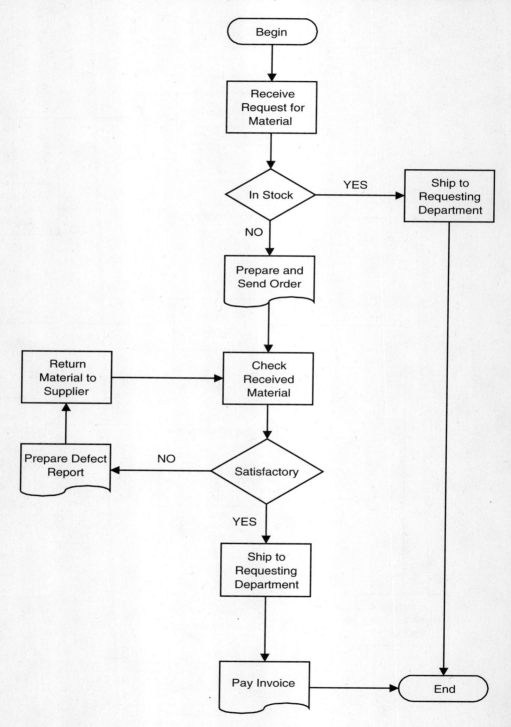

FIGURE 4–4.   The Consensus Process:

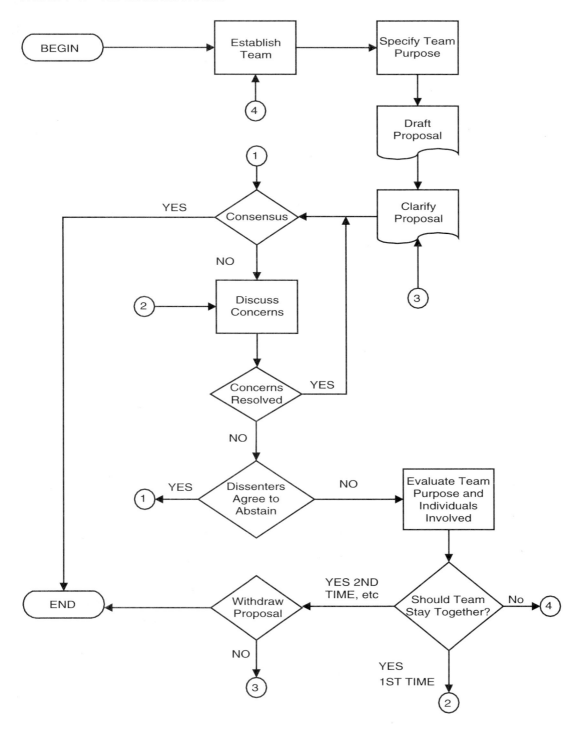

fect (positive or negative) being charted represents the head of the fish. A central backbone (horizontal line) is attached to the head, and bones (branched lines) representing major categories such as materials, equipment, resources, policies, or other appropriate categories are attached to the backbone. Causes related to each of the major categories are grouped along these lines. The branched lines can, in turn, be subdivided to depict minor causes. The success of the cause-and-effect chart depends on identifying as many potential causes as possible.

This type of chart has many applications; it is particularly effective for brainstorming sessions. It lends itself easily to identifying the environmental impact (effects) associated with the environmental aspects (causes) of different processes. It can also be used to analyze and develop solutions for a process that is out of control. Figure 4–5 shows an example that identifies the opportunities for reduced costs that can result from implementing an EMS. In this example, implementation of the EMS results in many organizational and operational changes that contribute to lower costs. For example, the installation of a high-efficiency heating system, improved vehicle maintenance, and efficient light bulbs all increase energy efficiency and lower overall operating costs. Similar benefits can result from the causes related to the prevention of pollution, worker's compensation, and management of human resources.

FIGURE 4–5.　Cause-and-Effect Chart Showing Cost Reduction Opportunities with the Implementation of an EMS:

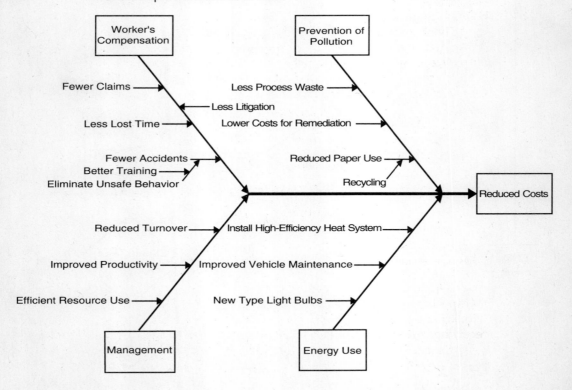

IDENTIFYING ENVIRONMENTAL ASPECTS AND IMPACTS

The standard (ISO 14001 Subsection 4.3.1) requires the organization to establish and maintain an up-to-date procedure for the identification of environmental aspects. An *environmental aspect* is any "element of an organization's activity, product or services that can interact with the environment" (ISO 14004, Subsection 3.3). (Appendix C contains an example procedure.) The *aspect* is the cause and the *impact* the effect. The aspects and significant impacts are then used to set objectives and targets.

ASPECTS

Aspects (and impacts) can be identified for the operations at the facility, the products or services produced, and contractors providing raw materials. However, the organization is responsible for only those aspects over which it can be expected to have control. For example, a pesticide product could have an impact on the environment if it is disposed of improperly after use, but it is not possible for the manufacturer to control the customer's behavior. It is possible, however, to provide warnings and labels that explain proper disposal practices. Additionally, the standard does not require a life cycle assessment or an evaluation of each individual product, component, or raw material input. The intention is to identify significant aspects and impacts so it may be possible to select categories of activities, products, or services for evaluation or to prioritize evaluation based on their potential significance.

There are many types of environmental aspects, and they can have beneficial or negative impact on the environment. Some examples are emissions, waste production, energy use, work practices, and the use of hazardous materials. The task of identifying environmental aspects is relatively easy, once the processes have been charted. Asking a series of questions about the inputs and outputs involved in making the product or in providing a service can help to clarify associated aspects:

Are hazardous raw materials used in the process?

Is there a potential for employee exposure to these substances?

Is there the possibility of an environmental release of these substances?

Is a hazardous by-product or waste produced from these raw materials?

Are nonhazardous waste materials produced?

Are there discharges to the air, water, or land?

Do discharges vary with changing conditions such as start-up, shut-down, or normal operation?

Asking these and other specific questions will help to identify environmental aspects related to material inputs and outputs, but aspects are also associated with other parts of an operation such as the warehouse, energy use, shipping and receiving, administration, housekeeping, and so forth. Ask the appropriate questions for each of these processes to identify as many environmental aspects as possible. A first look will rarely give a complete picture of the aspects associated with each process in an operation, and some processes may require recharting to clarify and fill in any gaps. One of the benefits of examining aspects associated with different processes is that employees begin to view their jobs and the organization in a broader environmental context and to see that they do have significant opportunities to minimize impact.

IMPACTS

Once the aspects are identified, the environmental impact associated with each aspect can be listed and evaluated. The cataloging of impacts should not be limited to current impacts, but should consider the potential for past and future impacts. Examples of impacts include degradation in groundwater, improvements in air quality, increases in worker illness, conservation of water, increased energy use, loss of habitat, and other ecologic effects.

Information resources that can be used to evaluate impacts include monitoring data, regulatory requirements, risk assessments, audits, and previously collected data from unplanned releases. The level of analysis will depend on the facility and the type of impacts. In some cases, a full-fledged risk assessment may be needed to evaluate and prioritize the impacts. In others, the evaluation might be limited to identifying the potential for health or safety impacts, ecosystem impacts, or noncompliance with regulations. The book by Louvar provides helpful discussions for identifying and analyzing hazards and risks associated with manufacturing processes, and the book by McBean discusses statistical tools for analyzing environmental data. Some questions that will help prioritize the impacts are:

- What is the scale of the impact?
  —Is it within the facility boundaries, or does it extend to local, regional, or even global scales?
- What is the severity of the impact?
  —Are the effects acute or chronic? Reversible or irreversible? Minimal or life-threatening to people or other organisms? Will evacuations be required? Will property be damaged? Will transportation be hindered?
  —Are sensitive environments such as wetlands, endangered habitats, areas of recreational or cultural significance, or drinking water supplies at risk?
  —What will the impact of a process failure or control mechanism failure be?
- How likely is the impact?
  —Could it occur as a result of normal operations or does it require a special set of circumstances? Does it require acute exposures or will it be triggered by chronic releases?
  —Does it occur continually? Daily, monthly, annually? Only in emergency circumstances?
- What is the duration of the impact?
  —Less than a day? A week, a month, a year?

The information gathered from delineating the scale, severity, probability, and duration of impacts can be used to prioritize the environmental significance of each impact. Additional interests that might be factored into the evaluation include liability, regulatory concerns, costs and efforts to change the impact, potential problems as a result of corrective actions, and public relations.

Finally, new activities present an opportunity to design a process with fewer or reduced impacts. Life cycle analysis, which is the subject of ISO 14040, is a method of examining a product from its inception to ultimate disposal. Raw materials, operational impact, potential for recycling, and final disposal are evaluated. Life cycle analysis is a complicated process, but its principles can be incorporated into product development or redesigned to reduce time to market, costs, and, ultimately, environmental effects.

LEGAL AND OTHER REQUIREMENTS

The next step in process analysis and planning is to evaluate legal and other requirements that apply to the organization. ISO 14001 (Subsection 4.3.2) requires the organization to establish and maintain a procedure to identify and have access to these requirements (see Appendix C for example procedures). Regulatory requirements include permit limits, licensing requirements, activity-specific rules (for example, calibration of air monitors), product-specific rules (for example, standards for X-ray machines), and industry-specific rules (for example, standards for the coke oven industry). Other requirements are those that go beyond legal minimums; for example, corporate policies on energy efficiency, dust suppression, or waste minimization. The procedure for maintaining legal and other requirements should identify the method, who is responsible for updates, and how the information will be distributed.

Although the standard does not require actual copies to be maintained, as a practical matter, *current* copies of these requirements should be available on site or easily accessible through a computerized database. A good practice is to organize the applicable requirements into a manual of regulations by process or activity area. Staying current of regulatory requirements has the added benefit of giving advance notice of new regulations that might affect the organization.

REGULATIONS

There are a number of ways to identify applicable legal requirements. An obvious starting point is to review the aspects and impacts of specific operations. For example, air releases will be governed by air pollution regulations; discharges to surface water will be governed by water pollution regulations; and so forth. Government agencies can provide lists of industries or activities and the regulations that apply. The use of governmental services, such as the *Federal Register*[3] (FR) in the United States, will keep the organization current on specific regulatory changes.

Another option is to subscribe to abstracting services such as the *Environmental Law Reporter*,[4] the *International Environmental Reporter*,[5] or *The National Reporter System*®.[6] Many abstracting services are available electronically, and some include international treaties and regulations in their databases. Ask questions prior to subscribing to electronic databases (most of these also apply to hard-copy services) to be sure that the service will meet the organization's needs:

- What is the regulatory coverage (state, national, international?)
- How frequently is the service updated? Ideally, updates should occur on a daily basis. Once every quarter is acceptable, but less frequent updates are not as helpful.
- Is the system easy to use? Ask for demonstrations and contact current users.
- Are searches easy to perform? Are they quick? It should be possible to search by key word, subject, citation number, acronyms; the ability to perform simultaneous searches across jurisdictions may be helpful.
- Can data be exported or printed with ease?
- What is the quality of on-line help? It should be uncomplicated and easy to access.

INTERNAL PERFORMANCE CRITERIA

As far-reaching as regulations often seem, they may not provide the desired degree of control, or they may be absent entirely. In these instances, additional performance criteria may help the organization to meet its objectives and targets. Examples of areas for which internal performance criteria can be developed include:

- Acquisition, property management, and divestiture. Acquiring new properties carries the risk of acquiring liability for existing pollution. Managing property in a way that minimizes liability helps reduce the costs associated with the transfer of property. Whether buying or selling, internal standards make sense.
- Suppliers and contractors. Many companies will wish to ensure that their suppliers and contractors comply with ISO 14001 to maintain quality and to reduce potential liability.
- Environmental communications. It is important to communicate both internally and externally. Effective communications within an organization will enhance everyone's understanding of programs and requirements. External communications should include regulators, customers, and the community.
- Environmental incident response and preparedness. It is always possible that the unexpected will happen; preparedness will limit the possibility of injury to employees or the public, damage to the environment, and liability.
- Environmental awareness and training. Informed and properly trained employees are more likely to achieve high productivity and high-quality work.
- Environmental measurement and improvement. Achieving environmental excellence requires continual evaluation. Performance can be evaluated using many different types of measurements, such as water quality monitoring, personal monitoring, or tracking of materials and wastes. Measurements can provide a base line that makes it easy to evaluate progress.
- Prevention of pollution and resource conservation. Preventing pollution and conserving resources is the best way to contribute to environmental improvement. These practices also have significant potential for reducing costs.
- Hazardous materials and waste management. Sound programs for managing hazardous materials and waste can reduce costs, prevent health and environmental effects, and minimize the potential for future liability.
- Energy and transportation management. This is an area of operations that provides opportunities for significant cost savings and beneficial environmental impact.

Organizations that view regulatory compliance as the end goal miss important chances to control their environmental management systems and improve performance. Opportunities for internal standards vary among organizations, but they will become apparent as the planning process advances, during implementation, or as part of the review process.

ESTABLISHING ENVIRONMENTAL OBJECTIVES AND TARGETS

The next step in the planning process is establishing objectives and targets. The standard (ISO 14001, Subsection 4.3.3) requires the organization to establish and maintain documented objectives and targets (see Appendix C for an example procedure). Objectives and targets provide a systematic way of implementing the commitments expressed in the environmental policy.

Objectives and targets should be developed in a systematic way that considers the potential benefits on the environment or health, potential costs, and personnel requirements. Objectives and targets should be established throughout the organization—for all functional areas and for staff and management. The process for setting objectives and targets should be inclusive, and it should take into account ongoing environmental initiatives. Those individuals who are accountable for the process will be in the best position to determine whether proposed objectives and targets can be reached. It is also important for objectives and targets to be consistent with other requirements of the organization.

OBJECTIVES

ISO 14001 defines an *environmental objective* as an "overall environmental goal, arising from the environmental policy, that an organization sets itself to achieve, and which is quantified where practicable." Objectives can be established for all aspects of the organization, including processing, raw materials and waste handling, emissions shipping and receiving, profit margins, training and communications, measurement, and continual improvement. The established objectives and targets should reflect a consideration of legal and other requirements, environmental aspects, views of interested parties, and technological options. The standard also requires objectives and targets to be established for pollution prevention activities. The annex to the specification standard clarifies that when technological options are considered, the organization may, but is not required to, consider the best available technology (BAT) when it is economically viable, cost-effective, and appropriate (Annex A.3.3). The use of the BAT in the standard may not carry the legal meaning prescribed by statutes such as the Clean Water Act or the Clean Air Act in the U.S. The organization should also consider its financial, operational, and business requirements when setting objectives and targets. This means it may, but is not required, to use total cost accounting methodologies to track the costs and benefits of environmental performance.

TARGETS

An *environmental target* is defined as a "detailed performance requirement, quantified where practicable, applicable to the organization or parts thereof, that arises from the environmental objectives and that needs to be set and met in order to achieve those objectives." In other words, objectives say what the organization will do over the long term, and targets specify how quickly the goals will be reached.

There are many ways of expressing targets, including percentages (30% increase in use of biodegradable cleaners), total quantities (create 4 hectares of wetlands), or quantities per unit of production (reduce solvent wastes by 1 kilogram per widget produced). Targets are typically used as performance indicators; that is, they are compared to actual performance to determine whether the organization is accomplishing its goals. It is important to make careful comparisons between actual performance and targeted performance. For example, when aggregate data are used for an entire facility (for example, quantities of waste reduction per unit of production), it is important to evaluate individual operating units because there may be a small portion of the organization that dominates. Also, a change in the method of measuring performance (for example, emissions stack testing versus in-stack continuous monitoring) may yield quite different results and interpretations. In the latter example, continuous in-stack monitoring is a more rigorous tracking program than using stack testing data, which may be collected only once every 2 years. For this reason, several years of data would be needed to interpret the significance of collected information. Some examples of objectives and targets are:

- Increase the use of recycled paper (objective) by 25% over 2 years (target).
- Install energy-efficient lighting in all buildings over an 18-month period.
- Achieve ISO 14001 certification within 20 months.
- Reduce the release of sulfur dioxide emissions by 25% within 2 years.
- Complete an inventory of hazardous materials within 6 months.
- Audit all operations within 1 year.
- Construct a state-of-the-art hazardous waste storage area by 2000.
- Replace the baghouse unit on the hospital incinerator within 20 months.

ESTABLISHING THE ENVIRONMENTAL MANAGEMENT SYSTEM

The standard requires the organization to establish and maintain a program to achieve its objectives and targets (ISO 14001, Subsection 4.3.4). The environmental management system is the action plan for reaching the objectives and targets. It should document and delineate what, who, when, and how the work will be accomplished. The overall program must be complete and must address all aspects of an operation, including processes, projects, products, services, and facilities.

Even though a good EMS will save money when implemented, resources must be available to start or modify a program. Needs will include technical and clerical staff, input from staff throughout the organization, data processing resources, supplies, and, possibly, contracts for consultants. It also must be clear who has overall responsibility for development of the EMS and who is responsible for each activity.

The responsibility and a time frame for achieving the objectives and targets should be assigned to key individuals at each relevant function and level of the organization. This requirement recognizes that key individuals may be located outside of the environmental department and could include staff from purchasing, engineering, processing, legal, or other departments. Individuals can be held accountable only after responsibilities, timetables, and resources are clearly identified.

SUMMARY

An efficient and effective EMS does not happen overnight. The initial phase involves analyzing all of the organization's processes to identify aspects, associated significant impacts, and legal and other requirements that might apply to the processes. This phase of planning provides information to establish objectives, targets, and performance criteria.

The most successful program will be integrated into the organization's overall management plan. The goal is to achieve excellence, but this does not happen overnight. A systematic approach that builds on existing components will decrease the time required to develop and implement the EMS. Recognize that there may be situations that require changes in the established EMS. A program of regular review will ensure that the EMS remains appropriate, relevant, and efficient.

ENDNOTES

[1] ABC FlowCharter® is available from Micrografx, Inc., 1303 Arapaho, Richardson, TX.

[2] Kaoru Ishikawa, *Guide to Quality Control* (Tokyo, Japan: Asian Productivity Organization, 1976).

[3] The *Federal Register* is printed by the U.S. Government Printing Office. It is available through libraries, state agencies, or it can be purchased from the U.S. National Technical Information Service, Washington, D.C. The index can be accessed on the Office of the Federal Register's electronic bulletin board "FREND" (202-275-1538). The table of contents, but not the text, can be viewed or downloaded free of charge. In some instances, copies of program-specific notices/rules can be obtained directly from the relevant EPA office.

[4] The *Environmental Law Reporter* is available from The Environmental Law Institute, P.O. Box 14175, Washington, D.C. 20044.

[5] The *International Environmental Reporter* is available from the Bureau of National Affairs, 1231 25th St. NW, Washington, D.C. 20007-1197.

[6] Information about *The National Reporter System*® can be obtained from West Publishing Company, P.O. Box 64526, St. Paul, MN 55164-9979.

CHAPTER 5

DOCUMENTATION

INTRODUCTION

ISO 14001 recognizes (and emphasizes) that documentation and document control are essential elements of an effective environmental management system. Well-written EMS documention increases employee and management awareness, aids implementation, and makes it easier to evaluate the EMS. One of the strengths of the standard is that it does not specify how the documentation will be written, organized, or maintained. The standard (ISO 14001, Subsections 4.4.4 and 4.4.5) requires the organization to document the core elements of the management system and interactions among the elements, to specify related documentation, and to control documentation (see Appendix C for an example procedure).

Some of the specific information that must be established and maintained is shown in Figure 5–1.

One of the criticisms of the ISO 9000 standard is that the documentation never ends! Before concluding that ISO 14001 has the same shortcoming, remember that the purpose of the standard is to improve the management of the environmental program and to build measures of quality into the system. This does not mean that a totally new documentation system must be created or that every action must be documented—using a lot of paper does not guarantee success. The documentation must be sufficient to provide the controls needed to ensure that the EMS works. Recognize that it will take time to identify organizational needs and to draft, evaluate, and revise any needed documentation. Work systematically and build on existing documentation. More than likely, many components are already in place.

This chapter outlines a documentation system patterned after ISO 9000 and provides additional examples in the appendixes. Appendix B contains an example Environmental Management Quality Manual, and Appendix C contains examples of procedures listed in Figure 5–1. These examples and others[1] can serve as guides, but it is best for an organization to resist the temptation to squeeze itself into someone else's documentation. Even though there are commonalities between organizations performing similar work, each organization's way of doing business (and, therefore, documentation) is unique. Simply filling in blanks without tailoring documentation to the organization is likely to result in inefficiencies. The intent of the standard is for each organization to develop documentation that fits the organization's individual needs. If existing documentation works, use it!

FIGURE 5–1.   Documentation Needed to Conform to ISO 14001:

- environmental policy
- procedure for identifying aspects
- procedure for identifying legal and other requirements
- procedure for periodically evaluating compliance with relevant environmental legislation and regulations
- procedure for establishing environmental objectives and targets
- procedure for documenting roles, responsibility, and authority
- procedures for training
- procedures for communication
- procedures for documentation and document control
- procedures for maintenance and operational control
- procedures related to identifiable significant environmental aspects of goods and services used by the organization
- procedures to identify potential for and respond to accidents and emergency situations
- procedures for preventing and mitigating environmental impact from accidents and emergency situations
- procedures for monitoring and measuring the key characteristics of the organization's operations and activities that can have a significant impact on the environment
- procedures for defining responsibility and authority for handling and investigating nonconformances, taking action to mitigate any impact caused by the organization, and initiating and completing corrective and preventive action
- procedures for the identification, maintenance, and disposition of environmental records (includes training records, audits, reviews, and so forth)
- procedures for periodic environmental management system audits and management review

PAPER OR ELECTRONIC DOCUMENTATION?

The standard allows either paper or electronic documentation systems. Electronic formats using software such as Folio Views® or Adobe Acrobat®² offer several advantages over paper systems. These tools create "infobases" and have many features that make them ideal for documentation. The programs have extensive indexing, linking, and search capabilities that allow users to locate information by topic, regulation number, key word, or other search parameters. For instance, if the word *training* were used as the key word for a search for training requirements, the software would locate every occurrence of the word. This capability allows all users on a network to be updated automatically, which, in turn, maximizes resources and eliminates the time required to photocopy new documents, retrieve outdated material, and destroy recalled documents. The software systems allow selected portions of the infobase to be printed or downloaded to a disk. When updates are made, it is possible to send out a network message informing people of the updated sections and instructing them to recycle any existing paper copies. A related feature is the ability to control access to documentation. For example, it is possible to authorize selected individuals to make changes in documentation and to limit the ability to print documents by creating read-only files. Some Folio Views® definitions are listed in Figure 5–2 to show the capability of this software.

Another advantage of electronic systems is the ability to perform needed tracking functions easily and accurately. This is especially important for tracking audit reports and the status of corrective actions, but it also allows data to be updated and reformatted easily for various internal and external reports.

Disadvantages include expense for the computers and related peripherals, software, maintenance of the system, and training needed to utilize the system. Computerized documentation without networking capability improves the ease of updating documents, but documentation control problems are similar to those for a paper system. Before making a commitment to an electronic system it is important to consider the organization's resources and other potential limitations such as ease of use and security needs. Ask for expert assistance and always check vendor references prior to purchasing.

FIGURE 5–2.   Definitions Used with Folio Views®:

Field—a searchable structure into which various types of information may be stored. Fields allow related bits of information to be collected on a subrecord level.

Group—a topical collection of records that users may define to create their own topical groups. Groups may be searched and displayed independent of the rest of the infobase; records may belong to multiple groups simultaneously.

Hierarchy Tree—a term that describes the relationship of levels in the infobase and any associated headings in the contents window. The hierarchy tree grows down; that is, its root is at the top and its branches extend down. The infobase itself is the root. The branches are the levels that are added to the infobase and any branches of those levels.

Infobase—a free-form collection of information (both text and graphics) with a comprehensive index. Text in an infobase may be searched, grouped by topic, linked, and edited. Functions similar to a database with a fully functional built-in word processor.

Jump Link—a link that goes from one point in the infobase to another specific point. A jump link is useful for cross-referencing and navigating the infobase.

Links—jumping points between one area in an infobase and another. For example, a link could go from an entry in the table of contents to the location of that entry in the main body of text. Links can also connect other infobases or other programs.

Master—an infobase that has been shadowed. The shadow file overlays the master infobase, allowing the addition or deletion of information in the shadow file without changing the master infobase. If the master infobase changes, the shadow file will have to be reconciled with the master to remain valid.

Object—nontextual information in the infobase. Objects can be, but are not limited to, graphics, animation, audio, and video.

Shadow Files—personal overlay files. A shadow file stores changes made to an infobase without actually changing the infobase. New shadow files may be created on any open infobase; only changes made after the new shadow file has been created are stored in the shadow file; other changes are stored in the infobase (if saved). Once a shadow file has been created, it may be opened and looks just like the original infobase, but it contains all changes made to the infobase. Additional changes may be made within the shadow file, once it is opened.

GETTING ORGANIZED—POLICIES, PROCEDURES, WORK INSTRUCTIONS, AND RECORDS

Each organization varies in its needs, but the starting point for determining which documents are needed is a preassessment and the environmental policy. An internal self-assessment and tools such as the questionnaire in Appendix A can help identify procedures that will be needed to develop or improve the EMS. Once the needed documents are identified, the nature and size of the documentation files will influence how the documents are constructed and organized. Because the standard does not specify how the documents should be developed or managed, any model that will control the EMS will work. One option that can help organize information is the documentation model that has evolved as best practice for ISO 9000.[3] This model arranges the organization's documents into four layers in a format that is easy to use, understand, and update. Based on an adaptation of this model, documents can be grouped into:

- environmental policy and supporting policies
- procedures
- work instructions
- records

Each layer of documentation becomes more detailed, from the environmental policy and supporting policies at the top to the records at the bottom. The remainder of this chapter explains each of these types of documents, followed by a discussion of how these documents can be organized into an easy-to-use, efficient documentation system.

ENVIRONMENTAL POLICY AND SUPPORTING POLICIES

At the completion of the planning activities, the organization will have a list or chart of processes that contribute to its successful operation. The overarching environmental policy provides the framework for developing supporting policies that might be needed for specific processes. For example, the overarching environmental policy may simply state that protection of the environment is the organization's policy; it may include general statements about the prevention of pollution, emergency response, or worker health and safety. In these instances, it may be important to elaborate on the general policy statement with a more specific statement about the organization's intentions in certain areas. Many organizations already have these policies in place, and they can be used as part of the documentation process. The environmental policy, in conjunction with the supporting policies, provides a nucleus around which carefully designed operational procedures, work instructions, and records can be developed. There are four items that should be evaluated when deciding which activities or operations require a specific supporting policy:

- vision, mission, and values
- environmental aspects

- objectives and targets
- guiding principles

The key is to be certain that supporting policies are developed that will control potentially negative environmental impacts identified in the process analysis. If needed, supporting policies should be approved by top management. To write a supporting policy, select an identified environmental aspect (objective, target) and identify what general actions are needed to prevent a problem or to maximize a benefit. Then, structure a policy that is in the spirit of the organization's vision, mission, and values, and one that is consistent with the organization's guiding principles. The policies should be clear, thoroughly documented, communicated throughout the organization, implemented at all levels, and updated on a regular basis. Supporting policies can be written in narrative form as a sentence or paragraph(s), as shown below, or in a more structured format, as shown in Figure 5–3, which is an operational policy for a wastewater treatment facility.

Some examples of supporting policy statements are:

Waste minimization procedures are implemented throughout the company.

All waste remaining is recycled or reused to the maximum extent possible.

Landfills or other disposal methods are a last resort when other means are not reasonably available.

All staff members are trained in the environmental aspects of our operation.

This training ensures that each employee understands the environmental management system.

ABC Corporation substitutes less hazardous materials as a hazard control measure whenever possible.

All monitoring instruments are calibrated according to the manufacturer's schedule and whenever there is deviation from the established control chart limits.

PROCEDURES

Procedures describe the activities that implement the environmental policy or supporting policies and document the flow of work needed to implement them. Each major activity related to the achievement of targets and objectives should be specified in a procedure. Procedures should also be consistent with the organization's vision, mission, values, and guiding principles to ensure integration and consistency across the EMS.

Procedures provide more information about what is to be done, who has the responsibility of ensuring that activities are conducted correctly, and where the activities will be performed. However, procedures should not be confused with work instructions, which are step-by-step instructions on how to perform an activity. For example, a procedure for maintaining and checking exhaust emissions for fleet vehicles should identify: 1) the reason for testing, 2) who is responsible for conducting the emissions testing and which vehicles are to be tested, 3) what

FIGURE 5–3.   Example of an Operational Policy for a Wastewater Treatment Facility:

XYZ, Inc.

Policy No.: 1
Title: Wastewater Treatment Facility Operations
Issue Date: 01/17/91
                                                                    Page 1 of 1

1.0   Objectives
       The objectives of this policy are to ensure proper operation of the wastewater treatment facility and to minimize discharges to water bodies.

2.0   Scope
       This policy applies to all locations of XYZ Corporation, all company-operated treatment facilities, and all related operations.

3.0   Responsibility
       All employees of treatment facilities are responsible for carrying out the work instructions associated with this policy. All managers responsible for the operation of those facilities are required to ensure that this policy is followed. Supervisors are responsible for ensuring that all required training is performed. Senior managers who have a treatment facility under their purview are required to ensure that this policy is followed.

4.0   Wastewater Treatment Facility Policy
       4.1   Each facility is operated in accordance with established procedures using design specifications as a guide to maximum performance.
       4.2   No facility is modified or operated in a manner different than that intended by the design engineers without an engineering evaluation to determine whether the modification in the physical plant or procedures will cause an increase in emissions to the environment.
       4.3   Only operators certified by the state are hired to work in any wastewater treatment facility.
       4.4   All operators are responsible for participating in required continuing education to keep their certifications current.
       4.5   New employees are trained in all aspects of facility operation. This must be done before a new employee is assigned responsibility for any operation or left alone at the facility.
       4.6   The supervisor and the Director of Environmental Health and Safety must be notified of any operational abnormalities that might cause an increase in environmental emissions.

5.0   Related Procedures
       (This section would include a list of procedures related to the implementation of the policy.)

6.0   Documentation
       (This section would list operational manuals, work instructions, records, and other documents related to the operation of the facility.)

7.0   Records
       (This section would identify where related records are filed and maintained, and for how long the records are retained.)

Revision No.: 2
Revision Date: 05/10/93
                                                            Approved by: MWC

---

testing is needed, and 4) the frequency of testing. A work instruction should describe in detail exactly how the test is performed, the needed equipment, and the required calibrations and maintenance activities.

The first step in writing a procedure is to create a flowchart or description of current practices. Next, identify all needed records related to the procedure. Compare the flowchart and related records to the requirements of ISO 14001. Make changes or adjustments as needed to conform to the requirements of the standard, obtain consensus, and prepare a written procedure that reflects the agreed-upon changes. Implement the procedure and allow it to be used for about 1 to 2 months (the length of time needed depends on the procedure and how frequently it is used). Next, refine it as needed and formally issue the procedure as Revision 0, and distribute it to those individuals and departments responsible for implementing the procedure. Distribute the procedures as they are completed, rather than waiting for all procedures and related documentation to be developed. Procedures must be revised when practices change; further revisions to the procedure are numbered consecutively.

Again, there are no requirements for using a specific structure for each procedure, but the procedures should have a standard format that includes identifying information, such as the procedure number, title, revision number, revision date, and the number of pages in the procedure. Figure 5–4 shows an example format that has numbered sections for each major category of information, and Figure 5–5 shows an example procedure using this format.

WORK INSTRUCTIONS

Work instructions are specific steps to be followed in accomplishing a task. For example, if the organization conducts environmental monitoring, the step-by-step instructions for operating, calibrating, maintaining, and decommissioning a sulfur dioxide monitor would be given in separate documents. A work instruction is needed whenever the lack of such an instruction could adversely affect the organization's ability to control the EMS. Instructions have a limited distribution, and they are typically housed in the departments that use them. Some instructions need to be detailed, others do not. All work instructions, whether they are produced inside or outside of the organization, must be controlled. Outdated instructions can have significant impact on the performance of the EMS. Work instructions should have identifying information, including a title, work instruction

FIGURE 5–4.   Example Format for Procedures:

1.0   Purpose or Objectives

2.0   Scope

3.0   Responsibilities

4.0   Procedure

5.0   Related Procedures

6.0   Documentation

7.0   Record Keeping

FIGURE 5–5.   Example of a Procedure for Performing Maintenance and Emissions Testing on Fleet Vehicles:

Procedure Number: 11

Title: Routine Maintenance and Emissions Testing of Fleet Vehicles

Issue Date: 10/15/95

Revision No.: 2.0   Approved By: FP

Revision Date: 01/20/96   Page 1 of 2

1.0   Objective

    1.1   To ensure that all fleet vehicles are in compliance with federal motor vehicle emission standards, that fuel use is minimized, and that emissions to the ambient air are minimized.

2.0   Scope

    2.1   All Company fleet vehicles.

3.0   Responsibilities

    3.1   The Fleet Manager has overall responsibility for this procedure.

    3.2   Individual Department Heads are responsible for ensuring that routine maintenance and emissions testing are performed on fleet vehicles within their departments.

    3.3   Individual Drivers are responsible for complying with the maintenance and testing schedule.

    3.4   The Vehicle Maintenance Supervisor is responsible for developing and notifying Drivers of the maintenance schedule. The Vehicle Maintenance Supervisor coordinates with the Environmental Manager to perform quarterly quality assurance tests on the emissions testing equipment by the Quality Assurance Unit, Environmental Management Department.

    3.5   The Vehicle Maintenance Technicians are responsible for performing needed maintenance and repairs to optimize vehicle performance and to minimize air pollution emissions.

4.0   Procedure

    4.1   Each vehicle is taken by the assigned Driver to the Vehicle Maintenance Section at the scheduled time. Drivers are assigned a relief vehicle for their shift by the Vehicle Maintenance Technician.

    4.2   The Vehicle Maintenance Technician logs each vehicle into the maintenance and vehicle emissions testing activites with the Vehicle Maintenance and Emissions Testing forms. These forms are generated on site by the Fleet Test software.

    4.3   The assigned Vehicle Maintenance Technician completes the needed routine maintenance and emissions testing. Each technician records the results of these activities onto the proper form.

(continued)

FIGURE 5–5.   (*Continued*)

> 4.4   The results of the maintenance and emissions testing activities are recorded by the Vehicle Maintenance Technician in the computerized maintenance and testing log, and the system is reset to flag the next testing period.
>
> 4.5   The Quality Assurance Unit performs quality control checks on the emissions testing equipment on a quarterly basis.
>
> 5.0   Related Procedures
>
> 5.1   Control of Vehicle Assignments, Procedure 12, Rev. 1
>
> 5.2   Compliance with Federal, State, and Local Environmental Regulations, Procedure 2, Rev. 0
>
> 5.3   Prevention of Pollution, Procedure 3, Rev. 1
>
> 5.4   Quality Assurance—Calibration of Vehicle Inspection and Maintenance Equipment, Procedure 13, Rev. 2
>
> 5.5   Record Retention, Procedure 4, Rev. 1
>
> 6.0   Documentation
>
> 6.1   Vehicle Maintenance Instructions and Form, Rev. 1
>
> 6.2   Vehicle Emissions Testing Instructions and Form, Rev. 0
>
> 6.3   Vehicle Log, Rev. 0
>
> 7.0   Record Keeping
>
> 7.1   The Vehicle Maintenance Technician adds the completed Vehicle Maintenance form and Vehicle Emissions Testing form to the Vehicle Log, which is maintained by the Vehicle Maintenance Supervisor. All records are retained for a period of 3 years by the Vehicle Maintenance Supervisor, according to the Record Retention Procedure.

number, revision number and date, authorizing name or initials, and the number of pages in the instruction. The process for developing work instructions mirrors the process for developing procedures (remember that work instructions should be developed by the individuals who do the work). Prepare a draft of the instruction that reflects the work as it is actually done, reach consensus among those affected, revise the draft as needed, issue the instruction, use it, and revise it as needed. Publish subsequent versions as Revision 0. Work instructions must be revised when practices change; additional revisions are numbered consecutively.

Examples of work instructions include regulatory standards that specify methods for collecting and analyzing groundwater samples at landfill monitoring sites; selecting, using, and maintaining personal protective equipment (PPE); or directions for logging samples into the laboratory. A work instruction can also be a form, blueprint, or process diagram.

An example of a work instruction for testing iron using a color comparison method is given in Figure 5–6.

FIGURE 5–6.   Example Work Instruction for Determining the Concentration of Iron in Water:

Laboratory Instruction No. 15
Analysis of Total Iron Using the Phenanthroline Method

Revision No.: 1.0   Issued by: D. Gray

Revision Date: 11/20/95   Page ___ of ___

I.   Required Equipment
    a.   graduated cylinder, 50 mL
    b.   Erlenmeyer flask, 125 mL
    c.   assorted pipettes
    d.   glass beads
    e.   Bunsen burner
    f.   tall form matched Nessler tubes, 100 mL
    g.   light comparator stand

II.   Reagents
    a.   concentrated hydrochloric acid, containing less than 0.00005% Fe
    b.   hydroxylamine reagent—dissolve 10 grams hydroxylamine hydrochloride ($NH_2OH \cdot HCl$) in 100 mL Fe-free distilled water.
    c.   ammonium acetate buffer solution—dissolve 250 grams of ammonium acetate ($NH_4C_2H_3O_2$) in 150 mL iron-free distilled water. Add 700 mL glacial acetic acid. (NOTE: prepare new reference standards with each preparation of buffer.)
    d.   Phenanthroline solution—dissolve 100 mg 1,10 phenanthroline monohydrate in 100 mL iron-free distilled water. Stir and heat to 80°C to speed dissolution, but *do not boil*. If the solution darkens when heated, discard it in the chemical waste collection receptacle marked, *Toxic Organic Wastes.*
    e.   Ten standard iron solutions prepared from stock iron solution (1.00 mL=200 µg Fe) in the appropriate concentrations for the levels being measured.

III.   Analytical Procedure
    a.   Wash all glassware with concentrated HCl and triple rinse with iron-free distilled water.
    b.   Mix the sample and measure 50 mL into the graduated cylinder.
    c.   Pour the 50 mL of sample from the graduated cylinder into the Erlenmeyer flask.
    d.   Add 2 mL concentrated HCl, 1 mL hydroxylamine hydrochloride, and a few glass beads to the flask.
    e.   Using a Bunsen burner, heat the sample solution to boiling and boil until 15–20 mL of solution remains in the flask.
    f.   Cool to room temperature and transfer to a 100 mL tall form Nessler tube.
    g.   Add 10 mL acetate buffer solution, 4 mL phenanthroline solution, and dilute to the mark on the Nessler tube with iron-free distilled water.
    h.   Mix thoroughly and allow 10–15 minutes for color development.

(continued)

FIGURE 5–6.   (*Continued*)

> i.   Compare the color against a series of ten standards (ranging from 1 to 100 µg Fe in the final 100 µL volume) that have been carried through the same procedure. Look vertically through the Nessler tube while it is held against an illuminated light comparator. Color standards are stable for up to 6 months if they are protected from evaporation.
>
> IV.   Calculations
>
> mg Fe/L sample=[µg Fe (in 100 mL final volume)]/mL sample

RECORDS

Records are documents that are produced throughout the day-to-day operation of the EMS. Records are important because they validate the performance of the EMS. In many instances, records are submitted to regulatory agencies to verify compliance with permits and regulatory standards. Examples include records of training, equipment calibration, complaint responses, malfunctions, and various reports. Records are typically stored and maintained by the departments that generate them, but some may be centralized in other locations for legal or internal purposes.

Records, whether written or computerized, must be accurate, accessible, and usable. It is important to update the contents of records when work practices change. Certain records must be secure to minimize the possibility of tampering or loss. Satisfying regulatory compliance requirements often requires the use of a standard format or prescribed forms for records and record keeping. All of these requirements must be considered when developing individual records and record-keeping systems.

Records should have identifying information, including a title, record number, revision level, and number of pages in the record. The record should also include an authorizing name or initials. Although individual records will differ, typical information would include the name or initial of the person who records the information; date and, perhaps, time; equipment name, model and serial numbers, if appropriate; and the required record. The process for developing records is the same as for procedures and work instructions. Figure 5–7 shows an example of a record used to track employee training.

TIPS FOR WRITING

Experience shows that documentation that does not reflect reality will not be used. For this reason, it is important for documents to be developed by those individuals who are responsible for the activities to be performed. This means that the person responsible for testing process water should be the person who writes the work instruction. If many individuals are involved in a procedure or if several work with the same instruction or record, one person can take the lead in

FIGURE 5–7.   Example of a Training Record:

XYZ Company
Training Record

Employee Name _____

Employee Number _____

Department _____

Course Name _____

Course Number _____

Date(s) of Training _____

Hours of Training _____

Location of Training _____

_____

Name of Instructor _____

Instructor's Affiliation _____

Exam Results _____

Results of Other Learning Evaluations _____

_____

_____

Instructor's Comments _____

_____

_____

Instructor:	Instructor's Signature	Date
Approved By:	Authorized Person	Date

Rev. 3 (5/12/90)

RKB

writing a draft, but all who are involved should review the document for accuracy and completeness. All documents must be modified when practices change. The consensus process is a useful tool for reaching agreement on documentation (see Figure 4–1 for a flowchart of the process). After revisions are made to a procedure, work instruction, or record, allow the revised document to be used for a while (for example, 30 days) before finalizing it. This provides enough time to determine whether the changes are adequate or whether additional work is needed before issuing the document.

The language for all documents should be clear and concise. Avoid flowery, bureaucratic, or technical jargon. Use the present tense rather the past or future tense. Use the third person rather than the first or second person. For example,

Preferred:

Monitoring equipment is calibrated by the environmental specialist on a periodic basis.

Not Recommended:

Monitoring equipment was calibrated by the environmental specialist on a periodic basis.

or

I will calibrate the monitoring equipment on a periodic basis.

Preferred:

Members of the Haz Mat Team receive training initially upon assignment to the team and on a periodic basis.

Not Recommended:

Members of the Haz Mat Team received training initially upon assignment to the team and on a periodic basis.

or

Jim Parks and Mary Smith were trained when they started working with us, and they will receive training every 6 months hereafter.

ORGANIZING THE DOCUMENTS

The ISO 14001 standard requires documents to be established and maintained, but it does not specify a particular organizational scheme. To some extent, the nature of the documents will dictate how they might be organized. Most documents can be collated into manuals, but drawings and paper format engineering diagrams might require maintenance in file drawers. A carefully constructed manual (or manuals) has several advantages. Primary documents can be organized and stored efficiently, and manuals provide an easy way to document related information, such as other procedures or records.

One method of organizing the documents is to group all the operational policies together and organize related targets, objectives, and procedures in a similar fashion. The manual would have separate sections for each—Policies, Objectives, Targets, and Procedures. Other manuals could be collated for work instructions and records. Larger companies are the most likely to use this system as they may have many policies and procedures, but it will work for anyone.

Another method that will work for most organizations is the documentation model developed for the ISO 9000 quality standards. Following this model makes sense if the organization already uses this documentation for a quality program. If the organization does not have a documentation system, the ISO 9000 model provides a clear organization that is easy to follow and implement. This model, which is discussed below, organizes the documents into four levels of documentation:

- Environmental Management Policy Manual
- Procedures Manual
- Work Instructions (Manual)
- Records (Manual)

THE ENVIRONMENTAL MANAGEMENT POLICY MANUAL

The Environmental Management Policy Manual (or any other name) is a document that provides an outline of the EMS and provides the framework for the remaining levels of documentation. The manual should present the organization's overall philosophy and goals for managing its environmental programs. The manual should be general; specifics are given in the operating procedures, work instructions, and records, which can, in turn, be organized into other documents. Excessive detail should be avoided so that the EMS manual does not require frequent updating. The manual should be circulated throughout the organization and made available at strategic locations.

One way of organizing the EMS manual is shown in Figure 5–8. The manual includes a table of contents, authorization page, environmental management policy, elements of the environmental management system, and the relationship of procedures to the standard. The authorization page establishes the applicability of the manual and the relationship of the manual to other documents in the organization. It also establishes document control by assigning a manual number and identifying the version (revision number) of the manual. The authorization page is a demonstration of the organization's support for the EMS, and it should be signed by the highest level of management, preferably the CEO. Figure 5–9 shows an example of an authorization page.

The authorization page is followed by the environmental management policy which should contain a statement stating that the policy has been disseminated to all employees within the organization. The policy statement is followed by a brief section that highlights the responsibilities and authorities of key positions with responsibility for the EMS. This section could be augmented by reference to an organizational chart, providing more detail on responsibilities and roles. However, it may be preferable to locate an organizational chart in a separate manual with procedures to minimize the need for updating the policy manual.

FIGURE 5–8.   Example Table of Contents for the Environmental Management
Policy Manual:

Table of Contents

1. Title and Authorization Page
2. Environmental Management Policy
3. Organizational Responsibilities for the EMS
4. Environmental Management System Elements
5. Relationship Between the Procedures Manual and ISO 14001

FIGURE 5–9.   Example of an Authorization Page:

ABC Company
Environmental Management System Manual

This Environmental Management System Manual is a statement of the organization, responsibility, procedures, and controls that have been implemented to conform with the requirements of the ISO 14004 standard.

The requirements of this manual apply to all personnel given responsibility for the implementation of the Environmental Management System and all the procedures given or referenced in this manual.

The contents of this manual are authorized and approved by the Chief Executive of the organization. The relationship between this manual, the ISO 14001 standard, and the Company Procedure Manual is given in Appendix A.

This manual is the property of ABC Company. The manual must not be copied in whole or part without written authorization. It must be returned to the Company when requested by the Quality Assurance Manager.

Katherine Culver
_____
Chief Executive Officer
January 1, 1994

Manual No. 16
Revision Date: April 1, 1996
Issued To: _____ *Paul Gilson* _____

Next, the elements of the EMS are summarized. This section includes brief statements about each of the elements of the management system. This material can be organized clearly and efficiently by including the numerical citation for each section of the standard and a brief description of the organization's policy related to that requirement. The manual concludes with a statement or matrix that shows how the procedures and the standard relate to one another. Alternatively, the section that briefly describes the elements of the EMS could contain a reference to the sections of the standard addressed by each element.

 Appendix B contains an example of an environmental management system quality manual. Because each organization differs, the manual is intended to serve only as a guide. This example includes supporting policies and shows how they might be incorporated into the manual.

THE PROCEDURES MANUAL

Individual procedures can be collated into a manual, according to procedure number or types of processes. The manual should be controlled and issued to those departments (and key staff positions) that are responsible for the documented activities. Controlling the manual will ensure that all personnel are working from the same basis and that no one is out of date. Issuing complete manuals rather than partial manuals is recommended because this simplifies the process of updating the manual. However, if the manual is extensive, it will probably be more efficient to issue a portion of the manual or a single procedure.

The manual should be organized into a logical format that is easy to use. Loose-leaf binders provide an easy way to update paper formats. The pages should not be numbered sequentially from start to finish because of the difficulty of renumbering pages when revisions are made to individual procedures. Replacing a section after revisions can be handled easily by numbering each procedure or section as a unit—for example, 1.1, 1.2, 1.3, 2.1, 2.11, 2.12, and so forth.

The manual should have a table of contents that includes a title/authorization page, organizational chart, and individual procedures. Additional sections that might be useful include lists of responsibilities by job title or position, a summary of records, and a summary of documents (see Figure 5–10 for an example table of contents).

The organizational chart identifies the roles within the organization. It is better to use position titles rather than the names of individuals because of personnel changes. Using titles will likely result in fewer revisions of the chart and make updating and controlling the manuals easier. Job titles in the organizational chart should match those in the procedures. Changes in the structure of the organization must trigger an update of the organizational chart.

The list of environmental management system responsibilities expands the organizational chart to identify the job title or position that is responsible for the procedure and a brief description of their overall responsibilities. Under each position, the list would contain a brief description of that position's responsibilities. All positions would be included. This list provides a quick reference of position responsibilities, and it allows information to be retrieved from the individual procedures very easily. The list should also identify back-up or substitute personnel, if any, and the person to whom the position reports. Figure 5–11 provides an example.

The procedure manual could also include a list that summarizes the organization's policies about the custody of records. This list should identify who has responsibility for the organization's records and the location and residence times for storing or archiving the records. Figure 5–12 provides an example format that could be used to establish this documentation.

FIGURE 5–10. Example Table of Contents for a Procedures Manual:

Table of Contents

1. Title/Authorization Page
2. Organizational Chart
3. Environmental Management System Responsibilities
4. Summary of Environmental Records
5. Summary of Documents
6. Procedures

FIGURE 5–11. Example of Documentation for EMS Responsibilities by Job Title or Position:

Title: Responsibilities for EMS	Page 1 of 2
Rev. No.: 1	January 1, 1996

Approved by: JD
Position Title: Quality Assurance (QA) Manager
Reports To: Vice President for Operations
General Responsibility:
  To assure that collected data are accurate, precise, and of high quality.

Procedure No. 5: Document Control
  The QA Manager has overall responsibility for document control. The QA Manager is responsible for coordinating document control activities with individual program managers.
  All employees are responsible for advising the QA Manager of any changes to procedures. The QA Manager is responsible for the contents and for authorizing all manuals, procedures, and work instructions.

Procedure No. 10: Calibration of Environmental Monitoring Equipment
  The QA Manager and the Environmental Surveillance (ES) Manager have overall responsibility for the calibration of environmental monitoring equipment. The QA manager is responsible for coordinating with the Environmental Surveillance Manager to train the Environmental Technicians in the routine calibration of equipment. The QA manager is responsible for QC checks on the calibrated equipment and for notifying the ES Manager of the results of the audits.
  (The QA Manager's responsibilities for other procedures would continue to be listed. Other positions would be handled in a similar manner.)

FIGURE 5–12. Example of Documentation for Controlling Records:

Title: Records Summary	Page 1 of 1
Revision No. 1	January 26, 1996

Approved by: TD

The following is a summary of policies related to organizational records.

Procedure No. 5:   Document Control
       The register of the Company manuals and procedures is kept permanently by the Quality Assurance Manager.
       The master copies of archived procedures are filed by the Quality Assurance Manager and retained for a minimum of 5 years.
Procedure No. 10: Calibration of Environmental Monitoring Equipment
       All calibration data remain in the calibration log that accompanies the equipment. These data are retained by the Quality Assurance Manager for the life of the equipment or a period of 3 years, whichever is longer.
       (The document listing would continue for each procedure.)

    Figure 5–13 shows an example of a document summary. This is a list of all documents that are required by each procedure and identifies the revision level for the document.

FIGURE 5–13.   Example of a Document Summary:

Title: Document Summary   Page 1 of 3
Revision No. 1   01/01/96
Approved by: JH

The document summary provides a list of documents that are required by each procedure and
identifies any revisions.
Procedure No. 5: Document Control
  Register of Manuals   Revision No. 1
  Chain-of-Custody Letter   Revision No. 1
Procedure No. 10: Calibration of Environmental Monitoring Equipment
  Equipment Calibration Log   Revision No. 2
  ABC Monitor Calibration Form   Revision No. 1
    (Additional documents would be listed for each procedure.)

WORK INSTRUCTIONS

Work instructions must be available in the area where the work is done. Although work instructions have limited distribution, they still require control. Collating work instructions into a manual helps keep these materials organized and readily available, but it may be appropriate to issue individual instructions rather than an entire manual. Depending on the number and type of instructions, several manuals or other storage devices may be needed. As work instructions can be lengthy, this is an area where computerization can be particularly helpful. A table of contents for work instruction manuals should include a title/authorization page, a summary of work instructions, and the individual work instructions.

RECORDS

Records can be issued and filed individually or a manual can be used to organize manuals. In practice, several manuals may be needed; for example, training, equipment calibration, complaint responses, malfunctions, equipment maintenance, and so forth. These records are typically kept by the departments that generate them, but some may be centralized in other locations if required or allowed by environmental regulations. A table of contents for a records manual should include a title/authorization page, a summary of records, and the individual records.

DOCUMENT CONTROL

To be useful, a document must be current, available, accessible, understandable, and easily maintained. The standard (ISO 14001, Subsection 4.4.5) requires the organization to establish procedures for controlling all documents required by the standard to ensure that:

- documents can be located;
- documents are periodically reviewed, revised as necessary, and approved for adequacy by authorized personnel;
- documents are available in current versions at all essential locations;
- obsolete documents are taken out of circulation promptly; and
- obsolete documents that are maintained for a legal or historical record are suitably identified.

Document control is achieved by identifying and controlling individual documents and collated manuals. As stated previously, it does not matter if documents are kept in electronic or paper formats. Either format requires a system for tracking and updating copies of policies, procedures, and work instructions. There are many options, but the basic requirements include: 1) the ability to identify and track all copies that have been distributed, 2) a structured system for recalling documents and reissuing them, and 3) a document format that facilitates revisions and replacement of pages (for example, document number, title, revision level and date, issue date, and page number according to section). Each individual should sign for the documents they receive. Refer to examples of documents in this chapter and Appendixes B and C for examples of document control information.

If individual documents are collated into a larger document or manual, it should have a table of contents that identifies the individual documents in the manual by revision level and an authorization page that identifies the issue date and number, and manual number. The manuals are then issued to specific individuals. The individual is required to sign for the manual that he or she is issued. It is the supervisors' responsibility to control documents in their areas; that is, they must be sure that the documents are complete, current, properly authorized, and properly maintained.

At first glance, this tracking system may seem to be excessive, but it allows outdated documents to be recalled and new issues to be released with minimal effort. The authorization page typically states the purpose of the manual and cautions against copying without permission, and it requires the manual to be returned when recalled.

SUMMARY

Documentation is a critical component of the EMS. Documentation can be in paper or electronic form, and it can be organized in any way that is logical, complete, easy to access, and controlled. Borrowing from the ISO 9000 experience, documents can be organized into records, work instructions, procedures, and policies. A concise overview of the EMS can be documented in an environmental management policy manual that contains the environmental policy, supporting policies (if any), and references to supporting documentation (procedures, work instructions, and records).

ENDNOTES

[1] W. Lee Kuhre, *ISO 14001 Certification* (Upper Saddle River, NJ: Prentice-Hall PTR, 1995). This book is a useful reference for operational procedures and contains a disk of templates.

[2] Information about Folio Views® can be obtained from Folio Corporation, 5072 North, 300 West, Provo, UT 84604. Adobe System, Inc. (345 Park Avenue, San Jose, CA 95110-2704) produces Adobe Acrobat®. An intranet can also be created using an HTML Editor.

[3] Some useful references for establishing ISO 9000 quality programs and documentation include: Patterson, James G. *ISO 9000: Worldwide Quality Standard* (Menlo Park, CA: Crisp Publications, Inc., 1995). Arnold, Kenneth L., *The Manager's Guide to ISO 9000* (New York, NY.: Macmillan, Inc., The Free Press). Owen, Bryan, Tom Cothran, and Peter Malkovich, *Achieving ISO 9000 Registration* (Knoxville, TN: SPC Press, Inc., 1994).

CHAPTER 6

IMPLEMENTATION— PERSONNEL, COMMUNICATIONS, AND DOCUMENTATION

INTRODUCTION

The previous chapters focused on the elements needed to plan the EMS: development of vision, mission, and values; identification of legal and other requirements; objectives and targets; process analysis; and formulation of policies, procedures, and work instructions. The next phase is implementation that integrates the planning activities into a cohesive program that will lead to the accomplishment of the organization's policies, objectives, and targets.

The time needed to implement the EMS depends on how much of the required documentation and support mechanisms are in place. An organization in the early stages of developing an EMS and/or constrained by resources can divide implementation into several steps. Prioritize the organization's needs and concentrate the initial efforts on those activities that will provide the greatest improvement in environmental performance. For many organizations in the initial stages of the EMS process, this may translate into complying with legal requirements. Stepwise implementation will help produce a smooth transition and minimize short-term resource demands. The ISO 14001 standard encourages the organization to focus on the alignment and integration of people, systems, strategy, resources, and structure. Appendix A (Appx A-2.c) contains a questionnaire that can help evaluate the implementation status of existing activities.

ENSURING CAPABILITY

Achieving environmental excellence requires money, human resources, and good management. In the past, many organizations underfunded and segregated their environmental programs from other functions because they were unaware of the benefits of integrated systems and because they believed "green" programs would make them less competitive. As noted earlier, more companies are aware that an effective EMS can result in significant benefits and opportunities for cost savings from reformulated products using less hazardous materials, increased worker health and safety, decreased liability, energy efficiency, waste minimization, and numerous other activities.

Ensuring capability means that an organization provides the human and financial resources to implement its EMS. To ensure capability, the standard (ISO 14001, Subsection 4.4.1) requires the organization to define, document, and communicate roles, responsibilities, and authorities, and to provide adequate resources to implement and control the EMS just as it would for any other management program (Appendix C contains an example procedure). It does not, however, specify the number or type of employees required or other resources for implementing the program. A system for allocating resources is especially important because without adequate resources, even excellent plans will fail. Other areas emphasized by the standard are accountability and responsibility; competence, training, and awareness; and integration and alignment.

ACCOUNTABILITY AND RESPONSIBILITY

To achieve the established targets and objectives, the standard requires the organization's top management to appoint a specific management representative who is accountable for the EMS. In smaller companies, one person may be designated; larger companies may assign this responsibility to several individuals. The standard allows the designated management representative(s) to have other duties, but it specifically assigns (ISO 14001, Subsection 4.4.1) overall responsibility for the EMS to this position(s), irrespective of other responsibilities. The assigned responsibility includes providing top management with reports on the performance of the EMS.

The standard provides flexibility, but the responsibility and accountability for the EMS must reside in the management level of the organization. The designated management representative could be the vice president of Environmental Programs, Quality Manager, Director of Engineering, Hazardous Materials Manager, owner of the company, or an environmental committee of managers representing different departments in the organization. The size and nature of each organization will determine the best way to handle this responsibility.

Designating a high-level, senior manager makes sense for several reasons. A senior manager will have the authority to ensure that the EMS is followed and will have access to needed resources and enough clout to be taken seriously. It is also important for the management representative to have the appropriate competence to evaluate the success of the EMS. However, this competence does not have to be in a technical area; good managers have the skills to utilize the expertise of technical professionals to arrive at sound decisions.

Other roles and responsibilities for the EMS must be clearly defined. There should be an organizational chart that clearly illustrates the chain of responsibility. Whenever the organization experiences change as a consequence of reengineering or downsizing projects, the chain of responsibility must be redefined. All personnel, from the production line to the corporate offices, must understand their role in implementing the EMS and must know they are accountable for their successes and failures. Figure 6–1 provides examples of typical environmental responsibilities and those who might be responsible for accomplishing the listed tasks.

RESOURCE ALLOCATION

It is management's responsibility to provide the human, technical, and financial resources needed to implement and control the EMS. However, this does not mean that management must do everything at once—a phased-in approach is acceptable. Deciding what resources to allocate requires having a reliable way to assess the costs and benefits of implementing the various components of the EMS. It is important for an organization to attempt to evaluate all of the costs associated with a product or activity.

Traditional cost accounting typically measures only the internal costs—what comes in and what goes out. For example, these costs include capital expenditures (buildings, equipment, utility connections, equipment installations, project engineering) and operating and maintenance expenses or revenue (raw materials, labor, waste disposal, utilities, water, sewage, energy, value of any recovered materials). The total or true cost, however, also includes indirect costs that are typically factored into overhead costs (for example, compliance costs, insurance, on-site waste management, pollution controls). Liability costs (penalties, personal injury and property damage, responding to emergencies) and intangible benefits (increased revenue from quality product, better company image, better worker productivity) are difficult to calculate, but they should be considered in the evaluation of the costs of developing an environmental management program, even if only in a qualitative sense. Another cost that is typically not measured in traditional cost accounting is the cost of lost opportunity associated with not produc-

FIGURE 6–1. Examples of Environmental Roles and Responsibilities:

Environmental Responsibility	Positions with Responsibility
Approve vision, mission, values, and policy	Vice President for Environmental Management; CEO, President
Develop operational/process-specific policies and procedures	Manager, process supervisor
Develop objectives and targets	Manager, supervisor, appropriate staff
Environmental monitoring and measurement	Quality Assurance Manager, Quality Control Officer, environmental staff
Regulatory compliance	Vice President of Environmental Management, operations/process manager
Ensure EMS compliance	Vice President of Environmental Management, all managers
Evaluate environmental aspects of business	Buyers, contract officer, finance manager, accounting officer
Total cost accounting	Manager, financial staff
Continual improvement	All management and staff

ing as a consequence of malfunctioning equipment, lack of available parts, spills, or other unplanned and uncontrolled events.[1]

Figure 6–2 shows an example of a case study that compares traditional costing to a total cost assessment for a white water/fiber reuse project at a coated fine paper mill. The project would permit fiber, filler, and water reuse on two paper machines at all times. The potential benefits included raw material and water use conservation, reductions in wastewater generation, and energy use. The energy savings, in particular, are substantially beneficial. This case did not involve hazardous materials or waste, so future liability costs were not calculated. Quantitative estimates of less tangible benefits, such as lower incidence of worker illness and injuries, were not included because the data were not adequate, but these benefits were addressed in a qualitative statement. Although the assessment in Figure 6–2 is not comprehensive, it does show the potential that the total cost accounting approach has for more clearly identifying financial indicators related to environmental projects, activities, and products. In this project, the internal rate of return increased from 21 to 48%, and the simple payback of 4.2 years decreased to 1.6 years. By excluding the savings associated with freshwater pumping, treatment, and heating and wastewater pumping, the company's traditional cost accounting makes the project appear substantially less profitable than it is. It should be noted that total cost accounting can also identify costs that make a project less attractive. In either case, a truer picture of costs and benefits that enables better decision making emerges.

One key area that an implementation plan should cover is resource allocation for new projects. It is much more efficient and cost-effective to design new projects in an environmentally sound manner than to correct problems after the project is completed. Environmental considerations such as the prevention of pollution and

FIGURE 6–2.   Example of a Traditional Cost Accounting and Total Cost Assessment for a White Water and Fiber Reuse Project:

	Traditional Cost Accounting	Total Cost Assessment
Total Capital Costs	$1,469,404	$1,469,404
Annual Savings (BIT)[a]	$ 350,670	$ 911,240
Financial Indicators		
Net Present Value, Years 1–10	$ 47,696	$2,073,607
Net Present Value, Years 1–15	$ 360,301	$2,851,834
Internal Rate of Return, Years 1–10	17%	46%
Internal Rate of Return, Years 1–15	21%	48%
Simple Payback	4.2 years	1.6 years

[a]Annual operating cash flow before interest and taxes.
Source: Allen L. White, D. Savage, and M. Becker. *Revised Executive Summary. Total Cost Assessment: Accelerating Industrial Pollution Prevention Through Innovative Project Financial Analysis with Applications to the Pulp and Paper Industry* (Washington, DC: U.S. Environmental Protection Agency, 1993).

resource conservation are often omitted from the decision-making process for capital projects. This is partly because environmental benefits can be difficult to assess in these projects. The guidance document (ISO 14004, Subsection 4.3.3.3) recommends the inclusion of these projects in the implementation of the EMS. This means that the organization should have policy, procedures, and controls for reducing environmental impact of capital projects. Some areas of consideration are:

- How are designs evaluated to ensure that the project will not cause undue environmental impact? Is there a standardized procedure for incorporating these considerations into the design process?
- How is return on investment calculated? Which of the following factors are considered—potential liability, penalties and fines, reduced staff time for monitoring, reduced cleanup costs, reduced disposal costs, reduced paperwork, reduced government interaction, improved community relations?
- Are other cost factors evaluated? Examples include: the cost benefits of prevention over control and cleanup; insurance savings from reduced liability, health and worker's compensation; potential problems with disclosure rules; better employee relations.

When these and other factors are evaluated, environmental considerations in project development often improve the financial attractiveness of a project. There will certainly be compromises along the way, but considering the total value of environmental aspects in product and product design will provide environmental and fiscal benefits.

Life cycle assessment (LCA) of products is a particularly useful tool for evaluating the material and energy impact associated with bringing a new product into production, but this tool can also be used for existing products. Figure 6–3 shows a model of life cycle stages, and Figure 6–4 shows a simplified LCA worksheet that could be used to identify environmental impact related to a product or activity. Life cycle assessment can be combined with cost accounting (Life Cycle Cost Assessment) to provide estimates of the total costs and benefits associated with a product or activity. Life cycle assessment has the potential of producing long-term cost savings for the manufacturer through the selection of less hazardous waste disposal and reduced liability.

The EPA has developed a method of total cost accounting for pollution prevention and waste minimization projects that takes into account usual costs (equipment, labor, and materials), hidden costs (compliance and permits), liability costs (penalties, fines, and future liabilities), and less tangible costs (consumer response and employee relations). The EPA documents[2] contain data collection sheets and profitability worksheets that range from relatively simple checklists to more complex evaluations. Appendix F lists helpful World Wide Web sites that can be accessed for more information.

OPPORTUNITIES FOR SMALL AND MEDIUM-SIZED BUSINESSES

One very positive strategy for small and medium-sized businesses is to work cooperatively with similar companies to reduce the total cost of implementing the EMS. In fact, cooperative strategies can often be developed with industries of any size if there is a common interest in sharing technology and expertise. Some areas

FIGURE 6–3.   Life Cycle Stages:

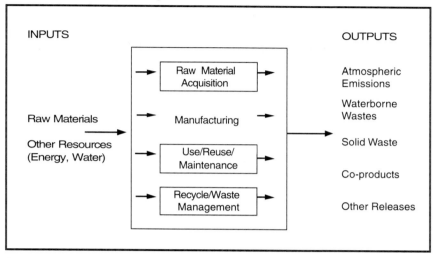

Source: U.S. Environmental Protection Agency, Office of Enforcement and Compliance Assurance, *Federal Facility Pollution Prevention Project Analysis: A Primer for Applying Life Cycle and Total Cost Assessment Concepts,* EPA 300-B-95-008 (Washington, D.C.: U.S. Environmental Protection Agency, 1995).

where industries can work together include identification of and planning for common issues, joint use of facilities, joint development of technology, and sharing of the cost of hiring consultants.

Significant help may be available from industry and professional associations, government agencies, chambers of commerce, or other groups. They often can provide resource materials and training, and can identify issues of concern. They may also help provide potential development partner contacts. Some government-sponsored organizations may provide on-site consultations for pollution prevention, worker health and safety, and management/financial issues. Attending or cosponsoring seminars and conferences can also provide helpful information that smaller businesses need to develop and implement the EMS.

Also, do not overlook the help that may be provided by local colleges and universities. Many universities have executive education programs and business development centers that work in partnership with the private sector. Most universities require faculty members to perform public service, which means that it may be possible to arrange free or low-cost consulting for various aspects of the EMS. More intensive assistance may be possible through a sabbatical arrangement in which the faculty member works on a specific problem (for example, engineering controls, cost accounting, or new product development).

COMPETENCY, TRAINING, AND AWARENESS

It is nearly impossible to achieve environmental objectives and targets if managers and staff do not have the correct set of knowledge, skills, and training. The

FIGURE 6–4. Life Cycle Assessment Worksheet:

1. PROCESS STEPS

2. INPUTS

2a. Raw Materials (units)

2b. Energy Usage
    Electricity (kW-hr)
    Natural Gas (cubic ft.)
    Fuel (gallons)
    Other

2c. Water Usage (gallons)

2d. Other Inputs (units)

3. OUTPUTS

Products, Useful By-Products (item and amount)

Releases to the Air (including gaseous wastes)

Releases to the Water (including liquid wastes)

Solid Wastes

Source: U.S. Environmental Protection Agency, Office of Enforcement and Compliance Assurance, *Federal Facility Pollution Prevention Project Analysis: A Primer for Applying Life Cycle and Total Cost Assessment Concepts,* EPA 300-B-95-008 (Washington, D.C.: U.S. Environmental Protection Agency, 1995).

ISO 14001 standard (Subsection 4.4.2) requires the organization to identify training needs and ensure that all personnel are properly trained if their work could result in a significant environmental impact. Each organization must also establish and maintain a program to ensure that requirements of the standard related to knowledge and competency are fulfilled at each relevant function and level of the organization. Appendix C contains an example training procedure.

COMPETENCY

Hiring competent individuals is an obvious good business practice that greatly minimizes the possibility of actions that would increase environmental impact and liability. Ensuring competency begins with a consideration of the type of operations that are performed at the facility and identification of the skills required for each position. Each position should be considered in terms of needed education, training, and/or experience. Written job descriptions should list the scope of the position, responsibilities, needed qualifications, and special skills. Job descriptions should be advertised in the appropriate newspapers, journals, or personnel agencies to maximize the pool of applicants. Candidates should be screened and interviewed using checklists to provide uniformity during the evaluation process. Interviews that involve problem-solving questions can give insight into the candidate's knowledge, judgment, and experience. A candidate's references and evidence of a criminal record (beyond traffic violations) should be checked prior to hiring. Although performing these checks seems obvious, there are many instances when organizations fail to conduct the proper background searches and are surprised when problems arise.

TRAINING[3]

The training program should begin with an initial orientation to the organization and its environmental policy, the EMS, and specific job duties and responsibilities. Refresher training should be given for specific job functions, as specified by legal requirements or at a level needed to control the EMS. All training programs should be relevant to the achievement of environmental policies, objectives, and targets. Job-specific training must include a consideration of significant environmental impacts. Training programs can be strengthened by incorporating environmental, health, and safety topics into basic job training, rather than teaching these topics as separate units. This increases employee awareness about interrelationships and provides another way to integrate the EMS with other aspects of the organization.

In-house training programs can be supplemented by offering continuing education opportunities for employees. Continuing education opportunities allow employees to sharpen their skills and to enrich their professional development, and they can lead to higher levels of awareness and motivation. It improves the organization by increasing overall competence and enables it to retain and develop employees for higher-level positions within the organization.

One area of the training program that is easy to overlook is the quality of training staff. Trainers must be competent in their area(s) of training, but they must also be able to convey information in a manner that is organized, easy to follow, and pitched at the correct level. The pace should vary, and audiovisual materials should be clear and relevant. There are significant differences in the way individuals learn new information. For example, a concise lecture will be adequate for some people; others require visual reinforcement or hands-on experiences. A good trainer recognizes that differences exist and provides training that includes a variety of techniques, including small group breakout discussions and problem solving, demonstrations, case studies based on real situations, and role-playing.

Typical training programs will have at least the following elements:

Needs Analysis. A needs analysis must focus on individual operations. Some questions to consider are: What technical and environmental training or certifications are required for each job? What environmental training is needed for nonenvironmental positions? What training is needed for senior managers? How often is it needed?

Training Implementation Plan. How will the initial and refresher training be provided? Can in-house staff conduct the training or should contractors be used? Are training programs of the required type being taught elsewhere in the community? Can the development and presentation costs be shared with other businesses? Are trainers competent and effective? How will learning be evaluated?

Training Documentation. Training records must be established and maintained for each employee. Computerized training records are not needed, but they offer easy tracking capabilities and are readily available for specific job functions.

Training Program Evaluation. The success of training can be measured by good evaluation techniques, and evaluation is legally required for many programs. Some questions to ask include: Is training reaching the right people? Is training meeting the needs of the individual and allowing the organization to conform to the EMS?

This portion of EMS implementation is critical to success. No environmental program is likely to be successful without a thorough and well-designed training program that is integrated into other types of job training. When money is tight, it is tempting to let training slide, but this is seldom a prudent course of action. A highly skilled, aware, and motivated workforce is more productive and less likely to make costly mistakes. The organization's training program is successfully implemented if employees are aware of:

- the importance of following the environmental policy, procedures, and requirements of the EMS;
- their roles and responsibilities in conforming with the environmental policy, procedures, and EMS;
- their roles and responsibilities in emergency preparedness and response activities;
- the significant environmental impacts, actual or potential, of their work activities;
- the environmental benefits of improved performance; and
- the consequences of not conforming to established policies and procedures.

ENVIRONMENTAL AWARENESS AND MOTIVATION

Companies with successful EMS programs are characterized by employees at all levels who have a high level of environmental awareness and are motivated to achieve the environmental goals of the organization. Promoting these virtues within the organization is not difficult but it requires an integrated, continual effort. A single program on environmental awareness is not likely to produce desired results over the long term. Hiring competent individuals, providing them with the right job training, and involving them in decision making are actions that contribute to employees who are aware and motivated. Other features of a good employee awareness and motivation program include:

- Training programs to explain and discuss the corporate environmental values and the commitment of senior management to these values.
- Communication that helps develop a set of shared environmental values and commitments. This should begin in employee orientation, and it should be reinforced by supervisors.
- Communication that reinforces the importance of achieving environmental objectives and targets. Explain why it makes good business sense to achieve these objectives and what each person can do to help.
- A performance system that recognizes and rewards environmental excellence and continual improvement.

INTEGRATION AND ALIGNMENT

The ISO 14001 standard repeatedly stresses the need to integrate the EMS into the organization's overall management program because there is a much higher chance of success when the program is part of the business plan. An integrated EMS means that the entire organization incorporates environmental concerns into its decision making. For example, the personnel department appraises and rewards environmental performance; the accounting department may become involved in cost accounting or cost-benefit analysis; the purchasing department may screen prospective vendors and contractors on the basis of a prequalification environmental checklist, and so forth.

There may be a temptation to downgrade the importance of environmental management and the EMS when resources become an issue. Even during lean times, the EMS should be viewed as necessary for the well-being of the organization, not as an option. Decisions for allocating resources can be expedited by instituting a process for balancing the competing interests and priorities of the organization.

An example of competing interests is demonstrated by a case in which a production piping system develops small leaks that release product or raw material to the environment or production area. There is potential for worker exposure and environmental damage, but production must be shut down to make repairs. Stephen Covey describes this type of dilemma as a decision of whether to maintain production or to improve production capacity (P/PC).[4] Not making the repairs results in costs for lost materials and potential worker's compensation or environmental cleanup. Production downtime costs money. How can this P/PC question be balanced? Although the answers are not always straightforward, a conflict resolution process for allocating resources can facilitate the enumeration of costs and benefits and can improve decision making within the organization.

Integration and alignment of the EMS can have other benefits, including:

- Clearer and stronger organizational policies. The initial assessment, policy development, and review and improvement activities will provide opportunities for closing gaps, eliminating duplicate efforts, and correcting existing weaknesses.
- More effective allocation of resources.
- More efficient operational control and process documentation.
- Improved information management systems among departments or facilities.
- Enhanced training and staff development that produces workers with higher skill levels who are more productive and have fewer accidents.
- Well-defined roles, responsibilities, and accountability that increase operational efficiency and facilitate continual improvement.
- Performance appraisal and reward systems that hold management accountable for environmental management should improve the entire organization.
- Improved documentation and record keeping throughout the organization.
- Improved communication and reporting.

Of course, it is much easier to say that the EMS must be integrated into other management systems than it is to do it. Other business systems may have been in place for some time, and employees may be comfortable with the status quo. They may resist changes, especially if they have been rewarded for performing well in the old system. These are all reasons why visible support for the EMS is essential from senior management and why a conflict resolution process is needed when environmental requirements are in conflict, or appear to be in conflict, with other business objectives.

SUPPORT ACTIVITIES

COMMUNICATION AND REPORTING

Many of the problems encountered in general business operations are related to communication failures, and these same problems often plague environmental programs. Effective communication provides the information interchange that is needed for the successful implementation of the EMS. The standard recognizes the importance of conveying information and requires the organization to establish and maintain procedures for communicating within the organization and with external interested parties (ISO 14001, Subsection 4.4.3). Appendix C contains an example procedure. Many types of environmental information can be shared, but the standard requires communication only about the organization's aspects and the EMS.

Significantly, the organization is not required to initiate communication with external stakeholders, but only to receive, document, and respond to relevant communications from interested parties. This requirement presumably leaves room for the organization to decide what and with whom to communicate. Further, the standard requires the organization to consider processes for communicating with external stakeholders about its significant environmental aspects and to document its decision on this issue. This section of the standard is considerably weaker than the external communication requirements expressed in the European Union's European Eco-Management Audit System (EMAS) regulation, which requires the organization to publish a verified public environmental statement. At the time the ISO standard was developed, proponents of stronger language believed that public exposure provided an incentive for improved performance, whereas opponents were concerned about disclosing sensitive information that could result in increased regulatory scrutiny or other unwanted attention.

Effective internal and external communications are based on the following key principles:

- All employees should recognize that communicating should be a two-way process.
- Information should be presented in an understandable manner. Avoid jargon, explain fully, be patient in answering questions, do not make people feel stupid or unimportant.
- Information presented as fact should be verifiable from other sources and not distorted to avoid unpleasant truth.
- An accurate picture of environmental performance should be presented. If there are initial problems, communicating those problems will often lead to useful suggestions on how to improve performance.
- Data should be presented in a consistent format so that performance figures can be compared from year to year.
- Structured mechanisms should be used to receive, evaluate, and respond to comments and inquiries for information.
- Do not be afraid to communicate. Informed stakeholders are more likely to respond positively to the organization's goals. Recognize that secrecy about environmental impacts is generally not a prudent course of action. Strike the appropriate balance between sharing information and maintaining privilege.

COMMUNICATING WITH EXTERNAL STAKEHOLDERS

The implementation of procedures for external communications differs slightly, depending on the stakeholder, but a basic procedure involves the following elements:

- Record the request. Needed information includes the name; sponsoring organization, if any; date; and nature of the inquiry.
- Route the request to the appropriate internal group for evaluation and response.
- Respond to the request.
- Document closure.

Media requests might involve a telephone call or request for a face-to-face interview on television or radio. All media requests should be sent immediately to the Public Relations Department or to a designated person in smaller organizations. Employees should not individually respond to media requests unless asked to do so. Spur-of-the-moment responses can result in the disclosure of information that is incomplete, incorrect, or sensitive. The training program should reinforce this policy and the reasons for it.

All media requests should receive a timely, but not rushed, response. If the answer cannot be given in a timely fashion, be sure to let the requestor know when they can expect the response. It is a good practice to have the technical expert and a representative from the public relations field requests together.

If a request is made for a television or radio interview, it is especially important to ask for questions to be submitted prior to the interview so that thoughtful and accurate responses can be prepared. Exert control over where the interview will be conducted. This is especially important during emergency situations. For example, do not allow an employee to be interviewed against the backdrop of a fire, chemical spill, or dead fish. Insist on a neutral background so the audience can focus on the factual information being presented rather than on provocative images. Insist on recording all interviews and request the ability to review responses to ensure accuracy and to prevent comments from being taken out of context. Ask for audible interviews to include complete responses to questions. If requests are not accommodated, evaluate the interview request and the potential for damaging the organization. Each situation must be evaluated separately. Sometimes it is better to offer a news release or printed statement rather than a face-to-face interview.

Keep all responses concise and factual. If the answer is unknown, say so. Do not speculate. Attempt to be as positive as circumstances allow. For example, if facility discharges result in a fish kill, acknowledge responsibility and damage, but reinforce the company's corrective actions, commitment to prevent recurrences, commitment to correct unknown but related problems, and any other positive aspects related to the situation.

Finally, attempt to maintain cordial relationships with the media. Temper outbursts, defensiveness, evasiveness, and other efforts to stonewall are likely to encourage closer scrutiny. Cultivating relationships before disaster strikes can help ensure fair and equitable treatment under trying circumstances. Whenever possible, provide positive news releases and accommodate interview requests. If an interview results in the dissemination of distorted or incorrect information, it is important to be assertive and to attempt to secure a retraction or correction.

Regulatory or legal requests typically are written and relate to permit applications or other regulatory requirements. These requests should be routed to the environmental, health, and safety department or to a regulatory affairs department. All legal and regulatory affairs requests should receive a timely response. If a mandated response time cannot be met, request the extension as soon as possible.

Customer requests should be routed to a designated group. Typically, this would be the environmental, health, and safety department or a regulatory affairs department. Requests from customers can vary significantly and may or may not require a response, depending on contractual stipulations and legal requirements. However, in most instances, responsiveness is the best policy to ensure customer satisfaction. If time delays or additional costs are expected, this information should be given to the customer.

Public information requests or comments can be handled on an individual basis or routed to a designated group. Depending on the size and nature of the organization, each method can offer advantages as long as follow-up and tracking occur. Requests or comments from the public may or may not require a response. Each should be evaluated, and the response (or lack of one) should be given to the requestor in a timely fashion.

COMMUNICATION OPTIONS

There are many options for communicating information with stakeholders. An effective tool for internal and external communications is a company newsletter. It can contain information about the EMS, a column from the CEO, a question and answer section, a forum for recognizing employee contributions, and numerous other topics.

Corporate reports, including an annual report, are also appropriate for all stakeholders. The corporate annual report is a good avenue for communicating the environmental achievements of the organization. It can include an organizational profile, the environmental policy statement, objectives and targets, a performance evaluation that includes both positive and negative performance, community outreach efforts, and other topics.

Information that is contained in government records, documents prepared for government agencies (such as permit applications), industry association publications, media coverage, or advertising can all be made available upon request. In many cases, legal and regulatory information is in the public domain, and it is probably in the best interest of the organization to help interested parties obtain it.

DOCUMENTATION

Chapter 5 provides a detailed discussion of how to organize, develop, and control documents to support the EMS, so this section will focus on the features of the documentation system that will fulfill the requirements of the standard.

Successful implementation of a documentation system means the organization has and maintains information that describes the core elements of the EMS and provides direction to related documentation (ISO 14001, Subsection 4.4.4). The documents can be in paper or electronic forms and organized in any fashion that facilitates their use. Organizing the documents into a manual(s) and maintaining a summary of the contents of the documentation simplifies tracking and produces a system that can easily demonstrate that the organization has implemented and maintained the EMS. Some of the major topics and one way of organizing them are:

- Environmental Management Policy Manual
  —organizational responsibilities for the EMS
  —environmental policy
  —supporting policies
  —objectives, targets
  —directions to related documentation

- Supporting documents
  —procedures
  —work instructions
  —records

All documents, regardless of the type or method of organization, should be clearly defined and easy to access and use, should support awareness of activities needed to meet objectives and targets, and should demonstrate that the EMS elements appropriate to the organization are implemented and controlled. EMS documents that are integrated into corporate documentation are another demonstration of support for the EMS and evidence that the EMS is an integral part of the organization's operations (for example, environmental policy training should be incorporated into job orientation and environmental procedures should be integrated into other departments as appropriate).

Document control is required by the standard (ISO 14001, Subsection 4.4.5) and is crucial to successful implementation. Procedures for developing, revising, and issuing documents must be established and maintained to control all documents required by the standard. If the procedures are adequate and implemented, the documents will be current, available where they are needed, and easy to locate. Outdated documents can have serious consequences and can increase adverse environmental impact. They must be removed or archived as obsolete if they are needed for historical or legal purposes. Individual documents must have identifying information, such as the organization, type of document, revision number and date, and contact person. Documents must be retained for the specified period of time.

An organization has implemented the documentation and document control requirements of the standard if:

- documentation required by the standard or needed to control the EMS is developed and maintained;
- documentation is organized and provides directions to related documents;
- documents demonstrate that the EMS elements appropriate to the organization are implemented and controlled; and
- documents are retained for specified periods of time.

As noted in Chapter 5, procedures are not needed for every operation or activity in the organization. The test is simple, ask: Is this procedure needed to meet the requirements established by the environmental policy, objectives, and targets? If the answer is yes, write and implement the procedure and related work instructions and records.

SUMMARY

Successful implementation of the EMS depends on having procedures in place to ensure that the organization has the capability and supporting activities to do the work. Ensuring capability means providing the human, capital, and financial resources needed to meet the organization's targets and objectives. All members of the organization should understand their roles and responsibilities and should be competent to accomplish their assigned tasks. Procedures must be in place to ensure that staff have the needed education and skills to perform their work. Continuing education and training are essential ingredients for developing a team that is capable, aware, and motivated. Communications and documentation are two supporting activities that are needed to help the organization do what it says it will do and demonstrate that it has done what it said it would do. Communications and documentation should involve everyone in the organization and outside interested parties.

Although there are many ways to implement these programs, success depends on identifying needs and a systematic approach to planning and implementation. Integration and alignment of all of these programs within the organization will help promote environmental awareness and maximize resources.

The next chapter continues the discussion of implementation of the EMS by providing an overview of program elements for suppliers and contractors, maintenance activities, waste minimization and emergency preparedness and response, management of chemicals, and property management. These areas are highlighted because they are operations/activities that can improve environmental management in any organization and because they are specifically required by the standard.

ENDNOTES

[1] Peter F. Drucker suggests these uncontrolled and unrecorded costs could be as high as traditional costs in some manufacturing plants. These costs can be captured in activity-based accounting. Peter F. Drucker, *Managing in a Time of Great Change* (New York, NY: Penguin Books U.S.A., Inc., Truman Talley Books/Dutton, 1995). Some managers use the term *activity-based accounting* to identify the actual and overhead costs, but not opportunity costs, associated with a product or activity.

[2] *Pollution Prevention Benefits Manual, Phase II,* vols. 1 and 2, EPA/230/R-89/100 (1989) and *Facility Pollution Prevention Guide,* EPA/600/R-92-088 (1992) are two useful documents with information on total cost accounting. These and other documents can be obtained from the EPA Pollution Prevention Information Clearinghouse [(202) 260-1023]. Examples of worksheets from the facility guide are given in Chapter 7.

[3] A detailed treatment of designing training programs can be found in: Sallie E. Gordon, *Systematic Training Program Design: maximizing effectiveness and minimizing liability* (Englewood Cliffs, NJ: Prentice-Hall, Inc., 1994).

[4] Stephen R. Covey, *The Seven Habits of Highly Effective People* (New York, NY: Fireside/Simon & Schuster, 1989), p. 59.

CHAPTER 7

OPERATIONAL CONTROL AND EMERGENCY PREPAREDNESS AND RESPONSE

INTRODUCTION

The implementation activities discussed in Chapter 6 support the processes that are associated with the significant environmental aspects identified through the planning process. The standard (Subsection 4.4.6) requires operational control of these processes to ensure that established objectives and targets are achieved. This chapter provides an overview of procedural and program requirements for procurement of goods and services, waste minimization, emergency preparedness and response, hazardous materials management, and property management and transfer. Chapter 6 discussed the use of resources in existing operations and for new capital projects, Part 4 will provide a discussion of daily management activities associated with sources of air, water, and land pollution.

The questionnaire in Appendix A (Appx A-2.Doc.) can help evaluate the status of implementation efforts related to operational control and emergency preparedness and response.

OPERATIONAL CONTROL

Operational control means that procedures, work instructions, and records (see Chapter 5) will be developed, implemented, and maintained for many different operations and activities that were identified during process analysis phase of planning (see Chapter 4). Examples of operations that might be included are: research and development, design, engineering, purchasing, contracting, material handling and storage, production, maintenance, laboratories, transportation, marketing, advertising, customer service, and property acquisition (an example operational control procedure is given in Appendix C). Activities within these operations that require documentation can be grouped as:

- prevention of pollution and resource conservation in new capital projects, process changes and resource management, property transfers and management, new products, and packaging;
- daily management activities; and
- strategic management activities.

As noted in Chapter 5, procedures are not needed for every operation or activity in the organization. The test is simple: Is the procedure needed to meet the requirements established by the environmental policy, objectives, and targets? If the answer is yes, write and implement the procedure and related work instructions and records.

PROCUREMENT OF GOODS AND SERVICES

If the organization uses suppliers, contractors, and subcontractors, consider implementing a procedure that monitors their environmental performance. Environmental performance should be a special consideration for goods and services directly related to the environment, but even nonenvironmental goods and services provide opportunities for environmental protection. Another, more practical reason for controlling suppliers and contractors is to reduce future liability from environmental impact caused by improper practices.

Procurement decisions provide another opportunity for integration of the EMS with other management systems in the organization. Expanding procurement decisions beyond the typical concerns over price, quality, and delivery will require additional training of procurement staff. A useful reference for understanding materials management, production planning, and inventory fundamentals is a book by Arnold.[1]

There are several ways to provide oversight for the purchase of goods and services, and it is likely that different degrees of control will be needed. The highest degree of control would be to require all suppliers and contractors to be ISO 14001- and ISO 9000-certified. This requirement, however, may not be practical and could eliminate many potential vendors. Another option is to audit all suppliers and contractors, but this could prove prohibitively expensive. A third option is to screen vendors to learn about their environmental and safety and health programs, and to make procurement decisions, based in part on their environmental performance.

Screening is affordable, easy to control, and provides a systematic tool for comparing proposals. Screening for environmental impact could range from a simple questionnaire about the vendors' policies and impacts to detailed questionnaires with attached permits and other documentation. Examples of vendors that require closer scrutiny include suppliers of hazardous chemicals; contractors who transport, store, or dispose of hazardous wastes; emergency response contractors; maintenance contractors who will handle hazardous materials such as asbestos; and so forth.

The control of suppliers and contractors follows the typical model for procurement of any good or service: 1) screen the qualifications of potential suppliers and contractors; 2) request proposals from qualified vendors; 3) review the submitted proposals; 4) award the contract; and 5) provide oversight for goods or services provided by the vendor.

Screening qualifications should include the experience and qualifications of personnel; history of compliance with applicable environmental, health, and safety laws and regulations; financial stability/resources; proof of insurance; and references. It might be helpful to request a table of contents of the prospective vendors' or contractors' environmental health and safety manuals to provide additional insight into their programs. It is especially important to validate the qualifications of personnel. Ask for proof of special certifications, such as asbestos removal, laboratory certifications, or proof of permits to conduct certain activities such as the storage, treatment, or disposal of hazardous wastes. Screening criteria can be evaluated on a simple "yes" or "no" basis, or points can be awarded and weighted for each category of performance. The important point is to treat each company in the same way. Depending on an organization's needs, it may be useful to develop a list of approved vendors for future reference. If so, measures should be in place to review and update their qualifications on a periodic basis.

After qualifications are screened, the organization can request proposals from qualified vendors and contractors. It is generally best to put the request for proposal (RFP) in writing and to ask for written responses. This must be done for complex or significant proposals. The RFP should be clearly written and should provide sufficient detail for the scope of work, responsibilities, special work practices, product specifications, and so forth. Copies of material safety data sheets (MSDSs) or product life cycle information might also be requested. Including an acceptance agreement with the RFP that provides a way of short-circuiting last minute surprises when the award is granted. Everyone should be given the same information; that is, the RFP, answers to questions, extensions of time, and site visits should be given to all potential vendors at the same time. The initial screening of proposals should be to determine whether they are complete, on time, and submitted in a specified format. The formal screening (and ranking) of the proposal should focus on the key elements of the request. A checklist provides an easy way to evaluate the proposals. Criteria should include responsiveness to the request, quality of the proposal, timeliness, and cost. Additional criteria might be helpful. In some cases, it might be appropriate to visit each vendor's facility to examine their environmental, health, and safety operations; or potential vendors can be invited to give a presentation of their proposals.

After the successful vendor is informed and accepts the award, the remaining bidders should be notified that they were not successful. A written contract should be executed *before* the work begins. The contract should be detailed, should delineate liabilities, and should specify penalties for nonconformance with the terms of the contract. After the contract is signed, the work can begin. Even at this stage, oversight is needed to ensure that all terms of the contract are met. Typically, the oversight responsibility is delegated to the environmental health and safety department for environmental contracts or to the appropriate department for other contracts. It is important for managers to receiving training for evaluating the environmental performance of contracts for which they have responsibility.

MAINTENANCE ACTIVITIES

All facilities require a maintenance program for a variety of equipment and structures. This is particularly important for any operations that produce or use hazardous materials. An effective maintenance program provides significant cost benefits to any organization. Preventive maintenance can help reduce operating costs and improve operational reliability by eliminating breakdowns and the need for corrective maintenance. It also reduces costs by increasing the useful life of equipment. Corrective maintenance returns malfunctioning equipment to operating condition. Both types of maintenance activities are needed to avoid or minimize possible regulatory compliance problems. Other areas that should be included in policies for maintenance are attaining and maintaining compliance, maintaining process controls, and quality control.

The maintenance program should include the following components: 1) training program, 2) planning and scheduling, 3) documentation system, 4) inventory control, and 5) maintenance cost control.

The operation of any facility requires well-trained staff to maintain process equipment, environmental controls, monitoring equipment, vehicles, buildings, and grounds. The inspection, repair, and preventive maintenance required for modern processes and equipment can be very complex and can require a variety of skills. Hiring competent individuals is the first step toward an effective maintenance program, but these incoming skills must be augmented with an ongoing and documented training program.

Planning and scheduling activities are essential for effective preventive and corrective maintenance programs. The maintenance supervisor is responsible for preparing work schedules. It is important for the work schedule to list job priorities, work assignments, available personnel, and timing for the completion of the work. Planning activities should also include developing a maintenance emergency plan.

A detailed documentation system is needed for any maintenance program. Records must be established and maintained to provide maintenance histories on equipment, to diagnose problems, and to avoid equipment failure. Other examples of maintenance records include as-built drawings, shop drawings, construction specifications, capital and equipment inventories, and maintenance costs. It is especially important for these documents to be controlled to prevent lapses in required maintenance or redundant work. Some examples of additional documentation include operating reports, work schedules, activity reports, performance reports, expenditure reports, cost analysis reports, emergency and complaint calls, and process control data.

An inventory of spare parts, equipment, and supplies must be maintained and controlled. The inventory should be based on recommendations from the manufacturer of the equipment and supplemented by experience gained through the operation and maintenance of the equipment. The inventory should be large enough to avoid downtime, but it should not contain an excess. The quality principle of "just-in-time" (JIT) delivery is helpful for inventory control.

The JIT philosophy is based on the principle of eliminating waste and that of continual improvement. *Waste* can be defined as any amount above the minimum needed to produce quality, and it refers to equipment, materials, parts, workers' time, and even space. Just-in-time techniques were originally developed for repetitive manufacturing, but many of the basic concepts can be used in other types of organizations. Buyers who work in a JIT environment must plan for inventory needs. Planning typically requires answers to four questions: What are we going to make? What is needed to make it? What do we have? What is needed? Just-in-time delivery requires a supplier who can provide the quality needed to reduce or eliminate inspection for incoming inventories and one who can deliver at the required frequency. Cultivating a long-term relationship with suppliers that is based on mutual trust and cooperation is a critical aspect of a JIT inventory system.[1]

A maintenance cost control system should be integrated into all programs. Budgets should be developed from past cost records and current operating and maintenance conditions. They are typically categorized into preventive maintenance, corrective maintenance, and projected and actual major repair requirements. The maintenance department, like other departments in the organization, should receive periodic feedback on cost performance.

Figure 7–1 at the end of this chapter provides an example of a preassessment questionnaire for evaluating operations and maintenance controls. Although this questionnaire was designed for a wastewater treatment facility, the questions are generic and could apply to many different types of operations.

PREVENTION OF POLLUTION AND WASTE MINIMIZATION[2]

A commitment to the prevention of pollution is required by the standard (ISO 14001, Section 4.2) (Appendix C contains an example procedure). The standard defines the *prevention of pollution* as the "use of processes, practices, materials or products that avoid, reduce or control pollution, which may include recycling, treatment, process changes, control mechanisms, efficient use of resources and material substitution." All pollution emissions into the air, land, or water can be the object of the prevention of pollution and waste-minimization programs. There are many reasons that the prevention of pollution and waste minimization activities should be integrated into operational programs and the EMS. These activities can:

- save money by reducing waste treatment and disposal costs, raw material purchases, and other operating costs;
- reduce potential environmental liabilities;
- protect the environment, public health, and worker health and safety; and
- enhance corporate image.

Source reduction and recycling (Figure 7–2) provide two important opportunities for making significant reductions in emissions and waste. From an environmental perspective, source reduction is preferable to recycling because it elim-

FIGURE 7–2.   Pollution Prevention and Waste Minimization Techniques:

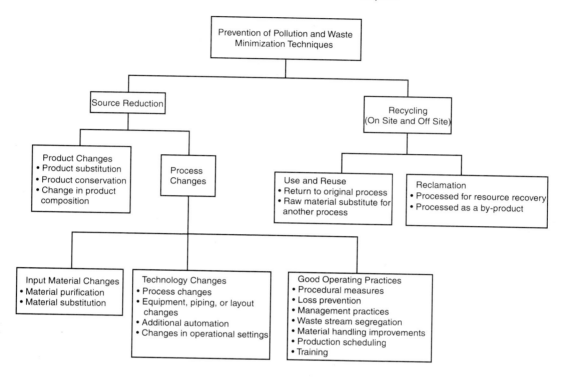

inates pollution and waste. Source reduction should be the starting point of a program, and treatment options should be considered only after source reduction options have been exhausted.

POLLUTION REDUCTION AND PREVENTION OPTIONS

SOURCE REDUCTION

Source reduction techniques include good operating practices, technology changes, input material changes, or product changes. Good operating practices are procedural, administrative, or institutional changes that an organization can use to minimize wastes. One advantage of implementing good operating practices is that they typically yield a high return on investment because they have minimal costs. Management and personnel practices such as employee training, incentives, and bonuses can be very effective ways to encourage employees to reduce pollution and wastes. Material handling and inventory practices include programs to reduce loss of materials due to mishandling, shelf-life expiration, and improper storage conditions. Loss prevention minimizes wastes by avoiding leaks from equipment and spills. Waste segregation practices reduce the volume of hazardous waste by preventing the mixing of hazardous and nonhazardous wastes. Cost accounting practices include programs to allocate waste treatment and disposal costs directly to the departments or groups that generate waste rather than charging these costs to general company overhead accounts. This makes these departments more aware of their treatment and disposal practices and provides a financial incentive to reduce wastes.

Technology changes are modifications of processes and equipment to reduce pollution and wastes. Examples include changes in the production process, equipment, layout, or piping and process operating conditions such as flow rates, temperatures, pressures, or residence times. The use of automation is another example of a technological change that can increase efficiency and reduce pollution and waste.

Changes in input materials are directed toward reducing or eliminating the hazardous materials that enter the production process or reducing the wastes generated during production. Examples include purifying materials and substituting less hazardous materials.

Product changes are directed toward reducing the waste generated from the use of the product. Examples include changing the composition of the product—substituting a more hazardous product with one that is less hazardous.

RECYCLING

Recycling allows hazardous and nonhazardous materials to be put to a beneficial use. These techniques—including use, reuse, and reclamation—can be performed on site or off site. At some point, however, even recycled materials may require eventual disposal.

Recycling could involve returning the material to the original process as a substitute for an input material, to another process as an input material, or to packaging. Reclamation is the recovery of a valuable material from a waste that can be sold or exchanged with another company.

Figure 7–3 provides examples of good operating practices and specific program areas that can reduce waste and pollution; Figure 7–4 at the end of this chapter provides a checklist for evaluating pollution prevention efforts.

IMPLEMENTING A PREVENTION OF POLLUTION AND WASTE MINIMIZATION PROGRAM

Implementation of a pollution prevention and waste minimization program requires a systematic process that includes planning and organization, assessment, feasibility analysis, implementation, and continual improvement (Figure 7–5). Appendix D contains worksheets that can be used to evaluate opportunities to prevent pollution.

PLANNING

Efforts to reduce pollution and waste begin with the development of a policy statement and top management support for the program. A team approach is especially important in developing a program to prevent pollution and to minimize waste because the entire organization is potentially affected by these activities.

FIGURE 7–3.   Waste Minimization Through Good Operating Practices:

Good Operating Practice	Program Ingredients
Waste minimization assessment	• Form a team of qualified individuals • Establish practical short-term and long-term goals • Allocate resources and budget for the program • Establish assessment targets • Identify and select options to minimize waste • Periodically monitor the program's effectiveness
Environmental audits/reviews	• Assemble pertinent documents • Conduct environmental process reviews • Carry out a site inspection • Report on and follow up on the findings
Loss prevention programs	• Establish spill prevention, control, countermeasures (SPCC) plans • Conduct hazard assessment in the design and operating phases
Waste segregation	• Prevent mixing hazardous wastes with nonhazardous wastes • Isolate hazardous wastes by contaminant • Isolate liquid wastes from solid wastes
Preventative maintenance programs	• Use equipment data cards on equipment location, characteristics, and maintenance • Maintain a master preventive maintenance (PM) schedule • Maintain PM reports on equipment • Maintain equipment history cards • Maintain equipment breakdown reports • Keep vendor maintenance manuals handy • Maintain manual computerized repair history file

Training /awareness-building programs	• Provide training for:
	  • Safe operation of the equipment
	  • Proper material handling
	  • Economic and environmental ramifications of hazardous waste generation and disposal
	  • Detection of hazardous material releases
	  • Emergency procedures
	  • Use of safety gear

Effective supervision	• Closer supervision may improve production efficiency and reduce inadvertent waste generation • Management by objectives (MBO), with goals for waste reduction
Employee participation	• "Quality circles" (free forums between employees and supervisors) can identify ways to reduce waste • Solicit employee suggestions for waste reduction ideas
Production scheduling/planning	• Maximize batch size • Dedicate equipment to a single product • Alter batch sequencing to minimize cleaning frequency (light-to-dark batch sequence, for example) • Schedule production to minimizing cleaning frequency
Cost accounting/allocation	• Perform cost accounting for all waste streams leaving the facility • Allocate waste treatment and disposal cost to the operations that generate the waste

Source: Adapted from U.S. Environmental Protection Agency, Hazardous Waste Engineering Laboratory, *Waste Minimization Opportunity Assessment Manual,* EPA/625/7-88003 (Cincinnati, OH: U.S. Environmental Protection Agency, 1988).

FIGURE 7–5. Prevention of Pollution and Waste Minimization Program Planning:

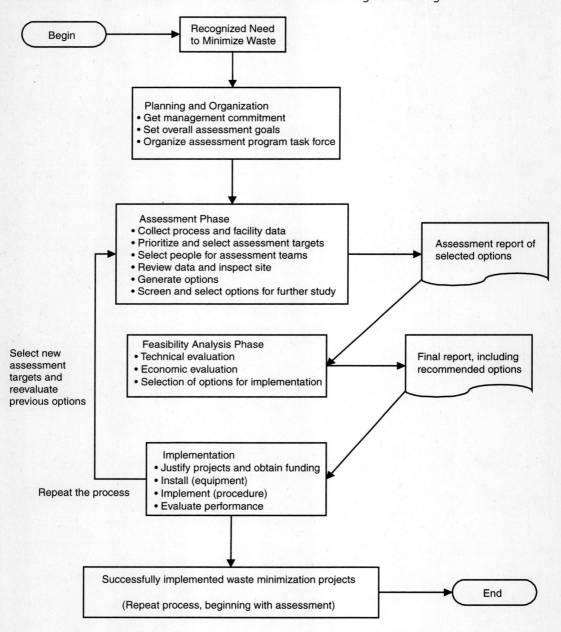

Source: Adapted from U.S. Environmental Protection Agency, Hazardous Waste Engineering Laboratory, *Waste Minimization Opportunity Assessment Manual,* EPA/625/7-88003 (Cincinnati, OH: U.S. Environmental Protection Agency, 1988).

Team members might be recruited from operations, engineering, maintenance, scheduling, materials control, procurement, shipping/receiving, facilities, quality control, environmental and health and safety, accounting, personnel, research and development, legal, management, and contractors/consultants.

ASSESSMENT

During the assessment phase, sources of waste and pollution streams are systematically identified. Typical sources of wastes from plant operations include materials receiving, raw material and product storage, production, and support services. Potential waste material from receiving includes packaging materials, damaged but not breached containers, spills from breached containers, transfer operations, and materials that do not meet specifications. Raw material and product storage sources include leaking pumps, valves, tanks, pipes and other containers, off-specification and excess materials, and spills into containment structures and tank bottoms. Production sources include off-specification products; leaking pipes, valves, hoses, tanks, drums, and process equipment; oils, additives, catalysts; water or solvent rinses; sweepings; filters; and a variety of other sources. Support service sources produce a variety of chemical and nonchemical wastes from cooling towers, boilers and power plants, maintenance, offices, cleaning, laboratories, and garages.

All potential sources of waste and pollution must be categorized and evaluated for potential minimization options. Information gathered during the identification of aspects will be useful, but additional data may need to be gathered. Questions to consider include:

- What are the pollution and waste streams? How much is generated?
- Which processes or operations create these contaminants?
- Which wastes are classified as hazardous and which are not? What makes the stream hazardous?
- What are the input materials that generate the pollution and waste streams?
- How much of a given input material enters the pollution and waste stream? How much is lost as fugitive releases?
- How efficient are the processes?
- Are unnecessary quantities of hazardous wastes generated by mixing hazardous wastes with other wastes? Are opportunities for recycling lost?
- What types of housekeeping practices are used to limit the generation of pollution and wastes?
- What types of process controls are used to improve process efficiency?

Figure 7–6 shows examples of information that might be helpful during the assessment phase.

Next, the pollution and waste streams and/or operations are assessed and prioritized. The best strategy is to focus on those streams that are the most important. Criteria for ranking wastes might include cost of treatment, potential environmental and safety liability, compliance status, quantity of pollution and waste, hazardous properties, ease of achieving reductions, potential for removing bottlenecks in production or waste treatment, potential recovery of valuable by-

FIGURE 7–6.   Input Information for Prevention of Pollution and Waste Minimization Assessments:

Design Information

- process flow diagrams
- material and heat balances for production and pollution control processes
- operating manuals and process descriptions
- equipment lists
- equipment specifications and data sheets
- piping and instrument diagrams
- plot and elevation plans
- equipment layouts and work flow diagrams

Environmental Information

- hazardous waste manifests and reports
- emission inventories
- waste analyses
- environmental audit reports
- environmental and occupational monitoring data
- permits/permit applications

Raw Material/Production Information

- product composition and batch sheets
- material application diagrams
- material safety data sheets
- product and raw material inventory sheets
- operator data logs
- operating procedures
- production schedules

Economic Information

- potential capital costs
  —purchased equipment
  —materials
  —utility systems and connections
  —site preparation (labor, supervision, and materials)
  —construction/installation (labor, supervision, and materials)
  —planning/engineering (labor, supervision, and materials)

(continued)

FIGURE 7–6.   (*Continued*)

       —start-up/training (labor, supervision, and materials)
       —regulatory/permitting (labor, supervision, and materials)
       —working capital
       —contingency (future compliance costs, remediation)
       —back-end (closure, inventory disposal, site survey)

- potential operating and maintenance costs
       —direct materials and labor
       —utilities
       —waste management labor, supervision, and materials
       —regulatory compliance labor, supervision, and materials
       —insurance
       —future liability
       —fines/penalities
       —cost of legal proceedings
       —personal injury
- revenues
- departmental cost accounting reports

Source: Adapted from U.S. Environmental Protection Agency, Hazardous Waste Engineering Laboratory, *Waste Minimization Opportunity Assessment Manual,* EPA/625/7-88003 (Cincinnati, OH: U.S. Environmental Protection Agency, 1988) and U.S. Environmental Protection Agency, Office of Enforcement and Compliance Assurance, *Federal Facility Pollution Prevention Project Analysis: A Primer for Applying Life Cycle and Total Cost Assessment Concepts,* EPA 300-B-95-008 (Washingtion, D.C.: U.S. Environmental Protection Agency, 1995).

products, and the available budget for minimization and prevention activities. The contaminant streams should be evaluated for potential options for reducing or eliminating the pollution or waste materials. This is best accomplished by a team comprised of individuals who have expertise related to the waste stream or pollution source. In some cases, the inclusion of an outside expert can facilitate the identification of potential solutions. The assessment team will develop a list of options for each pollution and waste stream, which can be screened for the potential for reducing costs and environmental impacts. The screening procedure should consider the following questions:

- What is the main benefit of the option?
- Does the necessary technology exist to develop the option?
- How much does it cost? Is it cost-effective?
- Can the option be implemented within a reasonable amount of time? Will production be interrupted?
- Does the option have a good chance of success?
- Are there other benefits of implementing the option?

Some options—such as housekeeping practices, procedural changes, and maintenance practices—can be implemented immediately with minimal costs. Others will require further study.

FEASIBILITY STUDY

Those options that appear attractive but require further study should be evaluated for their technical and economic feasibility. The technical feasibility study should involve everyone who will be affected by the changes. Sometimes outside consultants or pilot-scale testing will be needed to gather enough information to determine whether an option is feasible. A systematic and careful approach will avoid costly mistakes. Some technical evaluation criteria include:

- Is the option safe for workers?
- Will product quality be maintained?
- Are space, utilities, and additional labor available?
- Are the new equipment, materials, or procedures compatible with production, operating procedures, work flow, and production rates?
- How long will production be interrupted for implementation?
- Is special expertise needed to implement and operate the new system?
- Does the vendor provide acceptable service?
- Does the system create other environmental problems?

The economic criteria for selecting a given option will be different for each organization. The evaluation should consider standard measures of profitability, including payback period, return on investment, and net present value. The project should be broken down into both capital costs and operating costs and savings. Capital cost projections should include not only the fixed capital costs for designing, purchasing, and installing equipment, but also costs for working capital, permitting, training, start-up, and financing charges. In the United States, permitting costs associated with recycling options may make source reduction projects more attractive. Operating costs and savings should include reduced waste management costs, input material cost savings, insurance and liability savings, changes in the costs of utilities, changes in operating and maintenance costs for labor and supplies, changes in overhead costs, changes in revenues from increased or decreased productivity, costs or savings associated with quality, and increased revenues from by-products.

IMPLEMENTATION

A final report that lists the potential options and their associated technical and economic feasibility provides an important tool for making sound decisions. Everyone who was involved in the feasibility study should have an opportunity to review and comment on a draft report. This has the benefit of correcting potential errors but also helps foster employee commitment to selected options. Once the appropriate approval is gained for an option, implementation begins. Options that do not require additions or modifications to equipment can be implemented immediately. More complex projects will require traditional planning, design, procurement, construction, and operational phases. Additional funding sources (outside of the organization) can be explored (for example, governmental agencies, foundations, private sponsors). The final stage of implementation involves follow-up and documentation of waste reduction and prevention of pollution.

A program to prevent and minimize pollultion and waste should not be a one-time event, but an ongoing effort. After the highest-priority wastes and pollution sources are managed, efforts should continue to reduce lower-priority sources. Integration of waste minimization and prevention of pollution activities into an organization's way of life will produce the largest benefits. Appendix F contains useful World Wide Web addresses that can provide additional information on these topics.

EMERGENCY PREPAREDNESS AND RESPONSE[3]

Significant and often irreversible damage to the environment and health occurs as a result of accidents, spills, and other unplanned releases. The costs of cleanup and damage control range from thousands of dollars for modest releases to millions of dollars for significant accidents. An aggressive proactive management policy must be implemented to minimize damage. The need for an organization to integrate emergency preparedness and response into its operations is addressed in the standard (ISO 14001, Subsection 4.4.7). The basic steps in developing an effective program are identifying the need for procedures, implementing written procedures, conducting periodic tests, and continual improvement through review and revision.

Appendix C contains a sample emergency response procedure based on regulatory requirements in the U.S. (29 *Code of Federal Regulations* [(CFR) 1910.120]. This procedure is written in a more detailed manner than are the other procedures in the appendix to illustrate different ways of documenting procedures. Appendix C also contains an example spill and disaster management procedure.

Identifying the potential for emergency situations begins with the identification of aspects and process analysis activities (see Chapter 4). During this phase of planning potential accidental discharges to air, land, or water and related environmental and eco-system effects are identified as a consequence of abnormal operating conditions (such as reduced efficiency of pollution control equipment, start-up or shutdown procedures, maintenance failures, or changing process conditions), spills, or other unplanned circumstances.

After the potential for unplanned releases is identified, the organization is ready to develop procedures to respond to these situations. The core of the emergency preparedness and response procedures is a detailed plan that is designed to contain the accidental release and to protect people and the environment during the first few hours of the release. Document control is especially important for the procedures, work instructions, and records that are associated with emergency response because inconsistent or incorrect actions can be life-threatening and can cause significant environmental damage. Emergency response documentation should be distributed to the designated responders, facility, and corporate environmental health and safety representatives, fire department, hospital, police, and any other representatives required by organizational policies or legal requirements.

Emergency response activities can be grouped into planning activities, response actions, closure actions, and recovery actions. Key elements in each of these areas include:

- Planning Activities
  - —designate the planning team
  - —develop information on hazardous materials, including each material's potential impact on the environment and measures to be taken in the event of accidental release
  - —establish the emergency organization and responsibilities (availability)
  - —develop the emergency response plan, including actions to be taken in the event of different types of emergencies
  - —evaluate the medical fitness of designated responders
  - —purchase needed equipment including personal protective equipment (PPE); chemical, biological, and radiation cleanup supplies; communications equipment; fire extinguisher and related equipment; monitoring equipment; and so forth)
  - —establish needed agreements with fire, police, spill cleanup contractors, hospitals, and ambulance services
  - —establish internal and external communications
  - —prepare evacuation routes and procedures
  - —prepare medical treatment and first aid facilities
  - —prepare site security and control plan
  - —conduct needed training for responders
  - —conduct evacuation drills for employees
- Response Actions
  - —notify of emergency conditions
  - —evacuation
  - —employee accounting
  - —identify hazards and assess emergency
  - —remove injured employees
  - —initiate calls to external resources and agencies
  - —shut down utilities and services as needed
  - —secure area
  - —control the unplanned release
  - —clean up the released materials
- Closure Actions
  - —make needed reports
  - —postincident debriefing
  - —postincident analysis
  - —report on incident and needed improvements
- Recovery Actions
  - —certify for reentry
  - —restore operations
  - —restock supplies and equipment
  - —arrange for cleanup contractors
  - —make restitution

PLANNING

Detailed planning is the best way to prepare for the unexpected. Planning will be facilitated by forming a team to develop the emergency preparedness and response plan. This team should be interdisciplinary and should include representatives of process areas with potential for significant environmental impact and personnel from support functions within the organization. The planning team will identify tasks, assign responsibilities, and oversee the development of the plan.

One of the first steps is to conduct a hazard analysis: that is, assess the potential for unplanned releases from each process in the facility and evaluate the related health and environmental hazards. The MSDS (Figure 7–7) can provide important information about hazardous materials in a concise format including physical and chemical properties, health and toxicity hazards, fire protection measures, incompatibilities, spill containment, personnel protective equipment, and other special precautions. The MSDS is prepared by the manufacturer or importer of the hazardous material, and it should be included with the shipment of the material. If the MSDS is missing, request one from the manufacturer, importer, or distributor (Figure 7–8 shows an example of a form letter that can be used to obtain an MSDS). In addition to letting responders know what to expect, the hazard analysis has other benefits that include creating awareness of new hazards and identifying the need for preventive actions, such as monitoring systems or facility modifications. Information that is gathered during the hazard analysis includes identifying the nature of the hazards, locations and quantities of materials, types of containers, and the presence of any fixed suppression or detection systems. The chemical, physical, and biological hazards associated with each hazardous material must be evaluated and ranked according to risks. The potential for fire hazards and water reactivity are especially important. Identifying the potential for container breaches and explosions is critical, as these will result in the release of hazardous materials and/or energy. The next step is identifying possible interventions and the outcome associated with these actions. This planning exercise requires the behavior of the hazardous material and/or the container to be visualized under different scenarios. Questions to consider are:

- What is the hazardous material? Is it likely to be released as a gas, liquid, or solid?
- What hazards are related to the material? Is it flammable, water-reactive, toxic, or an oxidizer? In the event of a release, will heat or radiation be released? If the material is flammable, what extinguishing agents are needed? Will containers rupture? What path will the hazardous material follow? What dispersion pattern could result?
- What are the health concerns (hazardous exposure values, signs and symptoms of exposure)? What personal protective equipment is required? Are nearby communities likely to be affected?

The book by Louvar contains useful sections on fault-tree analysis, consequence analysis, exposure assessment, and risk management.

Some of the tools commonly used by safety professionals for performing a hazard analysis include:

- "What-if" analysis. This method is based on asking sequential questions, such as, What if the control equipment breaks down? or What if the pressure gauge malfunctions?
- Fault-tree analysis. This method is used to estimate the likelihood of events by working backward from a defined accident to identify and graphically display

FIGURE 7–7.   Material Safety Data Sheet Form:

Material Safety Data Sheet May be used to comply with OSHA's Hazard Communication Standard, 29 CFR 1910. 1200. Standard must be consulted for specific requirements.	**U.S. Department of Labor** Occupational Safety and Health Administration (Non-Mandatory Form) Form Approved OMB No. 1218-0072
IDENTITY *(As Used on label and List)*	Note: *Blank spaces are not permitted. If any item is not applicable, or no information is available, the space must be marked to indicate that.*

Section I

Manufacturer's Name	Emergency Telephone Number
Address *(Number, Street, City, State, and ZIP Code)*	Telephone Number for Information
	Date Prepared
	Signature of Preparer *(optional)*

Section II — Hazardous Ingredients/Identity Information

Hazardous Components (Specific Chemical Identity; Common Name(s))	OSHA PEL	ACGIH TLV	Other Limits Recommended	% *(optional)*

Section III — Physical/Chemical Characteristics

Boiling Point		Specific Gravity (H$_2$O = 1)	
Vapor Pressure (mm Hg.)		Melting Point	
Vapor Density (AIR = 1)		Evaporation Rate (Butyl Acetate = 1)	
Solubility in Water			
Appearance and Odor			

Section IV — Fire and Explosion Hazard Data

Flash Point (Method Used)		Flammable Limits	LEL	UEL
Extinguishing Media				
Special Fire Fighting Procedures				
Unusual Fire and Explosion Hazards				

(Reproduce locally)   OSHA 174, Sept. 1985

FIGURE 7–7. *(Continued)*

Section V — Reactivity Data

Stability	Unstable		Conditions to Avoid
	Stable		

Incompatibility *(Materials to Avoid)*

Hazardous Decomposition or Byproducts

Hazardous Polymerization	May Occur		Conditions to Avoid
	Will Not Occur		

Section VI — Health Hazard Data

Route(s) of Entry:	Inhalation?	Skin?	Ingestion?

Health Hazards *(Acute and Chronic)*

Carcinogenicity:	NTP?	IARC Monographs?	OSHA Regulated?

Signs and Symptoms of Exposure

Medical Conditions
Generally Aggravated by Exposure

Emergency and First Aid Procedures

Section VII — Precautions for Safe Handling and Use

Steps to Be Taken in Case Material Is Released or Spilled

Waste Disposal Method

Precautions to Be Taken in Handling and Storing

Other Precautions

Section VIII — Control Measures

Respiratory Protection *(Specify Type)*

Ventilation	Local Exhaust		Special	
	Mechanical *(General)*		Other	
Protective Gloves			Eye Protection	

Other Protective Clothing or Equipment

Work/Hygienic Practices

*U.S. Government Printing Office: 1987—181-504/64362

FIGURE 7–8.   Form Letter for Obtaining a Material Safety Data Sheet:

Letterhead

Date

Name and address of MSD supplier
(manufacturer, importer of distributor)

Dear Mr. or Ms.: _____

My company recently purchased your product *insert product name here,* and a
                                            (product identifier)
Material Safety Data (MSD) did not arrive with the first delivery.

Please send me an appropriate MSD sheet that will meet the requirements set forth in the OSHA
standards 29 CFR 1910.1200 and 29 CFR 1926.59.

Thank you for your cooperation.

Sincerely,

Employee name,
Job Title

Department
Company
Address
Telephone Number

Source: Adapted from *Regulatory Compliance Manual for Indiana Perchloroethylene Dry Cleaners* (Indianapolis, IN: Indiana Department of Environmental Management and Indiana Dry Cleaning and Laundry Association, 1995). This material may be used, in whole or in part, without permission. As such, the copyright on this book does not apply.

the combination of equipment failures and operational errors that caused the accident.

- Event-tree analysis. This method is the reverse of fault-tree analysis, and it is used to determine the likelihood of an incident occurrence. It works forward from specific events that could lead to an accident, to identify and graphically display the events that could result in hazards.
- Failure modes, effects, and criticality analysis (FMECA). This method ranks failure modes according to criticality to determine which are the most likely to cause a serious accident. Each system or unit of equipment is listed along with failure modes, effects of each failure on the system or unit, and determination of how critical each failure is to the integrity of the system.

After the hazard analysis is completed, planning for the response actions can begin. The emergency response team (ERT) should be designated and their areas of responsibility identified. Team members can come from any area within the organization, but they are typically selected from process areas or from the environmental health and safety department. They must have good judgment, the ability to act under pressure, and be in good physical condition. A team must be available for all shifts and trained to respond to the potential emergencies identified during the hazard analysis.

The emergency response plan should include actions to be taken in the event of different types of emergencies. Fortunately, not every accident presents a crisis situation. Minor spills and releases of hazardous materials in quantities less than 55 gallons pose limited problems that are not likely to spread beyond the work area. These incidents are often handled by the on-shift ERT without outside intervention. In the United States, releases greater than the reportable quantity (RQ) must be reported to the National Response Center [(800) 424-8802]. Larger releases of hazardous materials and fires pose moderate danger to life, health, and property. These releases have the potential for migrating off site and affecting the surrounding community. Serious incidents such as these require on-shift ERT response and some assistance from off-site resources. The most serious accidents pose a crisis situation. Examples include process unit fires or explosions and major releases of hazardous materials. These releases are very likely to go beyond the plant property, and they have the potential for affecting a large geographic area for an indefinite period of time. Each of these situations will require different levels of training, supplies, equipment, and response actions.

It is important to establish needed agreements with fire, police, spill cleanup contractors, hospitals, and ambulance services before an incident occurs. Contact persons and telephone numbers for these organizations and any required regulatory offices should be listed and readily available. Plans should be prepared for site security and control, medical treatment, first aid facilities, and for supplying information to the media and general public. The ERT members must undergo a physical examination by a physician to determine whether they are physically able to wear respirators and protective clothing. If possible, utilize a physician with experience in occupational medicine. The plan should also identify conditions under which team members must undergo periodic medical exams.

Each member of the team must be trained initially and on a periodic basis to conduct their assigned responsibilities in a safe manner. The team leader or on-scene incident commander is the person who assumes control of the incident and who has overall responsibility for the emergency response. This person must have the technical knowledge and leadership skills to provide overall control of the incident. Some members will receive awareness-level training that enables them to recognize hazards and to call the ERT into action. Operations-level training allows the responder to take initial actions to contain the release from a safe distance and to keep it from spreading, but they do not try to stop the release. Technician-level training allows the responder to take actions to contain the release. Responders with specialist-level training have the same capability of the technician, but they also have greater technical knowledge that allows them to provide support to the team and other groups involved in the incident. In the United States, training requirements for the ERT, which are given in 29 CFR 1910.120, are considered to be minimum; many organizations provide training in excess of these requirements. The emergency response procedure in Appendix C contains training requirements that go beyond the legally required amounts in the U.S.

The ERT members must have the equipment needed to carry out their assigned tasks. Examples include technical references, respirators, protective clothing, communications equipment, sampling equipment, direct-reading monitoring equipment and detection equipment, control and mitigation supplies and equipment (for example patches, plugs, booms, pumps, fire extinguishing agents, shovels, neutralizing agents and other chemicals, vacuum trucks), and decontamination supplies and equipment (for example, brushes, high-efficiency particulate air [HEPA] vacuum cleaner, bagging materials, chemical and biological degradation supplies, portable showers).

It is extremely important for the ERT and all employees to conduct drills prior to an emergency. The ERT members must be able to react quickly under physically and mentally challenging conditions. It requires practice to use protective clothing, respirators, and other equipment; to coordinate communications; and to perform needed activities smoothly and confidently. The ERT members should have drills as often as required to perform tasks safely. The type and freqency of drills will depend on the facility and expertise of the ERT members. However, a full-scale exercise that involves all outside agencies and aspects, such as medical services, should be conducted at least once per year. Individual scenarios tailored to specific types of actions (for example, confined space entry) may be required from two times per year in a facility that has limited hazardous materials to once per month for large, complex facilities with the potential for many different types of accidents. Tabletop exercises are an important component of emergency response training, and they can help to refine management's response to emergencies. These exercises should be coordinated with other agencies, in particular, the fire department. One of the weak areas in most emergency response actions is the lack of coordination among the responding organizations. Periodic training sessions on unified command are an important tool for ensuring that response actions are coordinated and will function smoothly.

Likewise, it is important for all members of the facility to practice evacuations at least once per year. Evacuation routes must be planned, and up-to-date maps showing the primary and backup evacuation routes, closest exits, assembly points, and emergency equipment locations should be posted throughout the facility. Ideally, these drills should include all members of the facility, but this is often impossible. During these drills, employees must practice evacuation routes and must assemble at designated locations for roll call. Disabled employees will require help and should have designated assistants. At the completion of all drills, reports should be prepared that identify weak areas and needed corrective actions. The implementation of corrective actions must be verified.

Emergency communications must be available and tested at least once per month and during the drills. A designated internal telephone number should be available throughout the facility. Calls should be routed to a designated dispatch post that will notify the ERT and/or other organizations as needed. Typically, emergency alarms with backup batteries are used. The alarm must be audible from every part of the facility and under the noise conditions present in the facility. Special provisions may be required for hearing-impaired individuals.

RESPONSE ACTIONS

Response actions include all those activities that occur from the initial report of the incident through the decontamination and equipment cleanup phases of the incident. The focus of response actions is to stabilize, confine, contain, and control the release of hazardous materials, transfer and recover materials, prevent unnecessary damage, and stabilize the situation for the final cleanup operations. The Incident Management System (IMS) is an effective model for managing the response to an unplanned release.[4] This model takes a highly structured approach to evaluating potential hazards, performing release control actions, controlling the scene, and for medical surveillance, decontamination, and other activities.

In the United States, if potential releases and impacts are confined to the facility, the On-Scene Incident Commander or leader of the facility's ERT coordinates the response. However, in the event off-site impacts are possible, the On-Scene Incident Commander is typically the local fire chief or his/her designated representative, and he/she makes the final strategic and tactical decisions about the incident, with the input from the senior facility response official and other participating organizations.

CLOSURE

Closure of the incident involves debriefing to let responders and any other key players know their potential exposures, signs and symptoms of delayed health effects, identification of damage to equipment, need for stress counseling, expended supplies, unsafe conditions that would affect final cleanup and recovery, and methods for assigning further information-gathering responsibilities for a postincident analysis and debriefing. Debriefing should be conducted before individuals leave the scene.

The postincident analysis is a continual improvement tool in which the incident is reconstructed to identify areas of strengths and weaknesses. Areas of consideration include determining whether the incident was properly documented and reported,[5] determining the cost of responding to the incident, and establishing the need for corrective actions for future emergencies. The postincident analysis will be conducted by the On-Scene Incident Commander if the response involves the community response system. Because the postincident analysis also provides a foundation for potential legal actions, there is some hesitance on the part of the affected organizations to produce an internal postincident analysis report. Although this is a legitimate concern, the analysis is a valuable exercise that contributes to continual improvement. Initiating the report as a privileged communication may adequately address these concerns.

RECOVERY ACTIONS

During the recovery phase, the facility will be certified for reentry, and operations will be restored. Supplies and equipment expended during the response

www

should be reordered. Arrangements should be made to conduct any needed recovery phase cleanup operations and restitution. Each organization will have different needs and legal requirements for emergency preparedness and response planning, so the details of the plans will differ. In the United States, emergency plans must comply with state and federal regulations, including at least 29 CFR 1910.38 and 29 CFR 1910.120. Appendix F contains useful World Wide Web addresses for more information on these topics.

MANAGEMENT OF HAZARDOUS MATERIALS

Hazardous materials include explosives; radioactive substances; poison gases, liquids, and solids; flammable and nonflammable gases; combustible liquids; flammable solids; spontaneously combustible materials; oxidizers and organic peroxides; water-reactive materials; magnetized materials; infectious substances; and wastes of all of these groups of materials. Proper management of hazardous materials begins with planning activities prior to purchase and does not end until all wastes and unused materials are properly disposed of. The benefits of a systematic and well-structured approach to the management of hazardous materials include cost savings resulting from a JIT inventory system, improved worker health and safety, lowered potential for accidents and spills, and the elimination of potential liability for improper disposal of wastes. Hazardous materials can be managed most economically by building best practices into the design phase of new production processes or activities (see Appendix F for World Wide Web addresses on this topic).

www

PLANNING

Careful planning prior to the acquisition of hazardous materials is an important first step to minimizing the cost and risks associated with these materials. Planning activities should be integrated into the pollution prevention program for the facility. Existing hazardous materials or chemical management procedures and practices should be reviewed using the following discussion as a guide.

Cost and inherent chemical and physical properties related to the application are two important factors that are used to evaluate materials for intended uses. An important and sometimes overlooked related consideration prior to purchasing is the hazard posed by the material during storage, transport, use, and disposal. The decision to purchase should include an evaluation of characteristics such as toxicity, reactivity, flammability, corrosiveness, and stability; environmental requirements for storage and handling; needed engineering controls and personal protective equipment to minimize exposures; and needed controls to reduce environmental impact. It is also important to determine the type and quantity of waste that will be produced as a result of using the material. The waste evaluation should include treatment and disposal costs, and the potential marketability of by-products and wastes.

The decision to purchase should be made only after determining that the material has the best properties for the proposed application combined with lowest cost and lowest hazard. A careful search for alternative materials may produce choices that are safer to use, easier to store, and less toxic. Although these materials may cost more, a consideration of waste management and pollution control costs could make some alternative materials more competitive.

Maintenance chemicals, pesticides, and other chemicals for routine housekeeping are often overlooked in the chemical management system. They generally present fewer problems, but it is still important to select the best and least hazardous material for the job. Careful evaluation of these materials will ensure that problems are less likely to arise from maintenance, cleaning, and housekeeping activities.

After a decision has been made to purchase a hazardous material, the procurement process should include an evaluation of suppliers and disposal contractors to ensure that they follow best management practices for hazardous materials and have permits required by regulatory authorities.

OPERATIONAL CONTROLS

Operational controls for hazardous materials include having procedures and related documentation for receiving and storage, materials tracking, and transportation, routine use of materials, handling of empty containers, and waste disposal.

SHIPPING, RECEIVING, AND STORAGE

Formalized procedures for shipping, receiving, and storage of bulk and nonbulk hazardous materials are essential to sound management practice. Materials to be shipped and received should be properly packaged, labeled, and accompanied by a properly executed shipping paper. Shipments of hazardous materials must comply with national and international shipping regulations.[6]

When materials are received, the shipping paper should be examined to verify the contents of the shipment. Containers should be inspected before they are accepted to ensure that they are of the correct type, are properly labeled, and are in good condition. When bulk chemicals are accepted, it is important to ensure that they are delivered to the correct receiving container and that any spills are recorded, reported, and cleaned up. The MSDS should accompany each product; missing MSDSs should be requested immediately (see Figure 7–7 for an example MSDS form and Figure 7–8 for an example letter requesting an MSDS).

When choosing a storage location for a hazardous material, the properties of the material and incompatibilities must be considered. The MSDS should provide information about special storage requirements and incompatibilities. Workers should be trained in the proper handling of hazardous materials; warning signs or instructions should be posted as appropriate.

Hazardous materials should not be stored alphabetically. Liquids, solids, and gases must be segregated from one another and stored according to chemical and physical hazards. Bulk liquids should have adequate secondary containment in the event of a leak or spill. Flammable materials require controlled temperatures and proper grounding of containers. Unstable materials such as organic peroxides and materials that polymerize may have additional storage requirements. The potential for the formation of hazardous decomposition products must also be considered. For example, ethers can decompose into unstable organic peroxides if stored improperly or if the product is stored beyond its acceptable shelf life. Special attention should be given to fire-extinguishing agents—Is the agent appropriate and available in adequate amounts? Some agents such as graphite and powders for reactive metals require dry storage.

TRACKING SYSTEM

All information about received hazardous materials should be entered into a hazardous material or chemical tracking system. The tracking system can be limited to basic information about the manufacturer/vendor and quantities, storage, and use locations or it can incorporate extensive information about hazards related to the material. The tracking system can easily be integrated into other data management systems and inventories that are needed for regulatory purposes. Examples of information that could be included in the tracking system include the basic information about the product: product name, chemical name(s), Chemical Abstract System (CAS) number, concentration of ingredients, manufacturer, supplier, pricing history, processes in which the chemical is used, and the storage location and amount. Information about the hazards related to the chemicals could include the entire MSDS or selected portions, such as hazard information and special storage requirements.

USE AND DISPOSAL OF HAZARDOUS MATERIALS

The safe use of hazardous materials in the workplace requires well-documented work instructions, record keeping, and adequate training of employees in the proper storage, use, and transport of hazardous materials. In the United States, the hazards of all chemicals produced or imported must be evaluated, and information about these hazards must be transmitted to employers and employees by means of a comprehensive hazard communication program.[7] Basic elements of hazard communication include preparing inventories of hazardous chemicals, providing access to the MSDS; using labels or other forms of warning about hazards; and training workers in the proper use and disposal of chemicals, hazards, personal protective equipment requirements, and emergency response actions. In addition, the empoyer is responsible for establishing procedures to maintain the currency of the program and to evaluate its effectiveness. Facilities that use chemicals in laboratories should develop a chemical hygiene plan.[8] General principles underlying both the hazard communication standard and the chemical hygiene plan are to minimize all chemical exposures through proper work practices; to avoid underestimation of risk; to provide adequate ventilation, engineering controls, and personal protective equipment; and to provide adequate and properly designed workplaces so that hazardous materials can be managed in a safe and efficient manner.

Careful tracking of the type and quantity of hazardous materials used can help reduce costs and can provide early detection of losses due to leaks in tanks or plumbing. All spills should be documented with the type and amount of material spilled, cleanup procedures, health effects, environmental damage, and corrective actions.

Proper handling of "empty" containers is another important component of the hazardous materials program. Disposal and safety problems can readily occur if empty containers are not handled properly. After the usable contents have been removed, residues or small quantities of material may remain. Production and maintenance employees must be given instructions regarding the proper cleaning and disposition of these containers. Residual chemicals can be removed by draining, washing, or scraping of the containers. In some instances, remaining material can be salvaged and recycled; in other cases, it must be considererd a hazardous waste. Wash water used in cleaning may contain enough of the hazardous material that the wash water becomes a hazardous waste. Containerizing and sampling the wash water may be necessary to determine the proper disposal methods. Sometimes, the cleaned containers can be reused or recycled in-house, returned to the supplier, or sold to a secondary user.

Records should be maintained for the quantity and type of wastes generated. It is important to segregate hazardous wastes from those that are not hazardous. Mixing these materials increases the total volume of hazardous waste and increases disposal costs. To minimize potential liability, use only waste handlers and facilities that have the legal authority to haul, treat, or dispose of hazardous wastes. Be sure to evaluate these providers for best management practices. Proper documentation and use of good providers can signficantly lower the risks of accidental releases and future liability.

PROPERTY MANAGEMENT AND TRANSFER

Poor property management can produce significant legal and financial problems. The best insurance against liability related to property and property transactions is knowledge about the environmental conditions affecting a property. Environmental problems can afflict building envelopes, the interior and exterior of buildings, and the property's surface and subsurface water and soil. Both currently owned property and potential acquisitions should be evaluated for environmental hazards. The need for environmental assessments prior to acquiring a property is self-evident. Buyers do not want to be surprised by hidden problems and attendant legal and corrective action costs. Similarly, sellers who know about environmental problems related to their properties can use this knowledge to make informed decisions prior to placing their properties on the market. Once hazards are identified, management options that range from removal of the hazard to management in place can be evaluated for cost and effectiveness.

The American Society for Testing and Materials (ASTM) supplies voluntary standards on environmental site assessments in the United States that provide a good foundation for evaluating properties. The ASTM standards (E1528-93 and E1527-94) provide protocols for evaluating environmental hazards in commercial real estate.[9] The ASTM's *1528-93: Standard Practice for Environmental Site Assessments: Transaction Screen Process* provides a standard questionnaire on environmental hazards that forms the basis of the evaluation of the property. This questionnaire is used as a guide by the environmental professional as he/she evaluates the property during a site visit, and it is administered to the owner and occupant (if applicable). The transaction screen also requires a limited review of government and historical records in an effort to identify existing environmental hazards at the property. A more complete evaluation can be obtained by following the guidance in ASTM's *E1527-94: Standard Practice for Environmental Site Assessments: Phase I Environmental Site Assessment Process*, which is discussed below.

PHASE I ENVIRONMENTAL ASSESSMENT PROCESS

The goal of the Phase I site assessment is to identify recognized environmental hazards associated with a property. The standard practice is designed to identify the range of contaminants within the scope of the Comprehensive Environmental Response, Compensation and Liability Act (CERCLA) and petroleum products. Hazardous substances under CERCLA include substances (and elements, compounds, mixtures, solutions) specifically identified in air, land, and water regulations. These include hazardous waste, hazardous air pollutants, toxic pollutants, and specific materials identified in the CERCLA regulations.

A properly conducted Phase I site assessment allows a user to satisfy one of the requirements to qualify for the *innocent landowner defense* under CERCLA. The Phase I assessment constitutes "all appropriate inquiry into legal ownership and uses of the property consistent with good commercial or customary practice" as defined in CERCLA. The utility of the Phase I site assessment practice, however, extends beyond establishing the innocent landowner defense. The standard practice developed by ASTM allows the environmental condition of any property to be identified in a systematic and complete manner.

The ASTM standard recognizes that there may be environmental conditions or issues related to commercial real estate that are outside of the scope of the standard (which is limited to CERCLA's definition of hazardous substances and CERCLA liability). Additional issues that may be important to a property evaluation include, but are not limited to, asbestos-containing materials, organic emissions from new or recently renovated construction, radon, lead-based paint, lead in drinking water, mold and moisture damage, and wetlands. A complete evaluation should consider these and any other factors related to the structures and grounds of the property.

The ASTM standard requires the Phase I site assessment to be performed by an "environmental professional." According to the standard, an environmental professional is a person who has the training and experience needed to conduct a site reconnaissance, interviews, and other related activities and who has the ability to interpret information and to develop conclusions about the environmental hazards related to the property. The standard does not require the environmental professional to have specific schooling or certifications, but certifications are developing for this specific expertise. For example, the National Association of Environmental Risk Auditors (NAERA) does provide a certification for Phase I site assessments that is based on the ASTM standard.[10]

The Phase I assessment begins with a review of records that will help identify recognized environmental conditions related to the property (Figure 7–9). At a minimum, certain environmental sources must be searched to evaluate the potential for existing contamination by underground storage tanks, solid waste disposal sites, hazardous waste disposal sites, and other sites where hazardous materials may be stored or leaking. Additional state and local records should be searched to provide a complete picture of the property's environmental condition. This record search should also include records of contaminated public wells, polychlorinated biphenyl (PCB) locations, and any records of spills or emergency releases. In addition, the environmental records search must include a review of the physical setting of the property. This search can be satisfied by reviewing a current 7.5-minute topographic map, but other sources may also be required to provide needed information about the geologic, hydrogeologic, hydrologic, or topographic characteristics of the subject property.

FIGURE 7–9. Types of Records Required for a Phase I Environmental Source and Physical Setting Review:

Type of Record	Approximate Distance to Be Searched from the Property Boundary
Federal- and State-Required Environmental Records	
Federal list of sites identified for remediation, National Priorities List (NPL)	1.0 mi (1.6 km)
Federal list of sites proposed for remediation, Comprehensive Environmental Response, Compensation and Liability Information System (CERCLIS)	0.5 mi (0.8 km)
Federal list of sites that treat, store, or dispose of hazarous wastes, Resource Conservation and Recovery Act (RCRA) treatment, disposal, and recovery (TSD) list	1.0 mi (1.6 km)
Federal list of generators of hazardous wastes, RCRA generators list	property and adjoining properties
Federal Emergency Response Notification System (ERNS) list	property only
State lists of hazardous waste sites identified for investigation and remediation (NPL and CERCLIS equivalents)	1.0 mi (1.6 km)
State landfill and/or solid waste disposal site lists	0.5 mi (0.8 km)
State leaking underground storage tank (UST) lists	0.5 mi (0.8 km)
State registered UST lists	property and adjoining properties
Additional State or Local Records	
Lists of landfill/solid waste disposal sites	distances at least equal to those specified above
Lists of hazardous waste/contaminated sites	
Lists of registered USTs	
Records of Emergency Release Reports	
Records of contaminated public wells	
Mandatory Standard Physical Setting Source	
Current 7.5-minute topographic map	distances sufficient to ascertain impact of sources on subject property
Additional and Nonstandard Physical Setting Sources	
Groundwater maps [United States Geological Survey (USGS)] and/or State Geological Survey	distances sufficient to ascertain impact of sources on subject property
Bedrock geology (USGS and/or State Geological Survey)	
Surficial geology (USGS and/or State Geological Survey)	
Soil maps (Soil Conservation Service)	
Other physical setting sources	

The next step in the Phase I assessment is a review of historical sources to identify any past uses that might have affected the current environmental condition of the property. Standard historical sources include:

- aerial photographs
- fire insurance maps
- property tax files
- recorded land title records
- 7.5-minute topographic maps
- local street directories
- building department records
- zoning/land use records
- other sources, such as newspaper archives or personal knowledge of occupants or owners

After gaining knowledge from environmental and historical records, the site can be evaluated by conducting a site visit in which the property and its buildings are physically and visually observed. The site visit includes observation of the interior and exterior of the property and its buildings. The evaluation should consider the following:

- *General Site Setting*
  — current and past uses of the property and adjoining properties
  — geologic, hydrogeologic, hydrologic, and topographic conditions
  — structures
  — roads within and adjoining the property
  — potable water supply
  — sewage disposal system
- *Interior and Exterior Observations*
  — current and past uses of the property
  — hazardous substances and petroleum products in connection with identified uses
  — storage tanks
  — odors
  — pools of liquid and sumps
  — drums
  — hazardous substance and petroleum products containers
  — unidentified substance containers
  — PCBs
- *Interior Observations*
  — heating and cooling system
  — stains or corrosion
  — drains and sumps
- *Exterior Observations*
  — pits, ponds, or lagoons
  — stained soil or pavement
  — stressed vegetation
  — solid waste
  — wastewater
  — wells
  — septic systems

The final step prior to report preparation is a series of interviews with owners, occupants, and local government officials to obtain information about uses and condition of the property. Interviews with owners and occupants, which are likely to be conducted concurrently with the site visit, will include questions related to documents collected during the environmental records search. These interviews will also include questions about safety and emergency response plans, and characteristics and properties of the hazardous and petroleum materials located and/or used at the site. An important component of the interview process is directed at obtaining knowledge of pending, threatened, or past litigation; administrative proceedings; or notices for any governmental entity regarding possible violation of environmental laws or possible liability related to hazardous substances or petroleum products at the property.

The last stage in the Phase I assessment is preparation of a report that details findings and conclusions related to the environmental condition of the property.

SUMMARY

Policies, procedures, work instructions, and records must be written and maintained to ensure operational control of the facility. The need for operational controls is identified through the process analysis phase of planning for the EMS. Only those processes that could have significant environmental impact require documentation and control. Although the standard does not require contractors to be monitored, it makes good business and environmental sense to monitor all procurement activities. Properly managed maintenance activities, waste minimization and prevention of pollution programs, property acquisition and transfer activities, and hazardous materials usage programs can produce significant savings, increased operational efficiency, and decreased environmental and health impacts.

ENDNOTES

[1] J.R. Tony Arnold. *Introduction to Materials Management*, 2d edition (Upper Saddle River, NJ: Prentice Hall, 1996).

[2] In the U.S., waste minimization is a policy that is specifically mandated by the 1984 HSWA to RCRA.

[3] In the U.S., emergency preparedness and response is regulated under several authorities, including regulations under the CAA Amendments of 1990 (40 CFR Part 68—*Risk Management Programs for Chemical Accidental Release Prevention*), the OSH Act [29 CFR 1910.119—*Process Safety Management of Highly Hazardous Chemical, Explosive and Blasting Agents*; 29 CFR 1910.120—*Hazardous Waste Operations and Emergency Response (Hazwoper)*], EPCRA, and the OPA of 1990.

[4] Gregory G. Noll, Michael S. Hilldebrand, and James G. Yvorra. *Hazardous Materials Managing the Incident* (Stillwater, OK: Fire Protection Publications, 1995). This reference is available from Oklahoma State University, Fire Protection Publications, Stillwater, OK

74078-0118. Two other useful references are the voluntary consensus standards: *NFPA 472—Standard for Professional Competence of Responders to Hazardous Materials Incidents* and *NFPA 471—Recommended Practice for Responding to Hazardous Materials Incidents*. These can be obtained from the National Fire Protection Association, 1 Batterymarch Park, P.O. Box 9101, Quincy, MA 02269-9101.

[5] In the U.S., reporting is required under Section 311 of the CWA, Section 304 of SARA Title III, Section 103 of CERCLA, Section 6 of TSCA, Subtitle I of RCRA, and under HWTA. See Chapter 13 for more information.

[6] In the U.S., the HMTA covers interstate, intrastate, and foreign commerce. The Department of Transportation is responsible for issuing detailed regulations, which can be found in 49 CFR Subchapter C, *Hazardous Materials Regulations.*

[7] In the U.S., the OSH Administration in the Department of Labor is responsible for developing regulations. The rule can be found in 29 CFR 1910.1200.

[8] A good reference for developing a Chemical Hygiene Plan for laboratories is 29 CFR 1910.1450, *Occupational Exposure to Hazardous Chemicals in Laboratories.*

[9] Copies of both standards can be obtained from the ASTM, 1916 Race Street, Philadelphia, PA 19103.

[10] Information on the NAERA certification course or a list of certified NAERA Phase I site assessors can be obtained from NAERA, 6646 Colerain Avenue, P.O. Box 53185, Cincinnati, OH 45253.

FIGURE 7–1.　Checklist for Evaluating Operations and Maintenance Controls:

YES	NO	N/A	Questions to Ask
Responses			**Policies and Procedures**
———	———	———	1. Is there a formal or an informal set of policies for facility operations?
———	———	———	2. Do policies address:
———	———	———	a. attaining and maintaining compliance?
———	———	———	b. maintaining process controls?
———	———	———	c. quality control?
———	———	———	d. preventive maintenance?
———	———	———	3. Is there a set of standard procedures to implement these policies?
———	———	———	4. Are the procedures written or informal?
			5. Do the procedures consider the following areas?
———	———	———	a. safety
———	———	———	b. emergency response
———	———	———	c. laboratory
———	———	———	d. process controls
———	———	———	e. operating procedures
———	———	———	f. monitoring
———	———	———	g. energy conservation
———	———	———	h. waste disposal
———	———	———	i. labor relations
———	———	———	j. equipment maintenance
———	———	———	k. work orders
———	———	———	l. inventory management
———	———	———	m. maintenance planning and scheduling
———	———	———	6. Are the procedures followed?
YES	**NO**	**N/A**	**Organization**
———	———	———	7. Is there an organizational plan for operations?
			8. Does the plan include:
———	———	———	a. delegation of responsibility and authority?
———	———	———	b. job descriptions?
———	———	———	c. interaction with other functions?
———	———	———	9. Is the plan formal or informal?
———	———	———	10. Is the plan available and understood by the staff?

(continued)

FIGURE 7–1. *(Continued)*

YES	NO	N/A	*Organization*
———	———	———	11. Is the plan followed?
———	———	———	12. Is the plan consistent with policies and procedures?
———	———	———	13. Is the plan flexible?
———	———	———	14. Does the plan include emergency procedures?
			15. Does the plan clearly define lines of authority and responsibility in subfunctional areas, such as:
———	———	———	a. laboratory?
———	———	———	b. monitoring practices?
———	———	———	c. process control?
———	———	———	d. mechanical?
———	———	———	e. instrumentation?
———	———	———	f. electrical?
———	———	———	g. waste disposal?
———	———	———	h. supplies and spare parts?

YES	NO	N/A	*Staffing*
———	———	———	16. Is there an adequate number of staff to achieve policies and procedures?
			17. Are staff members adequately qualified for their duties and responsibilities by demonstrating:
———	———	———	a. certification?
———	———	———	b. qualifications?
———	———	———	c. ability?
———	———	———	d. job performance?
———	———	———	e. understanding of processes and equipment?
———	———	———	18. Is staff effectively used?
———	———	———	19. Has the potential for borrowing personnel been considered?
			20. Are training procedures followed for:
———	———	———	a. orientation of new staff?
———	———	———	b. training of new supervisors?
———	———	———	c. continuing training of existing staff?
———	———	———	d. cross-training?

(continued)

FIGURE 7–1.   (*Continued*)

YES	NO	N/A	*Staffing*
			21. Which of the following training procedures are used?
————	————	————	a. formal classroom
————	————	————	b. home study
————	————	————	c. on-the-job training
————	————	————	d. participation in professional organizations
			22. Does the training program provide specific instructions for the following operation and maintenance activities?
————	————	————	a. safety
————	————	————	b. handling emergencies
————	————	————	c. laboratory procedures
————	————	————	d. mechanical
————	————	————	e. instrumentation
————	————	————	f. electrical
————	————	————	g. equipment troubleshooting
————	————	————	h. inventory control
————	————	————	i. handling personnel problems
————	————	————	j. monitoring practices
————	————	————	23. Does management encourage staff motivation?
————	————	————	24. Does management support in-line supervisors?
			25. Is staff motivation maintained with:
————	————	————	a. encouragement for training?
————	————	————	b. job recognition?
————	————	————	c. salary incentives?
————	————	————	d. job security?
————	————	————	e. working environment?

YES	NO	N/A	*Operations*
————	————	————	26. Are operating schedules established and maintained? If so, how?_____
————	————	————	27. Do schedules attempt to attain optimum staff utilization?

(continued)

FIGURE 7–1.   (*Continued*)

YES	NO	N/A	*Operations*
_____	_____	_____	28. Are line supervisors included in personnel scheduling?
_____	_____	_____	29. Is staff involved in and/or informed of personnel planning?
_____	_____	_____	30. Is there sufficient long-term planning for staff replacement and system changes?
_____	_____	_____	31. Are there procedures for manpower staffing in emergency situations?
_____	_____	_____	32. Do process control changes interact with management controls? If so, how?_____ _____ _____
_____	_____	_____	33. Are laboratory results effectively used in process controls? If so, how?_____ _____ _____
			34. How are process controls initiated? _____ _____ _____
_____	_____	_____	35. Are there emergency plans for treatment control?
_____	_____	_____	36. Is there an effective energy management plan? Is the plan used?
_____	_____	_____	37. Are operations personnel involved in the budget process?
_____	_____	_____	38. Do budgets adequately identify and justify the cost components of operations?
_____	_____	_____	39. Are future budgets based on current and anticipated operating conditions?
_____	_____	_____	40. Do operating and capital budget limit constrain operations?
_____	_____	_____	41. Can budget line items be adjusted to reflect actual operating conditions?
YES	**NO**	**N/A**	*Maintenance*
_____	_____	_____	42. Are maintenance activities planned? Is the planning formal or informal?

(*continued*)

FIGURE 7–1.   (*Continued*)

YES	NO	N/A	*Maintenance*
———	———	———	43. Does the facility have management controls sufficient to affect realistic planning and scheduling? If the controls exist, are they used?
———	———	———	44. Are operating variables exploited to simplify maintenance efforts?
———	———	———	45. Are the supply and spare part inventories planned in conjunction with maintenance activities?
———	———	———	46. Have minimum and maximum levels been established for all inventory items?
———	———	———	47. Does the facility have a maintenance emergency plan?
———	———	———	48. Is the maintenance emergency plan current?
———	———	———	49. Is the staff knowledgeable about emergency procedures?
———	———	———	50. Does a plan exist for returning to the preventive maintenance mode following an emergency?
———	———	———	51. Are preventive maintenance tasks scheduled in accordance with manufacturer's recommendations?
———	———	———	52. Is adequate time allowed for corrective maintenance?
———	———	———	53. Are basic maintenance practices (preventive and corrective) and frequencies reviewed for cost-effectiveness?
———	———	———	54. Do the management controls provide sufficient information for accurate budget preparation?
———	———	———	55. Does the maintenance department receive feedback on cost performance to facilitate future budget preparation?
———	———	———	56. Are maintenance personnel involved in the budget process?
———	———	———	57. Do budgets adequately identify and justify the cost components of maintenance?
———	———	———	58. Are future budgets based on current and anticipated operating and maintenance conditions?

(*continued*)

FIGURE 7–1.   *(Continued)*

YES	NO	N/A	*Maintenance*
_____	_____	_____	59. Do maintenance and capital budget limits constrain preventive maintenance?
_____	_____	_____	60. Does the maintenance department receive adequate feedback on cost performance?
_____	_____	_____	61. Can budget line items be adjusted to reflect actual maintenance conditions?

YES	NO	N/A	*Management Controls*
			62. Are current versions of the following documents maintained?
_____	_____	_____	a. operating reports
_____	_____	_____	b. work schedules
_____	_____	_____	c. activity reports
_____	_____	_____	d. performance reports (labor, supplies, energy)
_____	_____	_____	e. expenditure reports (labor, supplies, energy)
_____	_____	_____	f. cost analysis reports
_____	_____	_____	g. emergency and complaint calls
_____	_____	_____	h. process control data, including effluent quality
			63. Do the reports contain sufficient information to support their intended purpose?
_____	_____	_____	64. Are the reports usable and accepted by the staff?
_____	_____	_____	65. Are the reports being completed as required?
_____	_____	_____	66. Are the reports consistent among themselves?
_____	_____	_____	67. Are the reports used directly in process control?
_____	_____	_____	68. Are the reports reviewed and discussed with operating staff?
_____	_____	_____	69. Are summary reports required? If so, what type(s)?_____
			_____
_____	_____	_____	70. Are reports distributed in a timely fashion?
_____	_____	_____	71. Are the proper individuals on a formal distribution list?

(continued)

FIGURE 7–1. *(Continued)*

YES	NO	N/A	*Management Controls (Maintenance)*
———	———	———	72. Does a maintenance record system exist?
			73. Does the maintenance record system include:
———	———	———	a. as-built drawings?
———	———	———	b. shop drawings?
———	———	———	c. construction specifications?
———	———	———	d. capital and equipment inventory?
———	———	———	e. maintenance history (preventive and corrective)?
———	———	———	f. maintenance costs?
———	———	———	74. Is the base record system kept current as part of daily maintenance activities?
———	———	———	75. Is there a work order system for scheduling maintenance? Is it explicit or implicit?
			76. Do work orders contain:
———	———	———	a. date, work order number, location, nature of problem?
———	———	———	b. work requirements, time requirements, assigned personnel?
———	———	———	c. space for reporting work performed, required supplies, time required, cost summary?
———	———	———	d. responsible staff member and supervisory signature requirements?
———	———	———	77. When emergency work must be performed without a work order, is one completed afterward?
———	———	———	78. Are work orders usable and acceptable by staff as essential to the maintenance program? Are they actually completed?
———	———	———	79. Is work order information transferred to a maintenance record system?
———	———	———	80. Does a catalog or index system exist for controlling items in inventory?
———	———	———	81. Are withdrawal tickets used for obtaining supplies from inventory?
———	———	———	82. Do the tickets contain cost information and interact well with inventory controls and the work order system?

(continued)

FIGURE 7–1.   (Continued)

YES	NO	N/A	Management Controls (Maintenance)
_____	_____	_____	83. Is the cost and activity information from work orders aggregated to provide management reports?
_____	_____	_____	84. Is the cost and activity information also used for budget preparation?
_____	_____	_____	85. Is the maintenance performance discussed regularly with staff?
_____	_____	_____	86. How is the cost of contract maintenance or the use of specialized assistance recorded?
_____	_____	_____	87. Are safeguards and penalties adequate to prevent maintenance cards from being returned without the work being done?
_____	_____	_____	88. Is the preventive maintenance record checked after an emergency equipment failure?

Source: Adapted from U.S. Environmental Protection Agency, Office of Water Enforcement and Permits, *NPDES Compliance Inspection Manual* (Washington, D.C.: U.S. Environmental Protection Agency, 1988).

FIGURE 7–4. Prevention of Pollution Checklist for all Industries:

Responses			**Questions to Ask**
YES	**NO**	**N/A**	***Material Receiving: Packaging Materials, Off-Spec Materials, Damaged Containers, Inadvertent Spills, Transfer Hose Emptying***
―――	―――	―――	1. Is a "Just-in-Time" ordering system used?
―――	―――	―――	2. Is the purchasing program centralized?
―――	―――	―――	3. Are package types and quantity selected to minimize packing wastes?
―――	―――	―――	4. Are reagent chemicals ordered in exact amounts?
―――	―――	―――	5. Are chemical suppliers encouraged to become responsible partners (for example, do they accept the return of outdated supplies)?
―――	―――	―――	6. Is there an inventory control program that can track chemicals from ordering to disposal?
―――	―――	―――	7. Is the chemical stock rotated?
―――	―――	―――	8. Is a running inventory of unused chemicals available for other departments' use?
―――	―――	―――	9. Are materials inspected before accepting a shipment?
―――	―――	―――	10. Are material specifications reviewed before accepting a shipment?
―――	―――	―――	11. Are shelf-life expiration dates tracked and validated?
―――	―――	―――	12. Is the effectiveness of outdated material tested?
―――	―――	―――	13. Have shelf-life requirements for stable compounds been eliminated?
―――	―――	―――	14. Are frequent inventory checks conducted?
―――	―――	―――	15. Is a computer-assisted plant inventory system used?
―――	―――	―――	16. Are materials tracked on a periodic basis?
―――	―――	―――	17. Are all containers properly labeled?
―――	―――	―――	18. Are staffed control points established to dispense chemicals and collect wastes?
―――	―――	―――	19. Are pure feeds purchased?
―――	―――	―――	20. Are less critical uses for off-spec material found for materials that would otherwise be disposed of?
―――	―――	―――	21. Are shipping containers reusable?
―――	―――	―――	22. Have attempts been made to substitute less hazardous materials for those that are more hazardous?
―――	―――	―――	23. Are rinsable/recyclable drums used?

(continued)

Figure 7-4. *(Continued)*

YES	NO	N/A	*Raw Material and Product Storage: Tank Bottoms; Off-Spec and Excess Materials; Spill Residues; Leaking Pumps, Valves, Tanks, and Pipes; Damaged Containers; Empty Containers*
____	____	____	24. Is there a Spill Prevention, Control, and Countermeasures (SPCC) plan?
____	____	____	25. Are properly designed tanks and vessels used only for their intended purpose?
____	____	____	26. Do all tanks and vessels have installed overflow alarms?
____	____	____	27. Are all tanks and vessels in good condition and properly maintained?
____	____	____	28. Are there written procedures for all loading/unloading and transfer operations?
____	____	____	29. Are secondary containment areas installed?
____	____	____	30. Are operators instructed not to bypass interlocks or alarms and not to significantly alter set points without authorization?
____	____	____	31. Are equipment or process lines that leak or are not in service properly isolated?
____	____	____	32. Are pumps without seals used?
____	____	____	33. Are bellows-seal valves used?
____	____	____	34. Are all spills and overflows documented?
____	____	____	35. Are overall material balances performed, along with estimates of the quantity and dollar value of all losses?
____	____	____	36. Are floating-roof tanks used for volatile organic compound (VOC) control?
____	____	____	37. Are conservation vents used on fixed-roof tanks?
____	____	____	38. Are vapor recovery systems used?
____	____	____	39. Are containers stored so they can be visually inspected for corrosion or leaks?
____	____	____	40. Are containers stacked so the chance of tipping, puncturing, or breaking is minimized?
____	____	____	41. Is concrete "sweating" prevented by raising the drum off storage pads?
____	____	____	42. Are Material Safety Data Sheets (MSDSs) maintained to ensure correct handling of spills?
____	____	____	43. Is adequate lighting provided in the storage area?
____	____	____	44. Are aisles kept clear of obstruction?

(continued)

FIGURE 7–4. (*Continued*)

YES	NO	N/A	**Raw Material and Product Storage: Tank Bottoms; Off-Spec and Excess Materials; Spill Residues; Leaking Pumps, Valves, Tanks, and Pipes; Damaged Containers; Empty Containers**
_____	_____	_____	45. Are proper separation distances maintained between incompatible chemicals?
_____	_____	_____	46. Are proper separation distances maintained between different types of chemicals to prevent cross-contamination?
_____	_____	_____	47. Are containers stacked against process equipment?
_____	_____	_____	48. Are manufacturers' suggestions followed for the storage and handling of all raw materials?
_____	_____	_____	49. Is the electric circuitry properly insulated and regularly inspected for corrosion and potential sparking?
_____	_____	_____	50. Are larger containers used for bulk storage whenever possible?
_____	_____	_____	51. Are containers sized to minimize potentially wetted areas (a height-to-diameter ratio equal to 1)?
_____	_____	_____	52. Are drums and containers emptied thoroughly before cleaning or disposal?
_____	_____	_____	53. Is scrap paper reused for note pads? Is paper recycled?
YES	NO	N/A	**Laboratories: Reagents, Off-Spec Chemicals, Samples, Empty Sample and Chemical Containers**
_____	_____	_____	54. Are micro or semimicro analytical techniques used?
_____	_____	_____	55. Have efforts been made to increase the use of instrumentation?
_____	_____	_____	56. Have efforts been made to reduce or eliminate the use of highly toxic chemicals in laboratory experiments?
_____	_____	_____	57. Are spent solvents reused/recycled?
_____	_____	_____	58. Are metals recovered from catalysts?
_____	_____	_____	59. Are hazardous waste products treated or destroyed as the last step in experiments?
_____	_____	_____	60. Are individual hazardous waste streams kept segregated? Are hazardous wastes segregated from nonhazardous waste?
_____	_____	_____	61. Are recyclable wastes segregated from nonrecyclable wastes?
_____	_____	_____	62. Are procedures in place to assure that the identity of all chemicals and wastes is clearly marked on all containers? *(continued)*

FIGURE 7–4.   (cont'd)

YES	NO	N/A	**Laboratories: Reagents, Off-Spec Chemicals, Samples, Empty Sample and Chemical Containers**
_____	_____	_____	63. Have mercury recovery and recycling been investigated?

YES	NO	N/A	**Operation and Process Changes: Solvents, Cleaning Agents, Degreasing Sludges, Sandblasting Waste, Caustics, Scrap Metal, Greases from Equipment Cleaning**
_____	_____	_____	64. Is the dedication of process equipment maximized?
_____	_____	_____	65. Are squeegees used to recover residual fluid on product prior to rinsing?
_____	_____	_____	66. Are closed storage and transfer systems used?
_____	_____	_____	67. Is sufficient drain time provided for liquids?
_____	_____	_____	68. Is line equipment used to reduce fluid hold-up?
_____	_____	_____	69. Is a cleaning system used that avoids or minimizes solvents, and is cleaning performed only when needed?
_____	_____	_____	70. Is countercurrent rinsing used?
_____	_____	_____	71. Are clean-in-place systems used?
_____	_____	_____	72. Is equipment cleaned immediately after use?
_____	_____	_____	73. Is cleanup solvent reused?
_____	_____	_____	74. Is cleanup solvent reprocessed into useful products?
_____	_____	_____	75. Are wastes segregated by solvent type?
_____	_____	_____	76. Is solvent usage standardized?
_____	_____	_____	77. Is solvent reclaimed by distillation?
_____	_____	_____	78. Is the production schedule adjusted to lower cleaning frequency?
_____	_____	_____	79. Are mechanical wipers used on mixing tanks?

YES	NO	N/A	**Operation and Process Changes: Sludge and Spent Acid from Heat Exchange Cleaning**
_____	_____	_____	80. Is bypass control used to maintain turbulence during turndown?
_____	_____	_____	81. Are smooth heat exchange surfaces used?
_____	_____	_____	82. Are on-stream cleaning techniques used?
_____	_____	_____	83. Has high-pressure water cleaning replaced chemical cleaning where possible?
_____	_____	_____	84. Has the use of lower-pressure steam been investigated?

Source: Adapted from U.S. Environmental Protection Agency, Solid Waste and Emergency Response, *RCRA Inspection Manual 1993 Edition,* OSWER Directive 9938.02B (Washington, D.C.: U.S. Environmental Protection Agency, 1993).

PART 3

EVALUATING AND IMPROVING PERFORMANCE

Part 3 discusses the different elements involved in continually improving the EMS. Chapter 8 focuses on making sure that collected data are of high quality. This chapter discusses quality assurance/quality control, sampling and analysis, laboratory needs, and corrective and preventive actions, as well as introducing the EMS audit. Chapter 9 provides an overview of the standard's requirements for management review and continual improvement, and different tools that can be used for evaluation and problem solving. Chapter 10 provides an introduction to auditing and selecting auditors, and Chapter 11 provides information that can be used to develop an occupational health and safety management system within the environmental program.

CHAPTER 8

CHECKING AND
CORRECTIVE ACTION

INTRODUCTION

The most important activities in implementing and improving an EMS are measuring, monitoring, and evaluating environmental performance. The standard recognizes the importance of these activities in determining the quality of an organization's environmental performance and establishes key requirements for the gathering and interpretation of a wide variety of data, corrective and preventive actions, record-keeping and information management, and periodic evaluations of the EMS.

REQUIREMENTS FOR MONITORING
AND MEASUREMENT

During the planning phase, considerable attention was given to defining objectives, targets, and performance indicators that were reasonable and measurable. This was a key exercise because performance cannot be evaluated if it cannot be measured. The reliability of generated data (and, therefore, the performance evaluation) depends on how well the monitoring and measurement activities are controlled. The standard (ISO 14001, Subsection 4.5.1) requires the organization "to establish and maintain documented procedures to monitor and measure on a regular basis, the key characteristics of its operations and activities that can have a significant impact on the environment." (Appendix C contains an example procedure for monitoring a measurement.) The wording of the standard does not require all activities and operations to be measured and monitored, only those that *significantly* impact the environment.

An organization should develop procedures for monitoring and measuring in the following areas:

- tracking performance, operational controls, and conformance with objectives and targets,
- evaluating compliance with environmental laws and regulations,
- calibrating and maintaining monitoring equipment, and
- retaining records for calibrating and maintaining monitoring equipment.

INFORMATION TO TRACK PERFORMANCE, OPERATIONAL CONTROLS, AND CONFORMANCE WITH OBJECTIVES AND TARGETS

Documentation is required throughout the standard. Once the organization's objectives, targets, performance indicators, and operational documentation are established and implemented, the information generated by the EMS must be recorded and tracked to determine whether the organization has, in fact, done what it said it would. Procedures are needed to ensure that appropriate records are kept and that data are evaluated to meet the needs of the organization. Statistical process control provides techniques that are often used to evaluate and improve management programs. Some of these methods are discussed in Chapters 4 and 9, and additional useful references for the statistical evaluation of data, control charts, and analysis of process problems are the books by Louvar and McBean. Taking a structured approach to data evaluation will help staff achieve a greater understanding of the dynamics of the organization's operations.

EVALUATING COMPLIANCE WITH ENVIRONMENTAL LAWS AND REGULATIONS

An organization should meet all applicable laws and regulatory requirements, and it should establish and maintain a procedure for periodically evaluating compliance with relevant legislation and regulations. The ISO 14001 standard requires regulatory compliance, but it does not specify how the tracking is to be accomplished.

In the U.S., the combination of permits and compliance auditing fulfills this requirement. Compliance auditing reports are detailed evaluations that seek to determine whether the technical and administrative requirements of laws and regulations have been met by the organization. The regulatory audit is a part of the EMS audit (required in Subsection 4.5.4), which is intended to determine whether the overall EMS conforms to the organization's environmental management plan and the requirements of ISO 14001.

PRODUCING GOOD DATA

The characterization of physical parameters, chemical constituents, and biological organisms is a cornerstone of environmental programs and process engineering. Regardless of the application, the collected data must be representative, accurate, precise, and reliable in order to provide useful information. The production of high-quality data can be ensured by a competent quality assurance/quality control (QA/QC) program, which is essential to preventing, detecting, and correcting problems in the measurement process and to demonstrating statistical control of the measurement process. *Quality assurance* is the management system that ensures that a QC program is in place and effective, whereas *quality control* refers to the day-to-day activities used to control and assess the quality of measurements.

The following discussion identifies the key elements needed to ensure that collected data can be used to evaluate performance. World Wide Web sites and other useful resources for producing quality data are found in Appendix F.

QUALITY ASSURANCE

Quality assurance is an all-encompassing program that utilizes administrative procedures and policies to maximize the performance of personnel, resources, and facilities involved with the collection, measurement, and analysis of data. An effective QA program is needed for all phases of data generation and includes the design of a project/study; sample collection and handling; laboratory or field analysis; sample storage and disposal; data validation, analysis, reduction, and reporting; and data storage systems.

In many instances, QA programs are specified by legal or regulatory requirements (for example, in the U.S. 40 CFR Code of Federal Register Section 122.41 requires laboratory QA for water programs). In the absence of a regulatory requirement, however, best practices and voluntary consensus guidelines and standards form the basis of effective QA programs. When laboratory and field services are contracted to another organization, the contract organization should be evaluated to ensure that it has implemented a QA program that is based on best practices and complies with regulatory requirements for all phases of data generation.

The basic elements of a QA program are:

- organization commitment to quality assurance,
- development of and strict use of the good data collection practices,
- consistent use of documented procedures, and
- established written protocols that are specific to each measurement program.

Implementing a QA program involves defining acceptable error rates, using QC to establish error rates at acceptable levels, conducting assessments to verify conformance to the established error bounds, reporting, auditing, and continual improvement. The entire QA program should be documented in a QA plan, and it should be the policy of the organization to require all personnel to adhere to the requirements of the plan. Typically, the organization would have only one QA plan that would be comprehensive enough to apply to most of the organization's activities. Each project area would then have its individual QA project plan. All policies and procedures should be reviewed periodically and revised as needed. Figure 8–1 shows an example of items that would be included in a comprehensive QA/QC plan.

The organization of the QA/QC program will depend in part on the size and nature of the organization's measurement activities. Larger operations will typically have an organization that includes the QA/QC Officer, Laboratory/ Field Supervisor, and analysts/technicians. Organizations with limited data collection needs may have one designated person for the entire program. The QA/QC Officer has overall responsibility for the program and should report directly to management. This should be a competent person who does not have di-

FIGURE 8–1.   Elements of a Comprehensive QA/QC Plan:

- Organization for laboratory/field activities
- Chain of custody; sample security
- QC targets for precision and accuracy
- Sampling procedures
- Analytical methods and procedures
- Calibration procedures and frequency
- Preventive maintenance and frequency
- QC checks and frequency
- Assessment procedures for evaluating precision and accuracy
- Data reduction, validation, and reporting procedures and frequency
- Performance and system audits and frequency
- Corrective actions
- QA reports

rect responsibilities for field/laboratory work. The QA/QC officer is responsible for developing and implementing the QA/QC program, including statistical procedures and techniques; evaluating data quality; conducting audits and making recommendations for corrective actions; checking corrective actions; scheduling; and writing reports. The Laboratory/Field Supervisor is responsible for supervising the QA/QC program activities, training technical staff, identifying inaccurate data, and developing and implementing corrective actions. The analyst/technician is responsible for adhering to policies, procedures, work instructions, and record-keeping requirements established in the QA/QC plan. This person is responsible for ensuring that field and/or laboratory work activities produce high-quality data.

The QA/QC program should be documented in a written QA/QC manual that identifies the individuals involved in QA/QC activities and their responsibilities, such as operation, maintenance, calibration, data reduction, and auditing activities. The QA manual should include the standard operating procedures, work instructions, and records that provide detailed requirements for specificity, completeness, precision, accuracy, representativeness, and comparability of the generated data.

All work assignments in the laboratory or for field monitoring should be clearly defined. The analysts and technicians should have the proper levels and amounts of education and training. All personnel should understand and follow the specified procedures.

QUALITY CONTROL

Quality control is the routine application of procedures within a sample analysis methodology to evaluate data. It includes the use of proper analytical procedures, audit checks, calibration procedures, and other control activities. The data generated by QC activities are also used to determine the need for corrective actions and to evaluate the effect of corrective measures.

The purpose of the QC procedures is to control the accuracy, precision, and bias of the measurement process. *Accuracy* is a measure of how close the sampled result is to the "true result," and it is reported in percent recovery. *Precision* is a measure of reproducibility of the results, and it can be reported in relative percent deviation, relative standard deviation, or standard deviation. The required levels of accuracy and precision are established in legally required or best practice methods or by the needs of the monitoring program. *Bias* is a systematic error in the measurement process. Examples of errors that could produce bias include improper storage temperatures, calibration errors, improper washing of containers, or even cigarette smoke or other contaminants in the laboratory environment.

Quality control procedures are needed for all field and laboratory activities to provide data that have a high degree of accuracy and precision, and low bias. Even simple procedures (for example, taking a temperature measurement or reading a pressure gauge) can produce invalid results if the specified methods are not followed exactly. Laboratory QC alone will not control the system because errors made in field sampling are not targeted by the laboratory QC program. A field QC program is also needed, and it would include the use of equipment blanks, field blanks, trip blanks, duplicate samples, split samples, and the calibration of field equipment. Examples of laboratory quality control checks include reagent blanks, method blanks, spike samples or matrix spikes, checks on standard and other laboratory calibrations, duplicate samples, laboratory duplicates, internal standards. Although there are different types of duplicates, split samples, spiked samples, blanks, and calibrations, these QC checks are all used to evaluate and improve the precision and accuracy of the laboratory/field data collection system.

Duplicate samples are samples that are taken separately from the same source at the same time. These samples provide a check on sampling equipment and precision. *Split* samples are samples that have been divided into two containers for analysis by separate laboratories. These samples provide a check on the analytical procedures and techniques. *Spiked* samples are samples to which a known quantity of the substance to be analyzed has been added. They provide a check on the accuracy of the analysis. *Blanks* are samples that are treated in the same way as the sample, but they are not exposed to the contaminant of interest. They provide a check on the potential contamination of the reagents and containers used to collect the sample. *Calibration* is the process by which the response of an analytical system is checked against standards with known composition. Calibration is used to check monitors, pressure and other gauges, thermometers, balances and scales, flow delivery, and any other measurement system that produces a quantifiable output.

The following activities are standard ways to evaluate and improve precision:

- Control samples are used as specified by standard or reference methods.
- Duplicate samples are analyzed with each batch of samples as specified by standard or reference methods (typically, 10% of the samples are duplicates).
- Precision control charts or other statistical techniques (such as the mean, standard deviation, variance) for each procedure are prepared and used.

- Corrective actions are taken when data fall outside the warning and control limits.
- The out-of-control data or situation and the corrective action taken are fully documented.

The following activities are standard ways to evaluate and improve accuracy:

- Control samples and calibrations are used as specified by standard or reference methods.
- Spiked samples are analyzed with each batch of samples as specified by standard or reference methods (typically, 10% of the samples are spiked).
- Accuracy control charts for each procedure are prepared and used.
- Corrective actions are taken when data fall outside the warning and control limits.
- The out-of-control data or situation and the corrective action taken are fully documented.

QUALITY CONTROL AND ASSESSMENT

Once the QA/QC program is established and used correctly and consistently, the quality of data can be maintained in a state of statistical control in which random error rates are kept at acceptable and known levels. Statistical control is the evidence that the system conforms to the established criteria, and the most common tool for demonstrating statistical control (or lack of it) is the control chart.

The QC measures discussed in the preceding section are the basis for establishing statistical control and for quality assessment. *Quality assessment* describes the techniques used to evaluate the performance of the field and laboratory activities on an ongoing basis. A basic tool for QC and quality assessment is the control chart. Control charts can help determine how much variability in a process is due to chance (and uncontrollable) and how much is inherent in the process (and controllable). The control chart is a visual tool that can be used to show the variability of duplicates and spike samples, reference materials, and other control measures.

Examples of control charts are shown in Figure 8–2. These charts show the individual data points (no units attached for illustrative purposes) plotted over time. The solid central line defines the best estimate of the variable that is plotted on the chart. The dashed lines are statistically determined upper and lower control limits within which the data must fall if the system is in control. The upper control limit (UCL) and lower control limit (LCL) are typically defined as ±3 standard deviations of the mean or the limits within which 99.7% of the data should fall. The system is out of control when data fall outside of the control limits.

Laboratory control charts also usually show an upper warning limit (UWL) and a lower warning limit (LWL), which are defined as ±2 standard deviations of the mean, or the limits within which 95% of the data should fall. Sometimes, control charts also show a ±1 standard deviation comparison line. Figure 8–2 does not show the ±1 or ±2 standard deviation lines; refer to Figure 9–5 for an example of these comparisons. Data that fall between the warning limits are satisfactory

FIGURE 8–2.   Example of a Quality Control Chart:

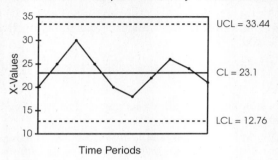

		1	2	3	4	5	6	7	8	9	10
X-Values		20	25	30	25	20	18	22	26	24	21

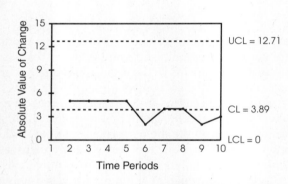

		1	2	3	4	5	6	7	8	9	10
Moving Ranges		0	5	5	5	5	2	4	4	2	3

and in control. Data that fall between the warning limits and the control limits in-dicate that the system is approaching an out-of-control (critical) situation and may require corrective action. When data fall outside of the control limits, the process (analyses) must be stopped immediately and corrective actions must be taken to bring the system into control once again.

The bottom graph in Figure 8–2 is a moving range chart that shows the ab-solute values of differences between subsequent readings plotted on a chart that also includes the upper and lower control limits. This chart provides useful infor-mation for spotting trends or repetitive patterns, and it also allows the user to de-termine whether the system is in control.

Auditing, which can include system, performance, and data audits, is another means of assessing quality. *System audits* are qualitative checks to ensure that the elements of the QA program are operational. *Performance audits* involve the analysis of performance evaluation samples, and they provide quantitative data assessment. *Data audits* are comprehensive random checks of all activities involving selected samples (from the time of sample container preparation to disposal of the sample). Selected samples would be checked for documentation, calculations, calibrations, transcriptions, report formats, and other parameters defined in the QA/QC plan. Auditing can also include interlaboratory and collaborative laboratory testing programs.

Any corrective actions that are recommended to bring the system back into control must be implemented, evaluated to ensure the system is in control, and reported. The corrective action may require revisions to documentation. Any changes should be controlled.

SAMPLE COLLECTION AND HANDLING

Properly designed and implemented sampling programs are essential to meeting the objectives and targets of the EMS. The type of sampling that is needed depends on the program objectives. Some sampling will involve surveys to gain a picture of existing conditions; others are directed at characterizing point sources. Samples can be collected at a single point in time (grab sampling), integrated over time (composite sampling), or collected on a continuous basis (continuous monitoring). Sample collection and analysis can be conducted manually or with automatic samplers and monitors. Regardless of the nature of the sampling, the sampling program must be planned and documented. The sample collection program should include:

- identification of the sampling site (including a site map)
- sample matrix (air, surface water, wastewater, soil, and so forth)
- number of samples in each matrix
- frequency of collection (daily, monthly, quarterly, and so forth) and duration of sampling program
- type of samples (grab, composite, continuous)
- method of sample collection and needed equipment
- needed analytical parameters (including method numbers and references)
- required field measurements and needed equipment
- transport requirements
- sampling and transport QC requirements
- required documentation

The sampling program is a written document that allows sample collection and handling to be conducted in a way that fulfills legal requirements and prevents the degradation of samples by improper collection or storage techniques. All documents, including records, should be readily available and understandable to those who use them.

The actual sample collection and handling procedures will vary, depending on the contaminant and collection medium, but common elements exist. These include:

- using proper sample collection techniques
- using appropriate containers/equipment and cleaning procedures for containers/equipment
- using properly calibrated equipment such as flowmeters, critical orifices, and others
- collecting QC checks such as blanks, duplicates, spikes at the proper level
- properly identifying samples
- using correct preservatives and holding times
- utilizing proper transport and storage temperatures
- properly packaging and shipping samples
- maintaining control over the sample through chain-of-custody procedures

Figure 8–3 incorporates these elements into a checklist that the U.S. Environmental Protection Agency (EPA) uses to evaluate the sampling program at a wastewater treatment facility. This checklist could easily be modified for other types of sampling.

Regardless of the type of sample, sampling methods must be those required by regulations (for example, in the U.S. 40 CFR Part 136 for water samples) or best practices. Further, the collected sample must be consistent with the reason for sampling. For instance, if the purpose of sampling is to evaluate the average concentrations of contaminants in a wastewater stream, composite sampling may be appropriate. However, composite sampling would not be appropriate if the sampling were conducted to determine worst-case concentrations. In many cases, sample locations are specified in regulatory permits. Proper sampling may re-

quire a knowledge of statistical methods such as those discussed in the book by McBean. A chain-of-custody record is required by many regulations, and it should be considered a best practice to reduce the chances of accidental or inten-

tional tampering of the sample. Proper chain-of-custody means that anyone who handles the sample, from collection to final disposition, must identify themselves and sign a document acknowledging receipt of the sample. Figure 8–4 shows an example chain-of-custody form for collecting a water sample.

LABORATORY REQUIREMENTS

The generation of high-quality data requires the use of proper analytical methods, equipment, and supplies in the appropriate environments. Analytical requirements will be specified in permits or as best practices developed by professional organizations. In some instances, government permits may allow alternative methods with the approval of the regulating agency. If so, the required written approvals should be available for inspection.

Because the laboratory uses and generates hazardous chemicals, worker health and safety is an important issue. Laboratory safety and hygiene programs should be an integral part of laboratory functions. Hazardous materials should be

FIGURE 8–3.   Checklist for Evaluating a Water Pollution Sampling Program:

Responses			Questions to Ask
YES	**NO**	**N/A**	
————	————	————	1. Are samples taken at sites specified in the permit?
————	————	————	2. Are locations adequate for the collection of representative samples?
————	————	————	3. Are flow-proportioned samples obtained when required by the permit?
————	————	————	4. Are sampling and analysis completed on parameters specified by the permit?
————	————	————	5. Are sampling and analysis done at the frequency specified by the permit?
————	————	————	6. Are the sampling methods as specified in the permit?
			Required method:_____
			If not, method used is ____ grab ____ manual ____ composite ____ automated composite
————	————	————	7. Are sample collection procedures adequate? Do they include:
————	————	————	a. samples refrigerated during transport?
————	————	————	b. proper preservation techniques used?
————	————	————	c. containers and sample holding times before analysis in conformance to regulatory or best management requirements?
————	————	————	d. samples analyzed in time frame needed?
————	————	————	e. chain-of-custody procedures employed?
————	————	————	f. adequacy of quality control?
————	————	————	8. Are monitoring and analyses performed more often than required by the permit? If so, are results reported in the self-monitoring report?
————	————	————	9. Do samples contain chlorine?
————	————	————	10. Is a contract laboratory used for sample analysis?
			If so, name: _____
————	————	————	11. Does the publicly owned treatment works (POTW) collect samples from industrial users in the pretreatment program?

Source: Adapted from U.S. Environmental Protection Agency, Office of Water Enforcement and Permits, *NPDES Compliance Inspection Manual* (Washington, DC: U.S. Environmental Protection Agency, 1988).

FIGURE 8–4.   Example Chain-of-Custody Form:

ABC Company

Water Sampling Record   Revision No.: _____
Sampling Record No.   Revision Date: _____
                                                               Page _____ of _____

Sample Collected by_____

Date _____   Time _____ am/pm

Facility Sampled _____

Facility Location_____

Sampling Location _____

Sample Type _____grab _____ composite

Observation/Comments_____

_____

_____

Sample Bottle ID Marking _____

Samples Split with Agency/Laboratory? _____ yes _____no

Name of Agency/Laboratory Representative _____

Title of Agency/Laboratory Representative _____

Sample _____

Time/Date _____

Received by _____ Affiliation, Title _____

Comments _____

Time/Date _____

Released to _____ Affiliation, Title _____

Comments _____

---

identified, stored, handled, and disposed of properly. Material safety data sheets or similar documentation must be available in a readily accessible location. It may seem obvious that adequate and properly designed physical space is needed to perform laboratory activities, but this is an area that is often poorly planned. There must be adequate fume hoods, general space ventilation and other environmental controls, emergency equipment, and special storage and disposal containers. It is especially important to provide separate rooms for eating. The laboratory must use approved methods; conduct calibration and performance checks on the analytical systems; properly manage laboratory instruments, equipment, and supplies; control samples and documentation; and follow the QA/QC plan. Requirements for the physical space, analytical methods, instruments, equipment, and supplies are enumerated below:

Requirements for Physical Space

- Adequate bench, instrumentation, storage, and record-keeping space, as well as space for secure and restricted access to samples
- Adequate environmental controls (temperature, humidity, lighting, ventilation, filtration, and other air-cleaning)
- Adequate and efficient fume hood system
- Electrical power and, if appropriate, voltage-regulated sources for delicate instruments
- Vibration-free area for accurate weighing and other procedures
- Adequate flammable and corrosive storage containers
- Adequate and appropriate containers for storing hazardous wastes until collected and transported by disposal company
- Adequate supplies of dry, uncontaminated compressed air
- Adequate supplies of laboratory pure water, free from chemical interferences and other undesirable contaminants
- Emergency equipment, fire extinguisher, eye wash station, shower, first aid kit, gloves, and goggles

Requirements for Analytical Methods, Instruments, and Equipment

- Use of analytical methods, instruments, and equipment that conforms to regulatory requirements or industry best practices
- Needed supporting equipment such as hot plate, refrigerator for samples, pH meter, thermometer, and balance
- Written documentation for the operation, maintenance, and calibration of instruments and equipment
- Proper laboratory custody (sample bottles, sample log-in, holding time and storage of samples, sample preparation for analysis, documentation, disposal of samples and products)
- Controlled documentation
- Written schedules for replacement, cleaning, operational checks, calibration, and other procedures

- Use of a QC system that conforms to regulatory requirements or industry best practices and that includes:
  — QC checks and records on purchase of equipment; instrument calibration and maintenance, and standard solutions and gases
  — Use of standard procedures for cleaning glassware and containers
  — Use of appropriate primary and secondary calibration standards (reagents, gases) that closely bracket actual sample levels
- Documentation of any regulatory-approved deviation from specified test methods

Requirements for Supplies

- Use of required reagent purity for specific analytical methods
- Standard reagents, solvents, and gases stored according to manufacturer's instructions
- Appropriate checks on calibration gases, stock and working solutions, and other calibration materials
- Method of tracking shelf-life of laboratory supplies and compliance with discard dates
- Proper labeling of standards and reagents

DATA HANDLING AND REPORTING

Laboratory and field measurements involve the use of numerous records and the generation of considerable data. Documentation begins with the preparation of sampling containers and reagents, and continues as samples are received, stored, analyzed, and disposed of.

Some data are recorded and reported without further manipulation. For example, a pressure gauge reading in psi is recorded directly onto the appropriate form. In other instances, mathematical or statistical calculations are used to convert the raw data into reportable data. All documentation related to raw and calculated data must be easy to locate, identifiable, and accessible. Examples of needed documentation and documentation that should be saved include: field and laboratory notebooks, worksheets, strip charts, chromatograms and other instrument outputs, computer-generated analytical output, QC charts, method detection limits, records concerning receipt of stock solutions and preparation of calibration standards, laboratory custody reports, sample preparation logs, calculations, maintenance and calibration logs, and final reports. All of these documents should be controlled and saved for appropriate periods of time, which might range from 3 years up to 10 years or more if the records are involved in a legal case.

The generated data must be reviewed and approved. This process involves all laboratory personnel, beginning with the analyst/technician and ending with the QA/QC Officer. When a QC error is identified, the source of the error must be established and corrected before testing can continue. All data collected since the last known acceptable QC check become suspect once an error is identified. The entire process of error detection and correction must be documented. The analyst/technician prepares the QC chart that shows out-of-control data or prepares other similar documentation. The Laboratory Supervisor issues a noncompliance report that includes corrective actions to the QA/QC Officer who, in turn, issues a written report to the Laboratory/Field Manager, defining a QC noncompliance event, along with any suspected sample data. Figure 8–5 shows an example of a noncompliance report form for a laboratory. This form identifies when the noncompliance occurred, what test method was used, and who did the test; and it also details information about the noncompliance, including the samples involved and the QC test results and performance limits.

FIGURE 8–5.   Example of a Noncompliance Report Form for Laboratory Data:

```
ABC Company
Laboratory Noncompliance Report

Date: _____

Noncompliance Test Method: _____

Date of Noncompliance: _____

Name of Analyst:_____


Suspect Sample ID    QC True Value      QC Measured Value     %Recovery      %Recovery Limit

_____      _____      _____       _____      _____

_____      _____      _____       _____      _____

_____      _____      _____       _____      _____

_____      _____      _____       _____      _____

Cause of Noncompliance:
_____
_____

Corrective Action:
_____
_____
_____


_____        _____
Laboratory Supervisor        Date
Rev. 1    09/96
RKB
```

The evaluation of data handling and reporting must be uniform and appropriate to the use of the data. Typical checks during an audit might include determining whether :

- Correct formulas are used to calculate final results
- Round-off rules are uniformly applied
- Significant figures are established for each analysis
- Provisions are available for cross-checking calculations
- Control chart approaches and statistical calculations have been determined for the purposes of QC and reporting
- The report forms provide complete data documentation and permanent records, and facilitate data processing

- Computer-based programs for data handling provide data in the form/units required for reporting
- Laboratory records are kept readily available and for periods of time specified by regulation or best practices
- Laboratory notebooks or preprinted data forms are bound permanently to provide good documentation, including the procedures performed and the details of the analysis, such as the original value recorded, correction factors applied, blanks used, and the reported data values; The notes are dated; They indicate who performed the tests and include any abnormalities that occurred during the testing procedure; The notes are retained as a permanent record
- All data handling documents are controlled

Figure 8–6 at the end of this chapter shows an example of a checklist that could be used to evaluate a QA program for a laboratory. The checklist include general questions about the laboratory and QC procedures, data handling and reporting, laboratory facilities and equipment, precision and accuracy, personnel, and safety.

CORRECTIVE AND PREVENTIVE ACTION

The standard (ISO 14001, Subsection 4.5.2) requires the organization to establish and maintain procedures for corrective and preventive actions. These actions should be integrated into monitoring activities, audits, and other reviews of the EMS. Findings from each of these activities should be documented, and the needed corrective and preventive actions should be identified. Management should ensure that the actions are implemented and that there is systematic follow-up to demonstrate effectiveness.

A corrective and preventive action procedure should:

- define who has the responsibility and authority for handling and investigating nonconformances
- identify and implement the corrective action,
- implement or modify controls needed to prevent recurrence of the nonconformance,
- record any changes in written procedures resulting from the corrective and preventive action, and
- document the actions.

The corrective or preventive action should be appropriate to the magnitude of the problem and resulting environmental impact. This means that the actions might be narrow or broad in scope, simple or complex, or completed in a single time frame or require extended periods for completion. For example, assume that an audit shows that monitoring equipment is repeatedly out of calibration. Further investigation shows that the calibration technician has not been formally trained in the given calibration techniques, and the monitoring equipment has not been maintained as required by the manufacturer. Two solutions include providing the technician with proper training now and periodically thereafter, and instituting a maintenance schedule. These corrective actions are relatively straightforward and short term. On the other hand, if a spill is caused by a leaking underground storage tank, the corrective action may require removal of the tank and replacement with a new tank. In addition, a corrective action plan will be needed for the contaminated soil. This situation is likely to require more time and effort to correct. Figure 8–7 shows an example of a corrective action form for correcting leaks in the clothes cleaning equipment at a dry cleaning facility that uses perchloroethylene, a hazardous material. This form provides information about the initial inspection that detected the problem (date, machine number, inspector, a description of the problem, and whether or not parts were needed to correct the problem). It also provides reminder guidance about repairing leaks within specified time periods and the need to complete the written log for entering repairs. The form makes it easy to know when parts were ordered, received, and installed. Tracking progress in correcting problems and understanding the repair history of equipment becomes easy with forms similar to this one.

FIGURE 8–7.   Corrective Action Form for a Dry Cleaning Establishment:

CORRECTIVE ACTION FORM

Date of Initial Inspection _____

Machine No _____

Inspecton _____

DESCRIBE PROBLEM:

ARE PARTS NEEDED?   ☐ YES   ☐ NO

  PARTS ORDERED _____

  DATE RECEIVED _____

  DATE INSTALLED _____

DATE PROBLEM CORRECTED _____

EXPLAIN:

If leaks are detected:
1. All leaks must be repaired within 24 hours.
2. If repairs are required, parts must be ordered within two (2) working days.
3. All parts received must be installed within five (5) working days.
4. A written log must be kept of all repairs made.
Rev.1   07/02/96

Source: Adapted from *Regulatory Compliance Manual for Indiana Perchloroethylene Dry Cleaners,* (Indianapolis, IN: Indiana Department of Environmental Management and Indiana Dry Cleaning and Laundry Association, 1995). This material may be used, in whole or in part, without permission. As such, the copyright on this book does not apply.

RECORDS

The standard requires the organization to "establish and maintain procedures for the identification, maintenance, and disposition of environmental records" (ISO 14001, Subsection 4.5.3). The records that must be maintained include any records that are needed to meet the requirement of the standard or the EMS. Examples of needed records include: inspection, maintenance, and calibration records; training records; audit records; incident reports; emergency response records; contractor and supplier records; and any records required by law or regulations. Figures 8–8 and 8–9 show two examples of records. Figure 8–8 is a form that is used to track the concentration of perchloroethylene (a hazardous material) that exits the control device intended to collect the material. This form requires the inspector to identify the machine number and the measured concentration of the perchloroethylene. If the concentration is greater than 100 ppm, corrective action is needed. Figure 8–9 is an example of a daily record that allows tracking of equipment downtimes for operators at a metal recycling facility. There are many potential formats for records, and Chapter 5 provides more details on developing and properly managing these important documents.

Records should be legible and should contain sufficient information to reconstruct the activity being documented. Records should be managed using a system that allows them to be retrieved easily. This requires records to be identifiable and traceable to the relevant activity, product, or process. The documentation system discussed in Chapter 5 provides one way to identify and track records. Although electronic data management systems are not required by the standard, they do facilitate retrieval and tracking of records. In many instances, electronic records can be created by the use of portable computers equipped with the requisite software.

Records should be protected against damage, deterioration, or loss. Backup records stored in a separate location are always a prudent idea. Records should be maintained in fire-rated rooms or cabinets that are locked. These requirements apply to both paper and electronic record-keeping systems. Records must be maintained for time periods that are specified in the EMS documentation. However, keeping records beyond needed times may not be a good idea as they take up space and may have legal implications in some situations.

The need to have secured access to some records and documents is a topic that requires careful consideration. Access to records should be limited, but some records may need special protection because they contain business information that is confidential or not appropriate for the courts or other audiences. Examples of information that may need special protection include any potential or pending legal action, preliminary findings of significant contamination, significant health and safety impact resulting from any work-related activity, and sensitive issues involving the public or problem employees. In the U.S., these records can be protected under the attorney-client privilege. Whenever special protection is needed, the attorney should be involved as soon as possible. In many instances, this means that the attorney will control the flow of information and initiate requests for documentation.

FIGURE 8–8. Weekly Concentration Log for a Dry Cleaning Establishment:

Carbon Adsorber Weekly Perc Concentration Log

Directions: 1. Measure the concentration of the perc in the exhaust duct after the carbon adsorber.
2. If concentration is greater than 100 ppm, attach a complete Corrective Action Report.

Date	Inspector's Initials	Machine Number	Concentration (ppm)	Concentration > 100 ppm?
				YES/NO
				YES/NO
				YES/NO
				YES/NO
				YES/NO
				YES/NO
				YES/NO
				YES/NO
				YES/NO
				YES/NO
				YES/NO
				YES/NO
				YES/NO
				YES/NO
				YES/NO

Rev.1   07/02/96

Source: Adapted from *Regulatory Compliance Manual for Indiana Perchloroethylene Dry Cleaners,* (Indianapolis, IN: Indiana Department of Environmental Management and Indiana Dry Cleaning and Laundry Association, 1995). This material may be used, in whole or in part, without permission. As such, the copyright on this book does not apply.

AUDITS OF THE EMS

The standard requires the organization to establish and maintain a program and procedures for periodic audits of the EMS (ISO 14001, Subsection 4.4.4). The purpose of the auditing program is to determine whether or not the EMS has been implemented and maintained so that it works as intended. If the organization seeks ISO certification, the EMS must conform to the requirements established in the standard. Chapter 10 provides an overview of auditing and qualifications of the audit team.

FIGURE 8–9.   Daily Record for an Equipment Operator at a Metal Recycling Facility:

Daily Readings Sheet for Fixed Operators

OPERATOR

DAYS OF WEEK

MON	TUE	WED	THURS	FRI	SAT

DATE   EQUIPMENT

METER READINGS   START

SHIFT

FIRST	SECOND	THIRD

ITEMS TO BE REPAIRED, IDEAS,
IMPROVEMENTS, REASON FOR
DOWNTIME—PLEASE BE SPECIFIC

DOWNTIME   HOURS

METER READINGS   DOWN

OPERATOR PM   COMPLETED

START TIME	END TIME	ACTIVITY COMPLETED	COMMODITY	CYCLES

25-Apr-96   CH
Rev. 1   Page 1 of 1

Source: Adapted, courtesy of American Iron, Minneapolis, MN.

SUMMARY

The purpose of the EMS is to ensure the achievement of the objectives and targets established during the planning phase. For the EMS to function properly, a system of checking and corrective action must be integrated throughout the EMS. The standard requires documented procedures for monitoring and measuring the key characteristics of the organization's activities and operations, including environmental performance indicators, regulatory compliance, and monitoring equipment. The organization must also develop procedures for handling nonconformances and auditing the EMS. An integral part of the checking and corrective action process is the operation of a comprehensive QA/QC program that can prevent, detect, and correct problems with laboratory, process, and field data.

FIGURE 8–6.   Checklist for Evaluating Laboratory Quality Assurance:

Responses			**Questions to Ask**
YES	**NO**	**N/A**	*General*
_____	_____	_____	1. Are EPA (regulatory)-approved analytical testing procedures used and on hand-in written form?
_____	_____	_____	2. If alternate analytical procedures are used, are proper approvals obtained?
_____	_____	_____	3. Are the calibration and maintenance of instruments and equipment satisfactory?
_____	_____	_____	4. Are QC procedures used?
_____	_____	_____	5. Are QC procedures adequate?
_____	_____	_____	6. Are duplicate samples analyzed? _____ % of time.
_____	_____	_____	7. Are spiked samples used? _____ % of time.
_____	_____	_____	8. Is a commercial laboratory used?
			Name _____
			Address _____
			_____
			Contact _____
			Telephone _____
			Certification No. _____
			Comments _____
			_____
			_____
YES	**NO**	**N/A**	*Laboratory Facilities and Equipment*
_____	_____	_____	9. Is the proper grade of laboratory pure water available for specific analyses?

FIGURE 8–6. (Continued)

YES	NO	N/A	*Laboratory Facilities and Equipment*
___	___	___	10. Is dry, uncontaminated, compressed air available?
___	___	___	11. Are an adequate number of properly operating fume hoods available?
___	___	___	12. Are adequate electrical sources available?
___	___	___	13. Are instruments/equipment in good condition?
___	___	___	14. Are written requirements for daily operation of instruments available?
___	___	___	15. Are written troubleshooting procedures for instruments available?
___	___	___	16. Are standards and appropriate blanks available to perform daily check procedures?
___	___	___	17. Are schedules for required maintenance available?
___	___	___	18. Is the proper volumetric glassware used?
___	___	___	19. Is glassware properly cleaned?
___	___	___	20. Are standard reagents and solvents properly stored?
___	___	___	21. Are working standards checked frequently?
___	___	___	22. Are standards discarded after the recommended shelf-life has expired?
___	___	___	23. Are background reagents and solvents run with every series of samples?
___	___	___	24. Do written procedures exist for cleanup and hazard response methods. Are cleanup response materials adequate and available?
___	___	___	25. Are gas cylinders replaced at 100–200 psi?

Comments_____

_____

_____

YES	NO	N/A	*Laboratory's Precision, Accuracy, and Control Procedures*
___	___	___	26. Are multiple replicates (blanks, duplicates, spikes, and splits) analyzed for each type of control check, and is the information recorded?
___	___	___	27. Are plotted precision and accuracy control methods used to determine whether valid, questionable, or invalid data are being generated from day to day?
___	___	___	28. Are control samples introduced into the train of actual samples to ensure that valid data are being generated?
___	___	___	29. Are the precision and accuracy of the analyses within acceptable bounds?

FIGURE 8–6.   (*Continued*)

Comments _____

_____

_____

YES	NO	N/A	*Data Handling and Reporting*
____	____	____	30. Are round-off rules uniformly applied?
____	____	____	31. Are significant figures established for each analysis?
____	____	____	32. Are provisions for cross-checking calculation used?
____	____	____	33. Are correct formulas used to calculate final results?
____	____	____	34. Are control chart approaches and statistical calculations for QC checks and reporting available and followed?
____	____	____	35. Are report forms developed to provide complete data documentation and permanent records to facilitate data processing?
____	____	____	36. Are data reported in the proper form and units?
____	____	____	37. Are laboratory records readily available and stored for required times?
____	____	____	38. Are laboratory notebooks or preprinted data forms bound permanently to provide good documentation?
____	____	____	39. Do efficient filing systems exist, enabling prompt routing of report copies?
____	____	____	40. Is all documentation properly controlled?

Comments _____

_____

_____

YES	NO	N/A	*Laboratory Personnel*
____	____	____	41. Are enough analysts present to perform the needed analyses
____	____	____	42. Do analysts have on hand the necessary references for the required or standard method procedures being used?
____	____	____	43. Are analysts competent by education?
____	____	____	44. Are analysts trained in procedures performed through formal or informal training or certification programs?

FIGURE 8–6.   (*Continued*)

Comments_____

_____

_____

YES	NO	N/A	*Laboratory Safety*
_____	_____	_____	45. Does the laboratory have a Laboratory Safety Program?
_____	_____	_____	46. Does the laboratory have a Chemical Hygiene Program?
_____	_____	_____	47. Are all chemicals, radioactive materials, biological agents, and gases properly labeled, segregated, and stored?
_____	_____	_____	48. Are an adequate number of fume hoods available?
_____	_____	_____	a. Are fume hood flows tested and acceptable?
_____	_____	_____	b. Are fume hoods free of chemicals when analysts are not working in the hoods?
_____	_____	_____	49. Is adequate flammable and corrosive material storage available?
_____	_____	_____	50. Are electrical outlets grounded?
_____	_____	_____	51. Are an adequate number of eyewash stations and shower stations available and functioning?
_____	_____	_____	52. Are wastes properly collected, stored, and disposed of?
_____	_____	_____	53. Are adequate first aid and fire extinguishing units available?
_____	_____	_____	54. Is there an adequate blood-borne pathogen program?

Comments_____

_____

_____

Source: Adapted from U.S. Environmental Protection Agency, Office of Water Enforcement and Permits, *NDPES Compliance Inspection Manual* (Washington, D.C.: U.S. Environmental Protection Agency, 1988).

CHAPTER 9

MANAGEMENT REVIEW AND CONTINUAL IMPROVEMENT

INTRODUCTION

Up to this point, the EMS planning and implementation phases have been discussed. If an organization follows all of these steps, it will have policies, procedures, work instructions, and a record-keeping system in place. Employees and management will be trained and aware of environmental issues affecting the organization. A system of measurement and evaluation will be operational. Also, everyone in the organization will have an increased awareness of the importance of achieving environmental excellence. However, even these significant accomplishments cannot ensure long-term success. Once the system is implemented, there must be periodic and continual review of the EMS to ensure that all parts are functioning efficiently. The requirements for review and improvement are discussed in the last section of the standard (ISO 14001, Section 4.6). This section, which is deceptively short, is crucial to the ongoing success of the EMS.

Once the EMS is established, it must be reviewed, evaluated, and modified as needed. The review process should cast a wide net and should examine all aspects, products, and services for environmental performance. The review process should cover both scientific data and management data. For example, the review should consider impact on financial performance, and marketplace position might even be considered. *Continual improvement* is defined by the standard as the "process of enhancing the environmental management system to achieve improvements in overall environmental performance in line with the organization's environmental policy" (ISO 14001, Section 3.1). This definition ties all of the elements in the EMS back to the environmental policy—the starting point of developing the EMS. Continual improvement provides an unending feedback loop through all parts of the EMS, and it ensures that the EMS will not be stagnant. The continual improvement program should reflect a dynamic process that routinely evaluates the organization's commitment to quality.

MANAGEMENT REVIEW

The standard requires top management to collect the needed information and to conduct the review at appropriate intervals. The review should be frequent enough to ensure the "continuing suitability, adequacy, and effectiveness" of the EMS (ISO 14001, Section 4.6). (Appendix C contains an example procedure.) The review goes beyond the narrow confines of determining regulatory compliance status. It should be broad enough to consider the impact of all environmentally related activities on the organization, yet it must be more than a superficial check. When performance failures are observed, a good review looks for interrelationships and causal chains to determine the root causes of problems. The review should be comprehensive, but it does not have to be completed at one time. Continual improvement of the EMS should be the organization's overall goal and the outcome of the review process.

The review should include the following elements:

- Objectives, targets, and performance indicators. Objectives and targets must be evaluated against the established performance indicators to determine whether the EMS is meeting the established goals. These reviews should be conducted by the managers responsible for defining them, with input from staff working in the areas where performance indicators are measured.
- Overarching environmental policy and supporting policies and procedures. These key elements must be reviewed by the managers who defined them to determine whether they are suitable and adequate based on performance indicators, changing conditions, feedback from internal or external audits, and other factors.
- Findings of internal and external audits. The EMS audit will determine whether the EMS conforms to the criteria specified in ISO 14001. For instance, are the required procedures in place? The EMS audit evaluates the performance of the EMS, not the organization's environmental performance. Other internal and external audits can provide additional information about environmental performance.
- Consideration of changing conditions and information. Organizations are constantly changing in response to pressures from the marketplace; changing regulatory and legal requirements; changing expectations and requirements of consumers, customers, and other stakeholders; and advances in science and technology. Some of these changes are relatively modest, though others—such as an environmental accident or the production of new products, services, or activities—may dramatically alter the way a business conducts its operations. The fact that organizations change requires the elements of change to be considered during the EMS review process. Changing conditions may require a modification of the organization's environmental policy.

All review activities should be documented, along with the conclusions and recommendations for necessary action. The review reports, which should be produced and communicated to the appropriate parties in a timely fashion, are the basis of the continual improvement activities.

To be successful, the review process cannot be accomplished in a vacuum. Even if consultants are used, there must be significant employee involvement to generate information needed to identify problems and to determine whether corrective actions are effective. It is also important to consider the views of interested parties during the review of the EMS. These outside perceptions can provide helpful insights into strengths and weaknessses of the EMS and can give the organization first-hand knowledge about the impact of the EMS on stakeholders.

Some basic questions to consider before conducting a review of the EMS are:

- Who will conduct the review? Are employees involved? If so, how?
- How frequently will the review be done?
- What elements of the EMS will be reviewed?
- What information will be included in the review process?
- Are the views of interested parties considered? If so, how?
- How will information be evaluated?
- How will the results of the review be documented and disseminated?
- How and when will corrective and preventive action plans be developed?
- Who is responsible for implementing and tracking corrective and preventive action plans?

CONTINUAL IMPROVEMENT

Once an organization makes the commitment to develop an EMS, it has started down the road to continual improvement. The standard institutionalizes continual improvement throughout the EMS process by requiring an explicit commitment to continual improvement in the environmental policy statement. This commitment, in turn, requires a process of checking and corrective action for all individual procedures and activities, and extends to the overall EMS through the management review and follow-up activities. Everyone in the organization should be aware of opportunities to improve performance—continual improvement should become part of the organization's way of life.

A comprehensive effort, such as the EMS, is seldom produced without flaws. It must be used over a period of time to identify what does and does not work. It is likely that several iterations will be necessary before major flaws (special causes of error) are corrected. Once these are controlled, attention can be given to the common causes that produce the majority of problems. Part of a program of continual improvement is evaluating progress toward objectives and targets. Is the organization progressing? Is the progress fast enough? Are there failures to reach objectives and targets that could result in significant liability or cleanup costs? The continual improvement process should also search for opportunities to improve the EMS that, in turn, will lead to improved environmental performance.

The process should determine the underlying reasons why the EMS does not conform to the environmental policy and the established objectives and targets. There are a number of statistical process control methods that can help identify the reasons for problems; some of these are discussed below. Once the root causes of problems are identified, a mechanism must exist to develop a plan to correct the problem and to prevent it from recurring in the future. The plan should emphasize preventive actions whenever possible—focus on preventing problems, not on correcting them. Once the corrective and preventive action plan is implemented, it must be verified. If the problem was not corrected or prevented, the plan must be modified or, in some cases, a new plan will be needed. A common problem at this stage of the process is a failure to update documents to reflect changes to the EMS. When it is determined that corrective and preventive actions are controlling those elements that they were designed to control, it is important to update documentation throughout the system. This includes policies (if necessary), procedures, work instructions, and records. All document changes must be controlled; that is, old documents must be removed and destroyed or archived and new documents must be issued.

Although continual improvement does not mean that all problems must be identified and corrected at once, it does require: 1) a process for identifying opportunities for improvement and 2) a process for verifying that corrective and preventive actions and improvement are effective and timely.

Some questions to help evaluate the continual improvement process are:

- How are corrective and preventive actions and opportunities for improvement identified?
- If weaknesses are identified, what is done about them?
- Who communicates needed actions? Are communications timely and effective?
- When changes are made, who checks to see whether they are effective? When?
- Are changes documented? Is the documentation controlled?

TOOLS FOR CONTINUAL IMPROVEMENT

There are many tools that can be used to solve performance problems. Some of these, such as the checklist, are very easy to use; others, such as quality control charts, require more sophisticated statistical techniques. All of these techniques, however, can be learned and applied to a variety of problems.

CREATING A CLIMATE FOR IMPROVEMENT

It is especially important for the organization to recognize that people who work directly with the processes and activities of the organization are the most knowledgeable about them. Seek input from managers and staff at all levels of the organization and from suppliers and vendors. Ask them for information about potential problems and for help in developing solutions to documented problems, but make sure that people believe they can offer suggestions and constructive criticism without retribution. Employees are often afraid they will lose their jobs or be punished if they reveal problems. Suppliers and vendors have a desire to please so that an organization will continue to buy from them or use their services. They may be afraid that any suggestions will be taken as criticism and hurt business opportunities. Take Deming's eighth point, "Drive out fear," to heart.[1] Encourage people to participate in the problem-solving process by fostering a climate of mutual respect, openness, and willingness to change. Suggestions, once offered, should be taken seriously.

The Plan, Do, Check, and Act Cycle

The Plan, Do, Check, Act (PDCA) cycle is a popular approach to solving problems in the quality movement. This form of analysis is similar to the Shewhart cycle, which was developed by Walter Shewhart during the 1930s for industrial applications; it is also known as the *Deming circle*.[2] Regardless of its name, this approach provides a set of systematic techniques that can improve decision making.

This process begins by assembling a team to consider a problem. The *Plan* phase allows the team to develop an understanding of reasons for a problem and to formulate a plan for correcting them. For example, assume that lead emissions at a secondary lead smelter were targeted for a 20% reduction, but that the measured reduction was only 10%. The team would analyze reasons for the gap between expected and measured performance, and would develop a list of corrective actions and a time for completing them. In the lead secondary smelter example, potential solutions might include additional seals for doors and windows, closer monitoring of the input materials and furnace conditions, replacement of control equipment, better housekeeping, and other corrective actions. During the plan stage, the potential list of problems would be prioritized, and solutions would be scheduled around the prioritized list. Someone in the team would be given the responsibility of tracking the problem and reporting on progress until the problem was solved.

The *Do* stage simply means implementing the plan. Depending on the problem, it may be preferable to make a test of proposed changes on a small scale. After the potential solutions are implemented, *Check* the results by measuring changes. This phase involves collecting data to determine whether the corrective actions were successful. Basic questions are: Has environmental performance improved as planned? If not, why? Was the cause correctly identified? Were there other reasons?

Finally, if the plan was successful, *Act*. Make the proposed changes in the system. Be sure to perform needed training, to document update and control, and to take other actions to integrate the changes into the EMS. Over a period of time, performance will improve as a result of going through the PDCA cycle repeatedly.

Brainstorm and Analyze

Putting the right group together and letting it brainstorm can have tremendous positive results, and there are many potential uses for this technique in the continual improvement program. Many quality programs use "quality circles" or "focus groups" to develop ideas for program development and improvement. These are similar in concept to brainstorming followed by analysis of data.

Brainstorming can be used for selected problems or as part of the day-to-day activities. Participants should be those people who are affected by the problem or who are working in a given area. At the end of the session, it is important for everyone to understand the problems and solutions that were discussed. Further, it is imperative for someone to be assigned the task of tracking the implementation of ideas and reporting back to the group by a given date.

Brainstorming sessions can take many forms, but a common strategy is to have a "free-form" idea session followed by analysis. At the outset of the session, participants are given the problem. Each person is free to contribute ideas, which are written on a flip chart. It is important for everyone to understand that there are no "bad" ideas. Equally, there are no "good" ideas at this juncture—participants should not make any value judgment comments, positive or negative. The pros and cons are not discussed, nor are the ideas organized by topic area. This initial phase of the session should be of adequate length to involve all participants and to create a relaxed atmosphere in which ideas can flow.

The brainstorming concept is similar to the free-association sessions used by psychologists. The basic notion is to get ideas to flow without excessive thought about their meaning or how they fit into the scheme of things. A good way to get discussions moving in the brainstorming session is to utilize flow charts and fishbone diagrams that were developed during the process analysis phase of establishing the EMS.

After the initial ideas are presented, there are several paths of analysis that will help the group to develop solutions. Depending on the nature of the problem, the SWOT (Strengths, Weaknesses, Opportunities, Threats) analysis may be a useful context for analyzing the ideas generated by brainstorming. SWOT analysis involves making a list of strengths and weaknesses and provides a context in which problem solving can occur. Opportunities and threats are both internal and external; examples include lists of resource availability, management support, customer desires, regulatory requirements, competition, skills development, and many other factors. An example SWOT analysis might appear as follows:

Strengths	Weaknesses	Opportunities	Threats
location	internal dissention	leadership improvement	foreign competition
market position	new leadership	new markets	market swings
skilled staff	turnover in key positions	possible new products	possible leveraged buyout
available capital	regulatory compliance problems	EMS development	new competition
developed programs	research and development problems	may buy out competitor	environmental regulatory court case
flat structure	vacant positions	possible legislative changes	slow markets

Another technique that can be used to build on the ideas generated during brainstorming is the FADE (Focus, Analyze, Develop, Execute) process, which was popularized by the Total Quality Management (TQM) movement. In this process, the information developed during brainstorming is organized and analyzed. The participants focus on specific topics, analyze these topics, develop solutions, and execute those solutions.

When used on a regular basis, brainstorming can become a powerful tool for creating environmental awareness and solving problems. Problems are identified and solutions discussed. Discussions can be tracked on a flip chart or board that remains in place for the next day's session, or they can be transcribed for distribution.

CREATING AND USING CHARTS

Graphical displays are ideal ways of presenting information. Properly constructed, they can be used to demonstrate patterns and to facilitate the understanding of complex data. To be useful, graphical displays must be clear. This means that the graph should have a meaningful title, legends that identify key symbols, and labels for the x-axis and y-axis. Likewise, tables should have a meaningful title, labels for columns and rows, and decimals that line up for numerical entries. Units of measurement in tables and charts should be identified. There are many different types of graphical displays, and this section briefly summarizes some of them.

FLOW CHART AND CAUSE-AND-EFFECT DIAGRAM

The flow chart and cause-and-effect diagram discussed in Chapter 4 (process analysis) are also useful tools for continual improvement documentation. Recall that the flow chart is used to provide a picture of process or information flows. By itself, a flow chart of the actual process may not pinpoint problem areas, but when the actual flow chart is compared to the ideal flow, it is often easy to determine where problems are likely. Similarly, the cause-and-effect diagram helps identify the possible causes of a problem, but it is not as useful for prioritizing them. However, both of these techniques, in conjunction with the Pareto chart (discussed below), will help prioritize causes, which allows solutions to be planned in a more cost-effective manner. It should be noted that sometimes the causes that occur more frequently (high incidence) are not always the highest priority. A small number of severe incidents may warrant a higher priority.

HISTOGRAM

Histograms are tools for showing how frequently an event occurs. They can be displayed as a bar chart or a combination bar chart and line graph. The height of the bars or points on a line represent how often the event occurs. Histograms can be used to track a variety of environmental performance measures. For example, assume that the particulate concentrations are measured on the roadways adjoining a facility that produces aggregate. The data from daily 24-hour measurements over a 30-day period are reported as follows:

Particulate Concentration (µg/m³)
85, 109, 100, 103, 112, 115, 98, 111, 90, 110, 110, 105, 105, 100, 98, 92, 94, 107, 106, 108, 105, 105, 97, 100, 94, 106, 104, 101, 106, 106

A histogram created from these data is shown in Figure 9–1. When these data are grouped according to how many times concentrations occurred in each category, the variability among the measurements becomes clearer. This variation can be examined to evaluate the possibility of underlying causes for the differences in concentration, or other statistical techniques could be performed to de-

FIGURE 9–1.   Example of a Histogram:

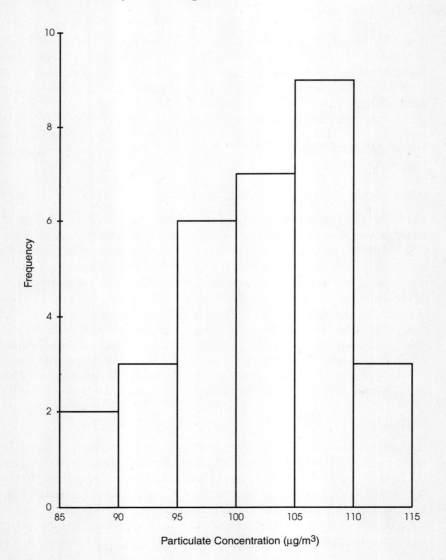

Particulate Concentration (µg/m³)

termine whether the concentration differences are significant. After possible causes for differences are examined, there may be opportunities to reduce the concentrations using dust suppression or other techniques, or the variability may simply be a function of meteorological conditions that cannot be controlled.

PARETO CHART

The Pareto chart is very useful for determining the hierarchy of causes that relate to an event. It is a type of bar chart that displays the frequency with which events occur. Pareto charts are often arranged with the highest frequency plotted on the left side of the chart, descending to the lowest frequency on the right. Sometimes, a line graph of cumulative percentage is plotted on the same chart. The Pareto chart is based on the principal that although a large number of causes may contribute to a problem, only a few of the causes are critical. In other words, a few causes account for most of the problems. The analysis of the Pareto chart involves evaluating the frequency of each cause against the cost of individual incidents to determine a strategy for corrective action.

Pareto charts can be constructed for any type of problem, ranging from improving a particular operation such as wastewater treatment to reducing the number of violations of standards or improving overall environmental performance. For example, assume that a facility was unable to meet the BOD (biochemical oxygen demand) limits for its wastewater discharges. A brainstorming session identified potential sources of the problem, and data were gathered to determine how many times each problem resulted in high BOD levels:

Cause of High BOD Discharges	Number of Incidents
Exceeding treatment plant capacity	12
Malfunction of activated sludge unit (ASU)	4
Introduction of concentrated wastes to treatment plant	9
Pump failure	3
Bypass for repairs	5
Power failure	2

A Pareto chart constructed from these data is shown in Figure 9–2. This chart shows the number of times each problem that caused high BOD occurred (referenced to left y-axis). The most common cause of high BOD was exceeding the plant's capacity. The least frequent cause was power failure. The line that is plotted on the graph shows the cumulative frequency (in percentage) for each of the causes (referenced to right y-axis). This plot shows that the most common cause, exceeding capacity, accounted for 34% of the high BOD discharges. Concentrated wastes and high BOD accounted for a total of 60% of the high BOD events. The cumulative total increases until it equals 100% for all of the causes together.

It should be noted that frequency of occurrence is not the only consideration when examining potential significance of the Pareto chart data. For example, there may be a problem or cause that does not occur very often, but each occurrence is very costly. If so, this problem or cause may require correction first.

FIGURE 9–2. Example of a Pareto Chart:

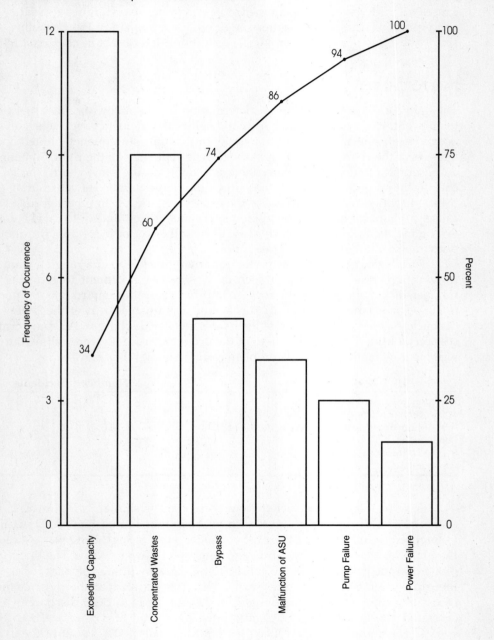

RUN CHART

A run chart is simply a plot of data over time to look for trends. If a linear or cyclic pattern occurs, underlying causes can be examined to determine whether corrective actions are needed. For example, assume a facility tracks its waste production each day over a 3-week period and generates the following data:

Day of Week	Waste Generated (tons)
M	156
T	144
W	138
T	136
F	147
S	123
S	118
M	162
T	151
W	143
T	135
F	149
S	105
S	111
M	158
T	154
W	122
T	131
F	153
S	122
S	109

A run chart created using these data is shown in Figure 9–3. This chart shows that waste production declines each weekend, followed by a jump on Mondays. There is also a midweek low on Wednesdays. Are these patterns due to reduced production on weekends or to different operations on high waste production days? Should (or can) these shifts be stabilized? How do these variations affect disposal costs? If the data were charted over time, it would be possible to determine whether the trend is long-term or temporary.

SCATTER DIAGRAM

Scatter diagrams chart the relationship between two variables. The scatter diagram is useful for evaluating cause-and-effect relationships. For example, assume that driving to a warehouse on a half-hour schedule produces the following trip information:

Period	Departure Time	Trip Time (minutes)
1	8:00–8:30	25
2	8:30–9:00	30
3	9:00–9:30	25
4	9:30–10:00	22
5	10:00–10:30	18
6	10:30–11:00	16
7	11:00–11:30	17

8	11:30–12:00	20
9	12:00–12:30	22
10	12:30–1:00	18
11	1:00–1:30	16
12	1:30–2:00	17
13	2:00–2:30	16
14	2:30–3:00	20
15	3:00–3:30	26
16	3:30–4:00	32

A scatter diagram generated from these data (Figure 9–4) shows the relationship between the starting time of a trip to the warehouse and the length of time needed to make the trip. Each 30-minute period of the day is numbered, and trips that begin during that period are averaged. The average trip time for each period is then plotted. On this chart, it is easy to see that periods 2 (8:30–9:00) and 16 (4:00–4:30) require more travel time, while periods 6 (10:30–11:00), 11

FIGURE 9–3. Example of a Run Chart:

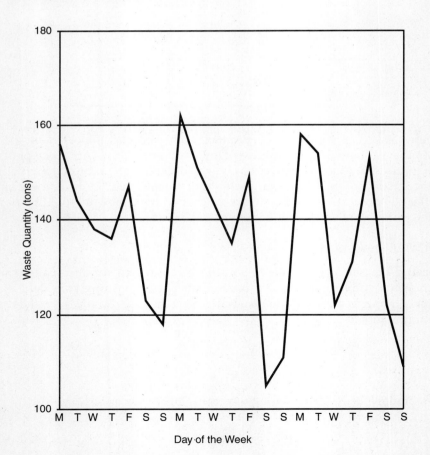

Day of the Week

FIGURE 9–4.   Example of a Scatter Diagram:

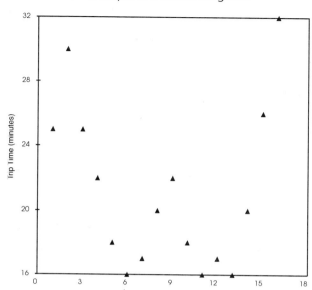

Time Period	1	2	3	4	5	6	7	8	9	10	11	12	13	14	15	16
Trip Time (minutes)	25	30	25	22	18	16	17	20	22	18	16	17	16	20	26	32

(1:00–1:30), and 13 (2:30–3:00) require the least. There is also an increase in travel time over the lunch hour. By examining the underlying causes of variability in trip time, it may be possible to reduce travel time and fuel use. Reductions in vehicle emissions can be of particular benefit to communities that experience carbon monoxide or ozone air pollution problems.

CONTROL CHART

Control charts are statistical quality tools (see Chapter 8) that are commonly used for process control. They are similar to run charts but have a statistically determined upper control limit (UCL) and lower control limit (LCL). Special causes of error (data that fall outside of the control limits) are typically more easily corrected than are common causes of error (data that fall within the control limits). Continuous improvement efforts should focus on shrinking the control limits (correcting common causes).

In the environmental field, control charts are commonly used to control the performance of environmental monitors and laboratory analyses. For example, assume that a sulfur dioxide monitor undergoes a daily span at 0.420 ppm as a calibration check, the following data resulting over a 16-day period:

Daily Span Value (ppm)
415, .422, .426, .431, .445, .436, .428, .422, .418, .410, .408, .415, .422, .420, .426, .438

Figure 9–5 shows a control chart for these data. If the monitor were operating within acceptable limits all of the data points should fall in a random pattern between the UCL and the LCL. These control limits are set at ±3 standard deviations from the mean of measurements for this monitor. The top two charts are the same, but the top chart shows only the control limits, whereas the middle chart shows the warning limits (±2 standard deviations) and the ±1 standard deviation

FIGURE 9–5.   Example of a Control Chart:

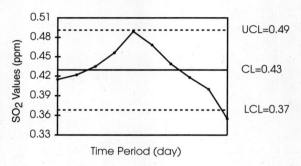

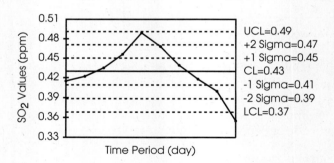

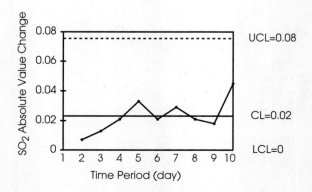

lines. These charts show that the daily span is drifting, and the monitor is out of control. The monitor may need maintenance, and it should be recalibrated. An example of a moving range chart is shown in Chapter 8, Figure 8–2.

BENCHMARKING

Benchmarking is a technique in which one operation is compared to an identical or similar operation that has better performance. This analysis can be done between facilities in multifacility operations, between top-performing departments within an organization, between local or regional companies, or between the best national and international companies. The objective of benchmarking is to find the best practices that can be reasonably applied to a given operation. Benchmarking can be used to compare performance in scientific or management activities. For example, an organization could benchmark its success in pollution prevention, waste management, training activities, environmental program costs, or other parameters.

The hope in benchmarking is that by copying (and ultimately improving) the best practices of others, the organization will improve its performance. Some leaders of environmental programs are 3M, Monsanto, IBM, Kraft General Foods, and others. Benchmarking does not have to be limited to similar industries.

SUMMARY

Management review and continual improvement are strategic, interrelated, and ongoing efforts that will ensure the success of the EMS over the long term. The performance of the EMS must be compared to the policies, objectives, and targets of the organization. Potential problems should be evaluated to determine root causes, taking care to explore all possible options. Once problems are linked to probable causes, a plan for solving them can be developed and implemented. Problem solving typically involves the collection of data in the planning and checking stages. Also, there are many tools that use the collected data to define problems, identify potential causes, and verify the success of solutions.

ENDNOTES

[1] W. Edwards Deming, *Out of Crisis* (Cambridge, MA: Massachusetts Institute of Technology, Center for Advanced Engineering Study, 1986) p. 59.

[2] W. Edwards Deming, *Out of Crisis.*

CHAPTER 10

AUDITING

INTRODUCTION

Auditing is a quality assurance tool that can verify whether management and technical practices exist, function properly, and are adequate to meet the organization's goals. Regulatory agencies routinely perform audits to determine whether an organization is meeting its legal responsibilities. However, auditing can be much more than an instrument for measuring regulatory compliance. Integrating audits into your organization's overall corporate decision-making process through the use of periodic and systematic audits can improve environmental performance by identifying strengths and weaknesses in the environmental program. Audits can be performed by outside consultants (third party) or internal staff (second party). ISO 14001 requires an audit of the EMS to demonstrate that the EMS is working properly and that the EMS conforms to the different elements of the standard. (Appendix C contains an example auditing procedure.) This EMS audit is required if you wish third-party certification or self-declaration.

The guidance for auditing is given in ISO 14010, which provides a guide on the general principles of environmental audits; ISO 14011, which provides specific audit procedures; and ISO 14012, which provides guidance on qualification criteria for environmental auditors.

REASONS FOR AUDITING

One important advantage of developing a program to audit environmental functions is that it is possible to be proactive and to identify and resolve problems rather than being thrown into crisis management when problems arise as a result of a regulatory inspection or unforeseen accidents and incidents. Although audits are not at the present time a substitute for regulatory oversight, they can complement regulatory oversight activities, provide an early warning system if problems exist and move the organization toward compliance with regulatory codes. Environmental audits can also help minimize liabilities related to property transfers, identify and exploit opportunities to turn waste and by-products into new products, use resources more efficiently, and evaluate risks from regulated and unregulated materials and practices.

Environmental system audits can help increase environmental awareness and evaluate whether or not the organization successfully:

- Develops organizational environmental policies that 1) implement regulatory and legal requirements and 2) provide management guidance for environmental hazards not specifically addressed in regulations.
- Trains and motivates facility personnel to work in an environmentally acceptable manner and to understand and comply with legal requirements and the organization's environmental policy.
- Communicates relevant environmental information with internal and external stakeholders.
- Requires third parties working for, with, or on behalf of the organization to follow its environmental procedures.
- Makes personnel proficient and available at all times to carry out routine environmental and emergency procedures.
- Applies best management practices and operating procedures, including "good housekeeping" techniques.
- Institutes preventive and corrective maintenance systems to minimize actual and potential environmental harm.
- Utilizes best available process and control technologies.
- Uses the most effective sampling and monitoring techniques, test methods, record-keeping systems, or reporting protocols (beyond minimum legal requirements).
- Evaluates causes behind any serious environmental incidents and establishes procedures to prevent recurrence.
- Exploits source reduction, recycle, and reuse potential, wherever practical.
- Substitutes materials or processes to allow use of the least hazardous substances feasible.
- Assesses environmental risks and uncertainties.

GENERAL REQUIREMENTS FOR EFFECTIVE EMS AUDITS

Successful environmental audit programs have certain characteristics in common, regardless of the reasons for the audit, type of organization, or type of auditor. One of the first requirements is explicit top management support for the auditing program. Management support for the auditing program should be demonstrated by a written commitment to follow up on audit findings to correct identified problems and to prevent their recurrence.

The environmental audit function must be independent of the audited activities to ensure objective and unobstructed inquiry, observation, and testing. Auditor objectivity should not be impaired by personal relationships, financial or other conflicts of interest, fear of potential retribution, or interference with free inquiry or judgment. This does not mean, however, that external auditors (third party) must be used. Internal auditors have the benefit of in-depth knowledge about the operations to be audited. They are often in the best position to evaluate an organization, but they must have management support and must be able to remain objective and neutral. Sometimes a combination of internal and external auditors is appropriate.

The audit team, regardless of composition, must be adequately staffed and trained to accomplish the audit objectives.. Training in-house auditors can be accomplished through external training courses, by hiring consultants to train staff, or by experienced personnel (if available) training junior staff. Training should include technical, regulatory, and management issues; mock audits; and the "how-to" aspects of conducting the audit, report writing, and communications skills. The list of references and resources list (Appendix F) give more information about resources for training and conducting audits.

The audit program is the foundation for the organization's auditing activities. It must be carefully constructed to ensure that the audit objectives are met through the collection, analysis, and documentation of useful information. The audit process must incorporate quality assurance procedures to assure the accuracy and thoroughness of the environmental audits. Quality assurance may be accomplished through supervision, independent internal reviews, external reviews, or a combination of these approaches. This will prevent problems before, during, or after the audit and will ensure that the goals of the audit are met. The organizational procedures for auditing should identify the program objectives, scope, resources, responsibilities, and the frequency of auditing. Written audit procedures should be used for planning audits, establishing the scope of the audit, examining and evaluating audit findings, communicating audit results, and conducting the follow-up activities.

The audit protocol should be developed and agreed upon by the involved parties prior to the audit. If needed, the protocol can be modified as the audit progresses, providing that all parties agree to the changes. The audit protocol should include specific procedures for collecting data, analyzing and interpreting data, and communicating results. The collected information should be factual, adequate, and convincing.

Information collected during the audit must be consistent with the objectives for the audit and must support the audit findings and recommendations. In the case of the ISO 14001 audit, the auditor will examine the organization's EMS to determine whether the EMS conforms to the requirements of the standard. For example, is there an environmental policy statement? Are procedures established and maintained for identifying aspects and impact, checking and corrective action, emergency response, and so forth? The auditor will determine whether the *management system* allows the established objectives and targets to be met. The auditor will not judge the organization's performance for each of its activities. For instance, the auditor will not evaluate the adequacy of individual objectives and targets. The auditor will determine whether objectives and targets have been established and whether procedures exist and are adequate to meet the established objectives and targets. Other types of audits, such as the regulatory compliance audit, are conducted to establish the level of performance in certain activities. These audits focus on specific qualitative and quantitative performance criteria to determine whether the organization is meeting its responsibilities.

The protocol should ensure that procedures exist that will result in the preparation of prompt, candid, clear, and appropriate written reports on audit findings. In-house audits will include provisions for corrective actions and schedules for implementation and verification procedures. Procedures should be in place to ensure that audit information is communicated to managers, including facility and corporate management, who can evaluate the information and make sure that identified problems are corrected. Procedures should also be in place for determining what internal findings must be reported to satisfy legal and regulatory requirements.

DEPTH AND FREQUENCY OF AUDITS

The depth and frequency of audits will depend on the type of audit being conducted. Some audits are comprehensive whereas others target selected areas. Audits can be conducted on a single activity, an individual site, or the entire corporation. For example, a corporate audit, which evaluates the entire organization, might be conducted once every 2 years. A comprehensive site audit, which considers all activities in a single installation, might be scheduled once each year, but an activity audit might be conducted 2 or more times each year. Legal and regulatory requirements may delineate the type and frequency of audits.

The ISO 14001 standard incorporates auditing into checking and corrective action activities (Section 4.5.4). The organization is required to establish and maintain an audit program and procedures to audit the EMS on a periodic basis. The goals of the EMS audit are to determine whether the EMS has been properly implemented, maintained, and conforms to the requirements specified in the EMS and the standard. Once the EMS audit is completed, the results must be provided to management for review. The organization's internal audits must be conducted frequently enough to ensure that the EMS functions properly.

The required frequency of the EMS ISO 14001 certification audit will depend on the requirements of the certification body. The certificate might be valid for a fixed period (perhaps several years) or indefinitely, based on successful surveillance at prescribed intervals.[1]

THE AUDIT PROCESS

The organization has considerable flexibility in developing its audit program and procedures. It can use internal or external auditors, decide the frequency of the audit activities, and select the audit procedures. The organization must, however, consider the results of previous audits and the environmental importance of each audited activity when the audit program is developed. The audit program and procedures should identify the activities and areas to be audited, the frequency of the audits, who is responsible for managing and conducting the audits, how and to whom audit results will be communicated, and requirements for auditor competence.

Prior to the conduct of an internal or external audit, it is important to lay a foundation so the audit can be conducted in an atmosphere of mutual respect, cooperation, and confidence. As noted previously, top management support and commitment to the audit process is essential. Managers and employees should be informed about internal or external audits before they occur. It is important for everyone to understand that the audit is not a witch hunt. Rather, the audit will help the organization achieve and maintain its environmental management goals by identifying problem areas and allowing management to focus on corrective actions.

The ISO 14011 standard follows established protocols in the conduct of the audit:

1. establishing the audit objectives (Section 4.1)
2. identifying roles, responsibilities, and activities (Section 4.2)
3. initiating the audit (Section 5.1)
4. preparing the audit plan (Section 5.2)
5. conducting the audit (Section 5.3)
6. preparing the audit findings (Subsection 5.3.3)
7. writing the audit report (Section 5.4)
8. completing the audit (Section 6)

As mentioned these steps are not unique to the ISO process, but are the best management practices for auditing that have been developed over years of experience. As such, these steps can be applied to all types of auditing.

ESTABLISHING THE AUDIT OBJECTIVES

Audit objectives and goals should be established prior to the audit. Whoever commissions the audit should be responsible for developing the audit objectives. Typically, this will be the auditee (the organization being audited); however, sometimes an organization other than the one being audited (called *client* in the ISO terminology) will commission the audit.

Audit objectives might include 1) helping management identify areas for improvement, 2) determining whether the EMS has been properly implemented and maintained, or 3) determining whether the internal management review process is adequate. The audit can also evaluate the EMS for external clients who are considering contractual relationships with the auditee.

IDENTIFYING ROLES, RESPONSIBILITIES, AND ACTIVITIES

Everyone in the organization has a role to play in making sure that the organization's audit activities are successful. These efforts must be ongoing, not just a one-time event. All employees must adhere to the organization's policies and procedures, and must follow their assigned work instructions. All employees are responsible for being aware of problems and for correcting or reporting them as appropriate. Supervisors and managers are responsible for ensuring that procedures are followed, problems are reported, and corrective action plans are developed and implemented. Ongoing audit activities should be coordinated by a designated person, committee, or department. This will help ensure integration among program areas and continuity over time. If everyone in the organization meets his or her responsibilities for day-to-day operations and periodic auditing, a comprehensive audit of the organization's technical or management systems should produce favorable findings.

The EMS audits required by ISO 14001 (Subsection 4.5.4) can be conducted by internal or external auditors. Even if your organization decides to pursue a third-party certification, it should consider an internal audit to prepare for the third-party certification audit. An internal precertification audit has some advantages over an external precertification audit. Internal audits, if they are independent and objective, are less costly and can provide a greater understanding of the organization simply because of the auditing staff's familiarity with the organization.

The principals to the audit include the client, auditee, and auditing staff. The interactions among these principals may differ slightly, depending on whether the audit is commissioned by the auditee or by someone else.

The **client** determines the need for the audit, defines audit objectives, and initiates the audit. In many cases, the client and the auditee are the same. However, in some instances, they may be different. For example, corporate headquarters may initiate an audit of individual facilities; a company planning to acquire a facility may initiate an audit prior to purchase; a company may require an audit of a supplier or a treatment, storage, and disposal facility. If the client and the auditee are not the same, the client is responsible for contacting the auditee to obtain its full cooperation. Once the decision is made to conduct an audit, the client must provide appropriate authority and resources to conduct the audit. The client should select the lead auditor or auditing organization and approve the composition of the audit team. Other responsibilities of the client include approving the EMS (or other) audit criteria and scope, approving the audit plan, and receiving the audit report and determining its distribution.

The organization being audited (**auditee**) plays a key role in determining whether the audit will be successful. It is critically important for the auditee to cooperate with the auditors to permit the audit objectives to be achieved. Meeting this responsibility involves providing access to the facilities, personnel, and relevant information and records, as requested by the audit team. It is important for the auditee to appoint responsible and competent staff to accompany members of the audit team. Staff are needed to act as guides to the site and to ensure that auditors are aware of health, safety, and other requirements. The auditee should also provide adequate facilities for the audit team to use as a staging area for their work.

An audit team, consisting of a lead auditor and individuals with expertise in several areas, may be needed to audit a more complex environmental management system. In less complicated situations, an individual auditor may be adequate.

The **lead auditor** is responsible for the overall coordination of the audit. The lead auditor should consult with the client to determine the scope and objectives of the audit. Next, relevant background information should be obtained on the organization (products, services, activities, site information). If audits were previously conducted, these findings should be gathered and reviewed.

The lead auditor will determine whether the requirements for an ISO 14001 or other type of environmental audit have been met. If the lead auditor determines that the audit can proceed, he or she forms the audit team in consultation with the client. Potential conflicts of interest must be considered when team members are selected. This is especially important for internal auditors. For an ISO 14001 audit, the lead auditor is responsible for directing the activities of the audit team in accordance with ISO 14010 and ISO 14011 guidelines.

The lead auditor prepares the audit plan in consultation with the client, auditee, and audit team members, and communicates the final audit plan to these participants. The lead auditor also briefs the audit team and coordinates the preparation of the working documents and protocol for conducting the audit.

It is the lead auditor's responsibility to represent the audit team in discussions with the auditee before, during, and after the audit. Additional responsibilities include resolving any problems that arise during the audit, recognizing when audit objectives are unattainable, and reporting the reasons to the client and auditee. The lead auditor is also responsible for communicating the results of the audit. The audit report should be submitted to the client within the agreed-upon time. Finally, if agreed to in the audit scope, the lead auditor should communicate recommendations for improving the EMS.

Auditors work under the direction of and support the lead auditor. The auditor collects and analyzes the audit evidence, documents the individual audit findings, and prepares working documents under the direction of the audit leader. Auditors are responsible for safeguarding and returning any documents related to the audit. Auditors also assist in writing the audit report. The auditor is responsible for planning and carrying out the assigned task objectively, effectively, and efficiently.

INITIATING THE AUDIT

Once the decision is made to conduct the audit and the auditor team is engaged, the next step in the audit process is to initiate the audit. This step involves determining the scope of the audit and a preliminary review of documents.

The scope should be determined for every audit prior to the conduct of the audit. The scope should be developed by the lead auditor and whoever commissioned the audit. In the case of client-initiated audits, the auditee should be involved in establishing the scope. Any changes in scope or objectives must be negotiated and documented. Careful attention to detail early in the planning process can help avoid costly and time-consuming changes.

The scope sets the boundaries for the audit. Will the audit be conducted on an entire multisite facility or on a single site? Will the audit be conducted on a newly acquired property, a facility slated for closure, or an operating facility? Which activities will be included? Is the audit being conducted for ISO 14001 certification or compliance, or for some other reason? What evidence will be evaluated? How will the results be reported? Who will be on the distribution list for the report? All these questions must be answered prior to the start of the audit.

The complexity of the audit will be determined largely by the anticipated environmental impact. A facility that has significant environmental impact will have a more detailed, in-depth audit than will a facility with no or few anticipated aspects and impacts. Time, technical constraints, and finances will also influence audits conducted for purposes other than for ISO 14001 certification.

Prior to conducting the audit, the lead auditor should determine whether there is adequate relevant information to meet the goals of the audit (documentation such as environmental policy statements, procedures, records, or other appropriate information). The auditor will also determine whether the resources exist to conduct the audit and whether there will be adequate cooperation from the auditee.

PREPARING THE AUDIT

Key elements of preparing for the audit include developing an audit plan, making the audit team assignments, and preparing the working documents.

AUDIT PLAN

The audit plan or protocol is a detailed description of what will occur during the audit. The audit should be conducted using documented and well-defined methodologies and systematic procedures. At the same time, the process should be flexible so that appropriate changes can be made during the conduct of the audit in response to information gathered during the audit.

Checklists are important tools for organizing, conducting, and recording the results of an audit, but they should not be relied upon completely. The auditor should retain a degree of flexibility that allows him or her to deviate from the checklist when appropriate. Checklists are not a substitute for knowledge. Internal audits can be conducted using commercially available audits, self-prepared checklists, or a combination of the two. Often, the checklists used by regulating agencies to conduct compliance audits can be used alone or with modification.

It is important for the auditor and manager to agree on the procedures to be used and audit criteria so both parties will be confident that the audit will provide reliable and high-quality results. The lead auditor should communicate the audit plan to the client, the auditors, and the auditee. Everyone should review and approve the plan. If there are objections to some part of the plan, they should be communicated to the lead auditor. These problems should be resolved before the audit begins.

During the conduct of the audit, appropriate information is collected, analyzed, interpreted, and documented to determine whether the audit criteria are met. The ISO guidelines on auditing include a general guideline for quality stating that the evidence collected should be "of such a quality and quantity that competent environmental auditors working independently of each other would reach similar audit findings from evaluating the same audit evidence against the same audit criteria" (ISO 14010, Section 5.5).

Regardless of the type of audit the audit plan should include the following components:

- the audit objectives and scope
- identification of the audit team members
- identification of the auditee's organizational and functional units to be audited
- the audit criteria

- the audit procedures
- identification of reference documents
- the expected time and duration of major audit activities
- the dates and places where the audit is to be conducted
- the schedule of meetings to be held with the auditee's management
- confidentiality requirements
- report content, format and structure, expected date of issue, and distribution of the audit report
- documentation retention requirements

In addition to detailing the scope and other management aspects of the audit, the protocol will include procedures for the direct observation of the facility, interviews with personnel and outside parties, measurements, review of existing information and data, and completion of questionnaires and checklists.

AUDIT TEAM ASSIGNMENTS

Each member of the audit team should be given specific EMS elements, functions, or activities to audit. These assignments should be made by the lead auditor in consultation with the members of the audit team. During the execution of the audit, the lead auditor may make changes to the work assignments to optimize the audit process. Before the audit begins, each member of the team should be instructed on the audit procedure to be followed for their individual assignments.

DOCUMENTS

Working documents may include procedures and checklists developed for evaluating the elements of the EMS, forms for documenting supporting evidence and audit findings, and records of meetings. To ensure proper document control, each document should be given a serialized number and should be listed with the number in a project document inventory that is assembled upon completion of the audit. Waterproof ink should be used to record all data. Video cameras and photographs are useful adjuncts to written information. Photographs should be identified by date, time, number of the photo on the roll, location, general direction faced by the auditor when taking the photo, and other comments (such as weather conditions, film type, and so forth). Maps and drawings should be simple and free of extraneous detail. They should include compass points and a basic scale to aid in interpretation of information. Maps can also be used to show where photographs were taken by number coding with the exposure number and a direction arrow.

Audit team members should safeguard confidential or proprietary information as agreed upon during the planning stages of the audit.

CONDUCTING THE AUDIT

The audit begins with an opening meeting between the auditee's management representatives and the audit team. This meeting establishes the official communication links between the audit team and the auditee. Another important goal of the meeting should be to promote the auditee's active participation in the audit.

The meeting should open with introductions and a review of the scope, objectives, and audit plan. The audit team should provide a short summary of the methods and procedures to be used to conduct the audit and should confirm that the resources and facilities needed by the team are available. The participants should agree on an audit timetable and should confirm the time and date for the closing meeting. The auditee should review the relevant site safety and emergency procedures for the audit team.

The audit team will participate in an orientation tour so it can become acquainted with the layout of the facility. At the completion of this brief tour, the audit team is ready to begin an in-depth evaluation of the facility.

The audit team will gather a variety of evidence during the audit through interviews, examination of documents, and observation of activities and conditions. The audit team should collect enough evidence to achieve the goals of the audit. The audit team will review records and record-keeping systems, regulatory permits, and a variety of other documents. Information gathered through interviews should be verified by supporting information from independent sources such as observations, records, and results of existing measurements. When information gathered through interviews cannot be verified, it should be labeled as nonverified information. Any sampling programs conducted by the auditee should be examined, with particular attention on the procedures for ensuring effective quality control of the sampling and measurement processes. The auditor should attempt to obtain enough evidence to allow significant and less significant findings to be taken into account. At the end of each day, the team members will share information and discuss findings.

PREPARING THE AUDIT FINDINGS

The audit team should review all of the collected evidence and should identify instances of nonconformity. Whenever findings of nonconformity occur, the audit team should document these in a clear, concise manner that identifies the audit evidence for the nonconformity. Documented nonconformances should be reviewed with the responsible auditee manager, and the auditor should attempt to obtain acknowledgment of the factual basis of all findings of nonconformities. Audit findings of critical nonconformities must be communicated to the auditee without delay.

The auditing process should be designed to provide a desired level of confidence in the findings and conclusions. However, even under the most optimum conditions, it is important for all parties to the audit to recognize that there is uncertainty associated with the audit findings. Some of the uncertainty can be attributed to the fact that the audit is conducted during a short period of time and includes only a portion of the available evidence. The limitations of evidence and other sources of uncertainty should be recognized by the auditor and taken into account during the planning of the audit and in the report of the audit findings.

Documentation of conformances may or may not be included in the audit findings. Including conformances in a report of an internal audit provides staff with positive feedback and this recognition can be a motivating factor. The EMS audit findings may also include documentation of conformances if prior agreement was reached in the scope of the audit.

CLOSING MEETING

A closing meeting should be held between the audit team, the auditee's management, and those responsible for the functions audited. This meeting should be held after the audit evidence has been collected but before the audit report is prepared. The purpose of this meeting is to present the audit findings to the auditee so that the factual basis for the findings is understood and acknowledged.

If disagreements occur, they should be resolved, if possible, before the lead auditor issues the report. If resolution is not possible, the lead auditor has the final say.

WRITING THE AUDIT REPORT

The audit report should contain the audit findings or a summary of the findings, and it should reference the supporting evidence. The report should be dated and signed by the lead auditor. The lead auditor and the client together determine what information will be included in the report.

The audit report will include specific information about the organization audited, the audit plan, and the audit findings. In the case of the EMS audit, the audit conclusions will note whether or not the EMS conforms to the EMS audit criteria. The auditor may comment on whether the EMS is properly implemented and maintained, and whether the internal management review process can adequately assure the continuing effectiveness of the EMS. During internal and EMS audits, the auditor may make recommendations for corrective actions if there is a prior agreement to include this information. A typical audit report may include, but is not limited to, the following information:

- the identification of the organization audited and the client;
- the date the audit was conducted and the time period covered by the audit;
- the objectives, scope, and plan of the audit;
- the audit criteria and a list of reference documents used in the audit;
- the members of the audit team;
- the auditee's representatives who participated in the audit;
- a statement of the confidential nature of the contents in the audit report;
- the distribution list for the audit report;
- a summary of the audit process and any obstacles encountered;
- audit findings and conclusions

Because the audit report will be used by technical experts and others, it must be coherent, concise, and internally consistent. It should be clearly written and free of technical and bureaucratic jargon.

The audit report should be completed by the date agreed to in the audit plan. If this deadline cannot be met, the auditor should formally communicate a revised date and the reasons for the delay to both the client and the auditee.

The audit report is the sole property of the client. The auditors and all recipients are responsible for safeguarding the confidentiality of the report. The audit report should be sent to the client by the lead auditor and distributed according to the agreed-upon list enumerated in the audit plan. In those instances when the client and auditee are different entities, the auditee should receive a copy of the audit report unless the client has specifically excluded the auditee. Any further distribution of the report outside the auditee's organization requires the auditee's permission.

The lead auditor, the client, and the auditee should agree to the retention and disposition of all documents related to the audit, including working documents, the draft, and final reports.

COMPLETING THE AUDIT

When all of the activities enumerated in the audit plan are completed and the audit report is distributed, the audit is closed.

AFTER THE AUDIT

The audit—internal or external—is a reality check for the organization. The audit will confirm whether the organization meets its stated objectives and will identify its strengths and weaknesses. Completion of the audit does not signal the end of the activities initiated by the decision to perform the audit. In fact, the audit is the beginning of a long-term commitment to quality and improvement. For an initial audit, the final report is the beginning of a comprehensive environmental management program. For an organization with an existing program, the final report is a barometer of performance.

Once the audit is completed, it is important to foster an atmosphere of positive reinforcement and continual improvement and to obtain verification and feedback on an ongoing basis. Management should recognize areas of success and should develop and implement an action plan to improve areas that fell short during the audit. A plan should be developed to correct existing problems and to initiate or upgrade programs to prevent the recurrence of problems. For example, this might involve developing an asbestos compliance program for maintenance activities or a preventive maintenance program for pollution control devices. After the corrective action plan is implemented, a follow-up audit should be undertaken to determine its effectiveness and to prevent backsliding. Even if no problems are found during the audit, follow-up audits are essential to ensure a quality program.

SELECTING AN ENVIRONMENTAL AUDITOR— QUALIFICATION CRITERIA

Guidelines for selecting lead auditors and the audit team are given in ISO 14012. Internal auditors should have the same competence as do external auditors, although they may not meet all of the criteria given in the standard. ISO 14012 identifies minimum personal attributes, training, education, and work experience, that auditors should possess. Auditors are responsible for maintaining their competence and should participate in refresher training as needed.

PERSONAL ATTRIBUTES AND SKILLS

Organizing, conducting, and evaluating an ISO 14000 or other audit requires a variety of skills and personal attributes. The auditor's responsibility is to gather factual information, to determine whether the goals of the audit are met, and to report the findings. The auditor must possess good organizational, communication, and people skills.

Audits, regardless of type, are stressful events, and the auditor must be sensitive to this fact. During the conduct of the audit, the auditor will interact with a variety of people, from management to line workers—each with a different style of communication and a different set of concerns about the audit. The ability to interact successfully with representatives of the organization being audited requires an auditor to have good interpersonal skills, such as tact, the ability to listen, and diplomacy.

Auditors who can maintain an objective and professional demeanor typically are better able to control stress and gain the confidence of the client and auditee. The ability to maintain objectivity requires the auditor to be independent of the audited activities. This typically will not be a problem when an outside auditor is retained to conduct the audit. However, independence can become an issue during internal audits. This problem can be minimized by selecting an internal auditor or audit team from staff outside of the department being audited. Sometimes the argument is made that someone who does the job is in the best position to know if it is done correctly. However, when people are close to the work, they sometimes become oblivious to problems or accept them simply because they have become "routine." Under these circumstances, it becomes difficult to objectively self-evaluate and self-criticize. For these reasons, the internal audit team should be composed of workers outside of the department or activity being audited.

The auditor must possess good oral and written communication skills in order to communicate in a clear, neutral, and unbiased way during all phases of the audit. Finally, the auditor must exercise due professional care when executing the audit. The auditor should incorporate quality assurance into all phases of the audit process to ensure quality results, and the relationship between the audit team and the client should be one of confidentiality and discretion. The audit team is responsible for the integrity and confidentiality of information related to the audit. This means they should not disclose information, documents, or the audit report to any third party without the approval of the client and, where appropriate, the approval of the auditee, unless disclosure is required by law.

EDUCATION AND WORK EXPERIENCE

The ISO 14012 standard requires auditors to have at least secondary education or its equivalent, and appropriate work experience. The experiential requirement includes fundamental topic areas such as development of skills and understanding in some or all of the following: science and technology, facility operations, environmental laws and regulations, environmental management systems, and audit procedures.

Figure 10–1 shows acceptable combinations of education and work experience as listed in the standard. Postsecondary education is considered appropriate if it addresses some or all of the above areas.

AUDITOR TRAINING

In the standard, the requirement for formal training in some or all of these appropriate work experience areas can be waived if the individual can demonstrate competence. The requirement for competence can be met through accredited examinations or relevant professional qualifications.

In addition to the formal training requirement, auditors should have completed on-the-job training that involves them in the entire audit process under the supervision and guidance of the lead auditor. The training should be completed within a period of 3 consecutive years and should include a total of 20 equivalent work days of environmental auditing and a minimum of 4 environmental audits.

Because of the complexity of many operations and the legal framework in which they operate, it is critically important to have the audit performed by individuals who have the needed expertise and people skills. A certification can help

FIGURE 10–1. ISO 14012 Minimum Recommended Education and Work Experience for Auditors:

Education	Required Work Experience	Comments
secondary education or its equivalent	5 years of appropriate work experience	The appropriate work experience criterion may be reduced by up to 1 year for appropriate postsecondary full- or part-time education.
university or postsecondary degree	4 years of appropriate work experience	The appropriate work experience criterion may be reduced by up to 2 years for appropriate postsecondary full- or part-time education.

ensure that auditors are prepared adequately; but checking credentials, past performance, and a personal interview always makes sense before hiring a third-party auditor.

LEAD AUDITOR

Under ISO 14012, the lead auditor must demonstrate the ability to provide the skills and leadership to manage the audit through additional training or demonstration of expertise. The additional criteria for lead auditor should be completed within a period of 3 consecutive years. Expertise can be demonstrated through interviews, observation, references, and/or assessments of environmental auditing performance made under quality assurance programs. The criteria for additional training include participation in the conduct of additional audits, and the auditor should participate as acting lead auditor under the supervision and guidance of a lead auditor.

ENVIRONMENTAL AUDITING POLICY OF THE U.S. ENVIRONMENTAL PROTECTION AGENCY[2]

In the U.S., the Environmental Protection Agency (EPA) has encouraged the use of environmental auditing since the mid-1980s. Drawing on its expertise and past experience with auditing, the EPA also participated in the development of the ISO standards.

The EPA believes that auditing should remain a voluntary activity, but it may seek to include environmental auditing as part of settlement agreements. Also, the EPA will take into account a facility's efforts to audit in setting inspection priorities and in fashioning enforcement responses to violations. For example, facilities that audit presumably will have a good compliance history, and the EPA may subject them to fewer inspections.

The EPA recognizes four types of assessments and audits as qualifying as Supplemental Environmental Projects (SEPs) in settlements of environmental enforcement cases. Inclusion of an SEP can mitigate the assessed civil penalty. The EPA defines *environmental auditing* as "a systematic, documented, periodic and objective review by regulated entities of facility operations and practices related to meeting environmental requirements." The types of assessments and audits that are permissible under the 1995 interim revised supplemental policy are:

- Pollution prevention assessments are systematic, internal reviews of specific process and operations designed to identify and provide information about opportunities to reduce the use, production, and generation of toxic and hazardous materials and other wastes. To be eligible for SEPs, the assessment must be conducted using a recognized procedure for pollution prevention or waste minimization.
- Site assessments are investigations of the environmental health status or threats to environment or health at a given site. Assessments could include characterization of contamination at a site or discharging to a site, health risk assessments, ecological surveys, or natural resource damage assessments. To be eligible for

SEPs, the assessment must be conducted using recognized protocols, if available.

- Environmental compliance audits are independent evaluations of compliance status with environmental requirements. Typically, the use of compliance audits as SEPs is limited to small businesses, and credit is given only for the costs associated with conducting the audit.

- Environmental management system audits are independent evaluations of an organization's environmental policies, practices, and controls. The evaluation may include the need for: 1) a formal corporate environmental compliance policy; 2) educational and training programs for employees; 3) equipment purchase, operation, and maintenance programs; 4) environmental compliance officer programs; 5) budgeting and planning systems for environmental compliance; 6) monitoring, record-keeping, and reporting systems; 7) in-plant and community emergency plans; 8) internal communications and control systems; and 9) hazard identification and risk assessment.

Additional information about media-specific audits and the U.S. EPA's audit policy can be found through the U.S. EPA World Wide Web sites given in Appendix F.

THE EPA'S USE OF AUDIT REPORTS

Recognizing the need for some measure of privacy within the regulated community, the EPA's 1986 policy stated that it would not routinely request environmental audit reports. However, the agency may request audit reports on a case-by-case basis when needed to accomplish a statutory mission or in criminal investigations. Audit reports may not be used to shield monitoring, compliance, or other information that would otherwise be reportable and/or accessible to the agency, even if there is no explicit requirement to generate that data. Examples of likely situations include: audits conducted under consent decrees or other settlement agreements; criminal investigations; or when the company uses the audit as a defense.

SUMMARY

Once the EMS is operational, it must be examined periodically to ensure that it is effective and moving the organization toward the goals identified in the environmental policy. Periodic audits that include technical and management systems allow the organization to identity areas of strength and weakness and to initiate corrective actions and follow-up. Audits, coupled with management review and a program of continual improvement, are powerful tools for achieving environmental excellence.

ENDNOTES

[1] There is currently a movement among ISO 9000-certified companies to modify the way some certifications are performed because there is a sense that mature companies that

have been certified/registered and have demonstrated the effectiveness of their quality systems should be able to qualify for a modified rectification scheme. The new methodology, called *Supplier Audit Confirmation* (SAC), places greater reliance on internal audits. This certification would fall between self-declaration and a full third-party certification. The Registration Accreditation Board has expressed interest in SAC and may allow it in the future. Though this is not directly applicable to ISO 14000 audits, changes in ISO 9000 are likely to influence future versions of ISO 14000.

[2] The EPA policies regarding auditing can be found in 59 FR 38455, July 28, 1994, *Restatement of Policies Related to Environmental Auditing* and 60 FR 24856, May 10, 1995, *Interim Revised EPA Supplemental Environmental Projects Policy*.

CHAPTER 11

OCCUPATIONAL HEALTH AND SAFETY MANAGEMENT SYSTEMS

INTRODUCTION

The climate created by ISO 9000 and ISO 14000 is the impetus behind several initiatives for occupational health and safety management systems (OHSMS). These initiatives are based partially on the recognition that occupational safety and health programs are logical components of quality systems, and they are often integral to the management of environmental programs. Several countries (the UK, Norway, and Australia) have proposed national OHSMS standards. The British Standards Institution's draft occupational health and safety (OHS) management systems standard, BS 8750, is intended only for guidance and not for use as a specification standard for third-party registration purposes. In the United States, the Occupational Safety and Health Administration (OSHA) addresses some management system issues through its Voluntary Protection Program (VPP). The American Industrial Hygiene Association (AIHA) has also developed a standard for OHSMS. At the present time, there are no international standards for an OHSMS,[1] but there are ongoing discussions about the need for OHSMS guidelines.

During the development of ISO 14000 standards, the ISO TC-207 assembled an ad hoc committee that explored advantages and disadvantages of ISO developing an OHSMS standard.

Advantages of an ISO standard are:

- It could lead to a unified standard that includes ISO 9000, 14000, and OHS. This could simplify implementation and reduce costs when compared with the cost of implementing three separate standards.
- An OHSMS could improve employee safety and health.
- Certification to an OHSMS could provide a competitive advantage in much the same way as ISO 14000.
- ISO may provide a good forum for building consensus on this issue.
- Having most companies certified as compliant with an OHSMS could help to avoid new regulations within the United States and other countries.

Disadvantages are:

- An international OHSMS could become a barrier to trade. There is fear that the European Union (EU) might use ISO as a forum to export its social agenda or to create a competitive advantage. Companies in less developed countries may not have the resources to invest in demonstrating conformance.
- It may not be possible to reach an international consensus on sensitive issues such as process safety, ergonomics, employee involvement, and child labor. Further, the ISO may not be an appropriate forum for resolving these issues.
- Occupational health and safety are integral issues in the labor/management relationship, which differs around the world. Achieving international consensus in different political, economic, and cultural contexts may be difficult.
- Significant resources could be required for implementation.
- If a standard became the measure of "due care," there could be a potential for increased liability from regulatory agencies and the courts.
- The role of labor organizations in developing the standards would be pivotal. Labor would view a standard more favorably if it included active employee participation in audits and development of corrective action plans, and if audit findings were made public.

Whether or not an international standard is developed in the future, it makes good business sense to include OHS in an environmental management program. The remainder of this chapter will highlight the U.S. VPP and summarize the AIHA guidance document.

Additional resources and World Wide Web sites can be found in Appendix F under "Occupational Safety and Health."

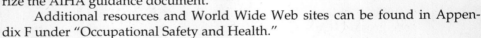

VOLUNTARY PROTECTION PROGRAM

The U.S. Department of OSHA developed the VPP to provide industries in the U.S. with a mechanism to enhance self-regulation and to reduce regulatory oversight. The certification process begins when an industry submits an application package to the regional OSHA office. The application must provide details of the hazards present in the workplace and the measures in place to deal with them. This effectively constitutes a written occupational safety and health program that, in its most complete form, is an OHSMS.

Next, the application is reviewed. If the written program is found to be satisfactory, OSHA schedules a site visit (typically 3 to 5 days in length) to conduct an in-depth evaluation of the facility. The adequacy of the program is evaluated in six areas:

- management commitment and planning,
- work site analysis,
- hazard prevention and control,
- safety and health training,
- employee involvement in program evaluation, and
- annual evaluation of the safety and health program.

The staff performing the OSHA review will evaluate all available information. The review would include employee interviews, records review, site inspections, and any other activities that might provide information on the adequacy and status of the occupational safety and health program.

The findings of the site visit team are then reviewed by OSHA management, who are generally not involved in the site visit. If the administrative review is satisfactory, the firm is designated as a VPP participant. To be awarded VPP status, the facility must have had a rate of injuries and lost work days below the national average for that type of industry. This comparison is based on rates published by the U.S. Bureau of Labor Statistics.

The final step in the program is official notification by the U.S. Assistant Secretary of Labor that the company has been awarded VPP status. These companies are then immune from preprogrammed OSHA inspections, but OSHA retains the right to inspect the facility in response to employee complaints, fatalities, chemical releases, or other significant incidents.

The VPP achievement hierarchy consists of three designations: Star, Merit, and Demonstration. Sites assigned *Star* status have safety and health programs recognized as comprehensive and have controlled hazards in the workplace. The *Merit* designation is for those sites that have only minor problems or that have a well-designed system that has not yet completely controlled hazards. *Demonstration* status is for companies that are not construction or general industry companies, such as agricultural industries.

The costs for being designated a VPP vary according to the current status of the program. More developed programs will have lower costs because they already have much of the necessary documentation and procedures in place to receive the designation.

There are many sources of help available to employers who wish to develop OHS programs and to achieve compliance with OSHA regulations.[2] Each state has an occupational safety and health organization that has been delegated certain powers by the federal government. States generally have a compliance assistance program that provides consultation services.

THE AIHA OHSMS GUIDANCE DOCUMENT

The AIHA guidance document (AIHA OHSMS 96/3/26) was developed and issued to assist employers and employees in achieving safe working conditions. The guidance document was peer reviewed by the AIHA OHSMS Task Force and the AIHA Board of Directors, which approved the management system. The guidance document is a controlled document and updated copies can be obtained from the AIHA.[3] The Guidance Document is intended to be used by health and safety professionals in the U.S. and in other countries as a basis for implementing an OHSMS. To promote harmonization with other standards, ISO 9001:1994 was used as a template for development of the OHSMS Guidance Document. The intention of the guidance document is to prevent, reduce, or eliminate workplace injury and illness, and it is applicable to an organization that wishes to:

1. demonstrate commitment to the health and safety of its employees,
2. implement a new OHSMS or maintain or improve an existing OHSMS,
3. assure itself or others of its conformance with its stated occupational health and safety policy, and
4. demonstrate the effectiveness of its OHSMS to an external organization.

Implementation of an OHSMS may be particularly valuable for small organizations because they have traditionally had little guidance in this field. Occupational health and safety management system conformance may also help smaller companies gain contracts as suppliers for larger organizations that have implemented an OHSMS, ISO 9000 quality system, or ISO 14001 environmental management system.

There may be legal, financial, or technical constraints that prevent immediate implementation of the complete OHSMS. In these cases, the guidance document encourages a risk-based prioritization of implementation activities that allows a phased-in implementation.

OVERVIEW OF THE GUIDANCE DOCUMENT

The guidance document contains both specifications and useful interpretations. As there are many similarities to the requirements of ISO 14001, the following sections summarize the main parts of the guidance document.

The first three sections present the scope, references, and definitions. Section 4.0 provides a discussion of the basic principles upon which an OHSMS should be established. The OHSMS should be specific to each organization and should build on existing programs. It should be implemented in the context of organizational policies, statutory and regulatory requirements, recommended guidelines, and labor agreements. The OHSMS should be based on the principle of prevention, and its implementation should improve the health and safety performance of the organization. Implementation requires a highly competent staff with the requisite training, experience, and professional judgment. Once implemented, continuous improvement should be an integral part of the OHSMS.

SECTION 4.1—OCCUPATIONAL HEALTH AND SAFETY MANAGEMENT RESPONSIBILITY

Section 4.1 of the guidance document covers OHS management responsibility. It requires development of an OHS policy similar to that developed for environmental management under ISO 14001. Senior management must develop a policy that has clear goals and objectives. The OHS policy must include a commitment to compliance with regulations or guidelines and to continuous improvement. It must also include a commitment to eliminating occupationally related fatalities, injuries, and illnesses. The policy must include a commitment to meaningful employee involvement in OHS planning, implementation, and evaluation. The OHS policy must be understood, implemented, and maintained throughout the organization.

Organizational responsibility and authority must be documented. The documentation must identify responsible persons and authority for implementation, and it must detail the interrelationships of personnel who manage the OHSMS, including those responsible for identification, evaluation, or control of work-related health and safety hazards.

A senior OHS management representative must have overall authority and responsibility for the establishment, maintenance, review, and documentation of the OHSMS. The senior manager should ensure that adequate resources are available to implement the OHSMS (for example, time, materials, technical support, financial support, and personnel).

SECTION 4.2—OHS MANAGEMENT SYSTEMS

Section 4.2 deals with specific requirements for the OHSMS: procedures, planning, and performance measures. A core requirement is that organizations shall "establish, document, and maintain the OHSMS." The organization is also required to prepare an OHSMS manual or equivalent documentation covering the requirements of the management system. The section on procedure development stresses that the procedures and their implementation should be integrated into each business function and operation. Procedures should be appropriate to the work being performed, and all operations should be documented.

Planning is essential to the effective functioning of an OHSMS. Good planning will enable the organization to develop procedures that describe how the organization will identify, evaluate, and control health and safety hazards.

Some issues to consider during planning activities include:

- utilizing prevention to control potential health and safety hazards;
- identifying and determining the compatibility of needed processes, resources, and controls;
- determining the feasibility of acquiring needed processes, resources, and controls;
- updating quality control, inspection, and testing techniques as necessary, including the development of new methods and instruments;
- emergency preparedness;
- medical surveillance and services; and
- preparation, dissemination, and storage of OHS records.

The organization should develop appropriate performance measures, collect and assess performance measure data, and use these data in evaluation and continuous improvement efforts. Some useful performance measures might be lost work days, number of occupationally related illnesses noted during physical exams, reduction in areas requiring hearing protection or other protective equipment, perception survey data, evaluation of unsafe behavior, number of near-miss incidents, employee comments or criticism, percentage of employees receiving required training, and average number of days taken to implement corrective action measures.

SECTION 4.3—OHS COMPLIANCE AND CONFORMANCE REVIEW

Section 4.3 contains requirements for documenting procedures for 1) assessing the applicability of legal requirements, 2) maintaining compliance and conformance with applicable requirements, and 3) documenting the performance of the OHS system.

This section requires the organization to develop and document OHS goals and measurable objectives, and to review and modify them as necessary. Employees should be involved in the development of the OHS goals and objectives. The OHS goals should be based on the concept of preventing work-related injuries. In addition, goals and objectives should be derived from and consistent with the organizational policy, and they should consider legal, technical, business, and ethical criteria. Once developed, the OHS goals and objectives should be documented and communicated to employees and other stakeholders.

The last part of this section requires a continuous improvement program. This program should be designed to ensure that goals and objectives are met or exceeded.

SECTION 4.4—OHS DESIGN CONTROL

The purpose of Section 4.4 is to ensure that OHS factors are considered during the design phase of new processes, products, or work stations. This section requires qualified OHS personnel to participate in decision making for process or operational changes, raw material changes, or purchase of new machinery. Ideally, their participation should occur early in the design phase to ensure that engineering or process controls are used to minimize potential hazards. It may be necessary to have a multidisciplinary team review processes designed to minimize health and safety impact. This team might include industrial hygienists, medical professionals, engineers, and other professionals.

Any new process design or modification to an existing process must consider a variety of inputs, including statutory or regulatory requirements, policies, and voluntary standards subscribed to by the organization. Process design changes and modifications must be approved by qualified OHS personnel prior to implementation. New designs and modifications must be documented and have outputs that can be verified. For example, if a vapor recovery system is specified by the design of a facility someone must verify that it was installed and functions properly to the specified recovery rate. The evaluation should also verify that the installed system is adequate to achieve its purpose.

SECTION 4.5—OHS DOCUMENT AND DATA CONTROL

Document and data control are specified in Section 4.5. Procedures must be developed that will control all OHS documents and data related to the requirements of the OHSMS, including laws, standards, regulations, and policies.

All documents must be reviewed and approved by the appropriate personnel before release. A master list of documents or other control mechanism must be developed. Changes in any of these documents will be reviewed and approved by the appropriate personnel (generally, the person who approved the initial version) before release.

Control of document versions is one of the most difficult tasks faced by those trying to maintain a current EMS or OHSMS. Computerization of documents helps greatly with document control. As recommended in the documentation section of this book, a software-based system such as Folio Views® reduces the problems associated with version control. Controlling printed copies then becomes the main difficulty.

SECTION 4.6—PURCHASING

This section identifies requirements for purchased products and services. The organization must have documented procedures for reviewing the health and safety aspects of products and for the evaluation and selection of contractors. These procedures must allow the organization to verify that purchased products and services conform to the OHS requirements, to purchase the least hazardous product feasible, to include OHS standards as performance criteria in contracts, and to maintain records of acceptable contractors.

The section also discusses factors involved with acquisition of goods and services, such as variability in toxicity, reviewing the qualifications and references of contractors, and other relevant considerations.

SECTION 4.7—OHS COMMUNICATION SYSTEMS

An effective communication system has many different aspects, some of which are enumerated in Section 4.7. The organization must make appropriate OHS documents available at all locations where operations essential to the effective functioning of the OHS quality system are performed. The documents must be current and controlled, and confidential documents must be secure. Invalid or obsolete documents must be removed from circulation, and appropriate archival techniques are needed to control these documents for historical or legal reasons. The communication of health and exposure hazards is also required in this section. The organization must communicate biological, medical, or environmental exposure information to employees promptly. Further, employees must be encouraged to report signs and symptoms, hazards, and/or hazardous conditions promptly and to make hazard control/prevention recommendations.

SECTION 4.8—OHS HAZARD IDENTIFICATION AND TRACEABILITY

This section requires the organization to develop documented procedures for identifying potential hazards through all stages of purchasing, production, delivery, and maintenance. The organization must have documented procedures and records to allow the unique identification of each production process or batch if it is required to do so by relevant legal, regulatory, or policy requirements. The guidance document requires materials to be traceable. This refers to the ability to determine the point of origin of raw materials, feed stock, or products that are used in other processes. This ensures that potential hazards are known or can be identified.

Section 4.9—Process Control For OHS

This section requires documented procedures that will control OHS hazards in production, installation, and servicing of equipment. The OHSMS should also ensure compliance with appropriate standards and regulations. Process parameters and process hazard characteristics must be monitored, and the organization must implement corrective actions whenever operations are out of control.

Organizations have many tools at their disposal for controlling employee health hazards. These are generally categorized as engineering (preferred), administrative, or personal protective equipment controls. In cases where it is impossible or prohibitively expensive to employ engineering controls, it is often possible to use administrative controls.

Equipment and processes should have an approval procedure that allows qualified individuals to evaluate the OHS impact from an operation or modification to the operation. If OHS impact or potential hazards become apparent only after the process is implemented, corrective actions should be initiated immediately to eliminate these potential hazards. The nonconforming process should be monitored by a qualified individual until the system is operating within specified quality parameters.

The organization must also maintain equipment properly. A poor maintenance program has the potential to increase overall costs and to create workplace hazards. Hazardous substances may leak into the workplace; excessive vibrations can cause accidents; product yields may drop; or there may be other negative consequences associated with specific processes.

Section 4.10—OHS Inspection and Evaluation

Many occupational illnesses and injuries are the result of employee practices, rather than conditions in the workplace. Periodic inspections and evaluations of the OHSMS can help eliminate behavioral and process problems that can adversely affect health and safety. Inspections and evaluations of the OHSMS should be conducted at intervals determined by the relative severity of the hazards present. The inspections should take a very proactive approach to the anticipation and prevention of hazards. Direct and indirect hazards should be considered. The type and frequency of inspections, as well as the records to be kept, should be detailed in the OHSMS. These records should be used during program evaluations to detect areas in need of improvement.

A preliminary assessment of the facility, which should be part of the OHSMS planning process, provides a helpful tool in planning inspections. It allows inspections to be targeted to the areas with the highest potential for worker illness or injury. Areas covered by the preliminary assessment include findings of previous inspections, material safety data sheets, employee health records, incident reports, corrective action plans, documented procedures, air monitoring records, and equipment calibration logs.

The organization should also establish and document procedures for evaluating purchases of personal protective equipment, monitoring and calibration equipment, contract services, hazard abatement equipment, or other products that may have an impact on OHS. Evaluation criteria should be established for each good or service. Factors to consider for the evaluation of equipment are the suitability for the intended task, condition, hazard data, and adequacy of operational manuals.

SECTION 4.11—CONTROL OF OHS INSPECTION, MEASURING, AND TEST EQUIPMENT

The evaluation of an OHS program requires accurate, precise, and reliable data to document routine exposures and exposures that could present a potential danger to life or health (for example, oxygen-deficient or -enriched atmospheres, toxic atmospheres, or explosive atmospheres). In cases where instrumentation is used or where biological or medical testing is done, it is important to ensure that sample collection and analytical methodologies meet the requirements of regulatory agencies or other applicable requirements. The guidance document requires the organization to document procedures for operating, maintaining, calibrating, and controlling equipment and methods used to collect data. Equipment should be used only in the manner specified by the manufacturer, and appropriate quality assurance/quality control measures must be taken. Data evaluation procedures must be documented, and procedures must be strictly enforced.

SECTION 4.12—OHS INSPECTION AND EVALUATION STATUS

This section requires the strict enforcement and documentation of the sample data evaluation and reporting mechanisms.

SECTION 4.13—CONTROL OF NONCONFORMING PROCESS OR DEVICE

Many things can cause a workplace fatality, injury, or illness. The two broad categories are unsafe behavior and failure of a control method. This section requires the organization to document and correct nonconformances. Corrective actions might include behavioral changes, modification or addition of engineering controls, or procedural changes. The nature of nonconformances and corrective actions should be communicated to all affected parties and appropriate managers. It is important to specify who has responsibility for assuring conformance.

Some methods of bringing the system into conformance may not be cost-effective. In these cases, other permanent measures must be implemented to control potential hazards. When designing these measures, remember that best management practices (and U.S. OSHA regulations) allow personal protective equipment to be used only when there are no other reasonable alternatives.

SECTION 4.14—OHS CORRECTIVE AND PREVENTIVE ACTION

This section requires the organization to investigate occupational injuries and illnesses in a systematic fashion. When nonconformances are found, procedures must be developed and maintained to implement corrective and preventive OHS actions. Any changes to existing procedures as a result of the corrective and preventive action must be documented.

The most important consideration in developing corrective and preventive measures is to ensure that the selected measures correct the root problem and not just the symptoms. For instance, assume that an employee walks in front of a forklift and sustains an injury. What cause should be eliminated? Is the employee at fault for not paying attention? This might be corrected by using signs and giving additional training. Was the accident caused because there was vibrating machinery nearby and the employee could not hear or see the forklift? This might be corrected by using shock mounts on the machinery and adding a mirror. The selected corrective and preventive actions should emphasize developing strategies to avoid future occurrences of the same type.

The procedure for corrective action begins with an investigation into the cause of the OHS nonconformance. Affected individuals must be notified, and the results of the investigation must be documented. The next step is to determine what corrective action is needed to eliminate the root cause of the OHS nonconformance. Finally, the implemented corrective action should be evaluated to determine whether it is effective and documented.

The guidance document also requires an organization to include the prevention of OHS hazards as an OHS policy priority. A number of factors should be considered when developing the prevention program. First, it should be based on a systematic approach to the anticipation, detection, assessment, and elimination of potential causes of OHS nonconformance. Second, an assessment method must be developed to verify the effectiveness of the prevention program. Third, management must be given relevant information so it can review the outcome of preventive actions. A clearly defined mechanism should be in place for accomplishing this portion of the standard. Fourth, the OHS must be integrated into the business functions of the organization. This last requirement allows proper planning for the preventive OHS actions and ensures the financial and human resources to accomplish the organization's goals.

SECTION 4.15—HANDLING, STORAGE, AND PACKAGING OF HAZARDOUS MATERIALS

This section recognizes the special hazards associated with flammable, combustible, corrosive, toxic, and other potentially hazardous materials. To minimize these hazards, the guidance document requires the organization to develop documented procedures for the packaging, use, storage, handling, and disposal of hazardous or potentially hazardous materials. The organization must establish a hazardous materials inventory system, must ensure proper labeling of hazardous material containers, and must maintain documentation of conformance with the OHS quality system as it relates to these materials. This section also encourages employee training and the communication of proper procedures to prevent improper disposal or release of hazardous substances into the environment. The highly regulated nature of these materials provides a structured framework for developing, documenting, and implementing measures to ensure worker health and safety. In the U.S., for example, these materials are regulated by the Department of Transportation, the Department of Energy, the Environmental Protection Agency, and OSHA. Other countries have equivalent agencies.

SECTION 4.16—CONTROL OF OHS RECORDS

The guidance document requires an organization to develop procedures for the identification, collection, indexing, access, filing, storage, maintenance, and disposition of OHS records. Records must be maintained that demonstrate conformance to specified OHS requirements, and records must be retained for the appropriate amount of time. For example, in the U.S., physical exam records for employees who wear respirators must be maintained for 30 years after the employee leaves the company.

Examples of needed records include exposure and training records, illness and injury investigation procedures, and medical files. The organization is also responsible for preserving the confidentiality of personnel medical records. Other needed records might include product and process information; inspection maintenance and calibration records; purchased product and service records; audit results; management reviews; emergency contingency plans; and valid employee OHS-related observations, complaints, and suggestions. The organization is required to maintain any other information that describes and documents compliance with applicable statutory and regulatory requirements.

SECTION 4.17—INTERNAL OHS MANAGEMENT SYSTEM AUDITS

Internal audits are essential to the successful implementation of an OHSMS. The audit requirement is a review of the program by internal staff, or an equivalent process. The purpose is to determine whether or not there are parts of the OHSMS that are nonconforming. Procedures for planning and executing the audit are necessary to a successful audit program. The audit plan should include: prioritized activities and areas to be considered in the audits, audit frequency, audit team make-up and required expertise, report content, communication of audit findings, and meaningful employee participation.

The audit should be comprehensive, should evaluate all elements of the OHSMS, and should be scheduled on the basis of risk severity and probability indices. The most severe and most likely problems should be given the most attention. The guidance document requires the auditors to be independent of the activity being audited, and worker involvement in the audit process is encouraged. This can help increase awareness and give employees a feeling of ownership of the OHSMS, which can aid greatly in achieving conformance. Deficiencies documented during the audit should be prioritized for corrective action. Follow-up OHSMS audit activities should verify that corrective action was taken and should record the results.

Audit findings should be organized into a report for the appropriate management staff. This report should identify specific OHS system deficiencies and the root causes of those deficiencies, and it should recommend corrective and preventive actions for the noted deficiencies. The report should also document the results of implementing corrective actions recommended in previous audit reports.

Section 4.18—OHS Training

This section requires all employees to be qualified for their tasks on the basis of appropriate education, training, certification, and/or experience. All employees who perform activities that place them at risk for occupational illness or injury should receive the appropriate training to deal safely with the hazards they may encounter. Training should comply with applicable regulations or best practices, and it should be fully documented.

The training should make employees aware of their health and safety roles and responsibilities; potentially hazardous conditions, including signs and symptoms of overexposure; hazard avoidance techniques; and the use and limitations of various control mechanisms. Workers should demonstrate comprehension and retention of training content.

Frequently, employees do not understand the importance of following the appropriate health and safety procedures. Training is one the most important tools that an employer has to increase employee awareness and to reduce the incidence of occupational illness and injury.

Section 4.19—Operations and Maintenance Service

The intent of this section is to ensure that contractors involved in operations and maintenance services make good-faith efforts to provide a safe and healthful environment for their employees.

This section requires the organization to establish and maintain documented procedures to ensure that contractors comply with specified OHS requirements. The documentation should include procedures to verify that contractors comply with the specified OHS requirements. These requirements can be fulfilled by requiring a contractor to comply with the organization's OHSMS or by requiring the contractor to establish an OHSMS for the contracted services.

SECTION 4.20—STATISTICAL TECHNIQUES

Collecting and analyzing OHS data provide an organization with opportunities to identify programs and systems that require improvement, to clarify resource allocation decisions, to identify areas where equipment or procedural changes may be useful, and to establish a causal link between incidents. The guideline encourages the organization to collect and use data for continuous improvement. It requires the organization to identify and use helpful statistical techniques to establish, control, and verify the effectiveness of the OHSMS. The use of statistical techniques must be documented in procedures and must be controlled.

Whenever statistical methods are used to analyze data, it is important to be aware of the strengths and limitations of these techniques. With careful application, the results of statistical work can be very valuable for process control and continuous improvement.

SUMMARY

Implementation of a thorough OHSMS should allow an organization to establish an OHS policy; to integrate OHS prevention and continuous improvement into the corporate culture; to identify and meet government requirements in a timely manner; to develop OHS goals and measurable objectives; and to prioritize the use of OHS resources and anticipate, recognize, evaluate, and control hazards.

Although an international standard has not yet been developed for an OHSMS, one may be developed in the future. The AIHA OHS management system establishes a useful framework for developing an effective OHSMS because it is closely associated to ISO 9001 and ISO 14001. Using the AIHA OHSMS for guidance should allow many industries to implement an OHSMS program with relative ease. The costs for implementation of this standard are not likely to be excessive for those companies with an existing OHS program.

ENDNOTES

[1] At the January, 1997 meeting of the ISO Technical Management Board, Resolution D/1997 was passed that decided that there was little support from major stakeholders for ISO to develop an OHSMS standard. The resolution also recognized that the need for an OHSMS standard might exist in the future.

[2] Sources of information include:
- *Guide to Developing Your Workplace Injury and Prevention Program*, which can be obtained from:

  Cal/OSHA Consultation Service Headquarters
  395 Oyster Point Boulevard, Room 325
  South San Francisco, CA 94080

- The American Industrial Hygiene Association has many useful publications, including: *Industrial Hygiene Auditing: A Manual for Practice*, which can be obtained from:

  American Industrial Hygiene Association
  1700 Prosperity Avenue
  Fairfax, VA 22031

- OSHA Training Institute
  155 Times Drive
  Des Plaines, IL 60018

- National Institute for Safety and Health
  4676 Columbia Parkway
  Cincinnati, OH 45226

[3] The address for AIHA is given in endnote 2. The AIHA can also be reached through *http://www.aiha.org*.

REGULATORY COMPLIANCE AUDITING

Part 4 highlights factors to consider when preparing for regulatory compliance audits for selected topics. The focus of this section is on regulatory compliance in the U.S., but this part will also be useful to others. Each of the chapters contains example forms and checklists that can be modified and adapted as needed. It should be noted that these chapters are not in-depth treatments of the audit process, but they do highlight factors that should be considered when evaluating programs. Chapter 12 provides an overview of the regulatory framework and selected regulations; Chapter 13 covers spill prevention, reporting, and response; Chapter 14, underground storage tanks; Chapter 15, water pollution; Chapter 16, hazardous wastes; and Chapter 17, air pollution.

CHAPTER 12

REGULATORY OVERVIEW

INTRODUCTION

One key aspect of the an environmental management program is ensuring compliance with applicable environmental regulations. Attaining and maintaining regulatory compliance should be one of the first goals of an organization as it moves toward achieving excellence in its environmental management program. Organizations, however, often encounter significant frustrations on the road to compliance. These include a complex regulatory environment, changing regulations, and the lack of uniformity in regulatory requirements at different levels of government.

In the U.S., the state and local governments must have regulations that are at least as stringent as those of the federal government. However, they can (and do) enact regulations that are more stringent. Because of the potential liability associated with noncompliance, it is critically important for an organization to have a mechanism in place so it can remain current on regulatory developments at the local, state, national, and international levels.

A lack of knowledge or understanding does not absolve an organization from the responsibility of complying with regulations. Furthermore, a lack of compliance can seriously affect the financial well-being of the organization and individuals within the organization. Administrative and civil penalties under the major environmental statutes in the U.S. can reach $25,000 per violation or day of violation. Administrative penalties are typically assigned for lesser problems or when the agency wants a quick resolution of the problem. When the Environmental Protection Agency (EPA) seeks civil or criminal penalties, the case is referred to the Department of Justice.

Administrative and civil penalties are typically assessed against the corporation rather than individuals, but the statutes also provide for criminal sanctions against organizations and responsible individuals for misrepresentation and knowing or negligent violation of the statutes. Criminal sanctions include fines and/or imprisonment. For example, under the Resource Conservation and Recovery Act (RCRA), criminal penalties include up to $50,000, 2 years imprisonment, or both for persons who "knowingly" commit certain violations. An individual who is convicted of "knowing endangerment" (a situation in which another person is placed in imminent danger of death or serious bodily injury) could be fined up to $250,000 and/or receive up to 15 years in prison under RCRA. Some penalties under the Clean Water Act (CWA) are stiffer. "Negligent" violations can result in a fine of up to $25,000 per day of violation and/or up to 1 year in prison. If the violation is "knowing," the penalties include a fine up to

$50,000 per day of violation and/or 3 years in prison. A violation of "knowing endangerment" triggers a fine of up to $250,000 and 15 years in prison for individuals and a fine of up to $1 million for corporations. In the event of a second conviction, the maximum fines and prison terms double.

It is obvious that gaining and maintaining an understanding of environmental laws and regulations, and achieving compliance is critical to the success of any organization. Regulatory updates can be acquired by subscribing to abstracting services (many are on-line), joining professional or trade groups, or reading the statutes and regulations (and interpretations of them). Reading the actual regulation is important, even if outside abstracting services are utilized. Some tools for researching legislation and regulations are given in Chapter 4; the EPA has many publications, on-line services, and hotlines that can provide additional help (see Appendix F). Access to these resources is also available through the *epa.gov* site on the WWW.

THE LEGISLATIVE AND RULE-MAKING PROCESS IN THE U.S.

At the federal level, laws begin with a bill that is introduced in either the House of Representatives or the Senate. Bills are assigned to a committee for consideration—a process that might include hearings, studies, or investigations. The committee issues a report and recommends whether or not the bill should pass. Differences that might exist between House and Senate versions are resolved in a conference of House and Senate representatives. The bill then goes to the full House of Representatives or Senate for a vote. If the bill passes in both the House of Representatives and the Senate, it becomes law when it is signed by the President or if it is not vetoed within 10 days. Figure 12–1 at the end of the chapter provides a list of major federal laws that cover environment.

Laws provide the framework within which administrative agencies such as the EPA or OSHA develop and promulgate regulations that carry out the wishes of the Congress. An agency may also be empowered to promulgate regulations through an executive order signed by the President.

The rule-making process involves publishing proposed regulations in the *Federal Register* (FR), providing for public comment through public hearings or written comments, and publishing final regulations in the FR. The final regulations become effective on the published dates, unless otherwise noted in the final rule. The FR also contains the regulatory history and explanations (including responses to public comments) for the agency's proposals. The FR has a monthly index and a monthly "List of Sections Affected," which allow the user to locate changes in the regulations since the FR was issued. The final rules are published once each year in the *Code of Federal Regulations* (CFR); individual sections of the code can be ordered from the National Technical Information Service (see Appendix F). The government also publishes a CFR "Index and Finding Aids" volume with the CFR. Because the code is published only once each year, it cannot be relied upon for the most current information. The FR (or an abstracting service) must be consulted for the most current requirements since the last edition of the

CFR was published. (The CFRs and FRs are available as reference documents in federal agency libraries, and CFRs can be purchased from the U.S. Government Printing Office.)[1]

KEY PROVISIONS OF SELECTED STATUTES

The following discussion provides an overview of key provisions of selected statutes. It is not intended to be an exhaustive treatment, but only to provide a brief survey of major sections of the statutes.

THE RESOURCE CONSERVATION AND RECOVERY ACT AND THE HAZARDOUS AND SOLID WASTE ACT

The Resource Conservation and Recovery Act of 1976 provides a "cradle-to-grave" management system for over 300 million metric tons of hazardous waste generated each year in the U.S.; it was significantly amended by the Hazardous and Solid Waste Amendments (HSWA) of 1984 to include small-quantity generators and a new program for underground storage tanks containing hazardous substances or petroleum (Subtitle I). The states have the authority (in Section 3006) to carry out the federal program, which is administered by the EPA, providing that the state program is consistent with and equivalent to the federal program. As noted earlier, the state program may be more stringent than the federal program. Compliance auditing for hazardous wastes is covered in Chapter 16; underground storage tanks are covered in Chapter 14.

The RCRA is divided into 10 subtitles. Subtitle A defines our national policy for managing hazardous waste and identifies objectives for reaching the policy, which includes the following elements: 1) the generation of hazardous waste is to be reduced or eliminated as expeditiously as possible, 2) land disposal should be the least favored method for managing hazardous waste, and 3) all waste that is generated must be handled so as to minimize the present and future threat to human health and the environment.

Subtitle C creates the national hazardous waste management system. The cornerstone of the law is Section 3001, which requires the EPA to develop regulations for the identification and listing of hazardous wastes. A waste is hazardous if it is contained in a list or if it meets certain characteristics. The first two lists include hazardous wastes from nonspecific sources (F list) and hazardous wastes from specific sources (K list). The third list, which is subdivided into two lists, identifies commercial chemical products, including off-specification species, containers, and spill residues, which, when discarded, must be treated as hazardous wastes. The U list contains chemicals that are considered to be toxic and are managed like other listed hazardous wastes. The P list consists of acutely hazardous wastes, which have more rigorous management requirements.

If a waste is not listed, it is regulated under RCRA if it exhibits one of four characteristics: ignitability, corrosivity, reactivity, or toxicity. The RCRA also governs wastes that are mixtures of hazardous and solid wastes, wastes that are derived from the treatment of hazardous wastes and debris, and environmental media that contain hazardous wastes. Section 3010 requires any person who manages a hazardous waste to file a notification with the EPA for each site at which hazardous waste is managed. Generators of hazardous waste are the critical link in the management program, and generators must comply with the standards authorized by Section 3002 of RCRA. These standards include proper handling and storage, and the preparation of the Uniform Hazardous Waste Manifest to track the shipment of the wastes. Transporters must comply with regulations authorized by Section 3003 for manifests, labeling, and the proper shipment and delivery of the waste to a treatment, storage, or disposal (TSD) facility. Section 3005 establishes the permit system for owners and operators of TSD facilities. Section 3004 requires TSD facilities to comply with performance standards, which include minimum technology requirements for treatment and disposal, groundwater monitoring requirements, and a ban on the land disposal of untreated hazardous wastes. The TSD facilities must comply with certain requirements that include performing detailed chemical and physical analysis of the wastes prior to acceptance, providing security to prevent unauthorized entry, training personnel, and developing a contingency plan in the event of accidents and spills.

Small-quantity generators that produce less than 1,000 kilograms (kg) (2,200 pounds) of hazardous waste per month must comply with special regulations. Although very small generators (less than 100 kg per month) are conditionally exempt from RCRA, they are subject to certain minimum standards. Regulations that apply to generators between 100 kg and 1,000 kg are similar to those for large generators.

Site inspections and federal enforcement of EPA's regulations are authorized in Sections 3007 and 3008.

The Hazardous and Solid Waste Amendments added Subtitle I to RCRA, requiring the EPA to develop regulations for new and existing underground storage tanks (USTs). The EPA has revised the UST regulations several times since the final adoption in 1988. The statute defines a UST as any one or combination of tanks (including underground pipes connected thereto) that is used to contain an accumulation of regulated substances, and whose volume (including the volume of underground pipes connected thereto) is 10% or more beneath the surface of the ground. Exceptions to the definition include farm or residential tanks of 1,100 gallons or less capacity, heating oil storage tanks, septic tanks, surface impoundments, pits, ponds, lagoons, and others. The regulated substances include "hazardous substances" regulated under the Comprehensive Environmental Response, Compensation and Liability Act (CERCLA), and petroleum and petroleum-based substances.

Owners and operators of existing tanks are required to meet certain responsibilities. An owner is any person who owns a UST that is used for the storage, use, or dispensing of regulated substances on or after November 8, 1984. A person who owned a UST immediately before the discontinuation of its use before November 8, 1984 is also considered an owner. Persons, such as lenders, who hold security interests in petroleum USTs but do not participate in the operation of such USTs are not considered to be owners. An operator is any person who controls or is responsible for the daily operation of the UST. The regulations for USTs identify requirements for notification of tanks, leak detection, record keeping, reporting, corrective action, closure, and financial responsibility.

Owners and operators of existing tanks are required, by December 22, 1998, to meet one of the following requirements: 1) new UST system performance standards, 2) tank upgrade requirements, or 3) closure and corrective action requirements. Owners and operators of existing tanks must comply with requirements for installation of tanks and piping, cathodic protection, financial responsibility, and release detection.

The EPA can assess civil penalties of not more than $10,000 per tank against owners and operators who knowingly fail to notify or who submit false information under the notification requirements [42 USC Section 6991e(d)]. A $10,000 penalty per tank per day of violation can be assessed for failure to comply with UST requirements relating to leak detection, record keeping, reporting, financial responsibility, closure, and corrective actions [42 USC Sections 6991b(c), 6991e(d)].

THE COMPREHENSIVE ENVIRONMENTAL RESPONSE, COMPENSATION AND LIABILITY ACT

The driving force for the enactment of the CERCLA in 1980 was the realization that the improper disposal of hazardous wastes, such as those at the Love Canal site, presented a great risk to public health and the environment. The CERCLA contains eight major sections that include the designation of hazardous substances, release reporting, remedial provisions, identifying and evaluating facilities (sites), the Superfund, liability and settlement provisions, EPA's information-gathering authority, and judicial actions.

The CERCLA, commonly known as the *Superfund law*, directed the EPA to effect cleanup at sites that release or threaten to release hazardous substances into the environment. A hazardous substance under CERCLA [Section 101(14)] is defined by reference to substances that are regulated under other environmental statutes. Hazardous substances under CERCLA include:

- "hazardous wastes" that are listed or have specific characteristics under RCRA,
- "hazardous substances" defined in the CWA (Section 311),
- "toxic pollutants" designated under Section 307 of the CWA,
- "hazardous air pollutants" listed in the Clean Air Act (CAA) (Section 112),
- imminently hazardous chemical substances or mixtures addressed in Section 7 of the Toxic Substances Control Act, and
- any element, compound, mixture, solution, or substance designated by the EPA as a hazardous substance pursuant to CERCLA.

A list of these hazardous substances can be found in 40 CFR Part 302. There is no minimum amount of hazardous substance that triggers a response action or liability for the release or threatened release of these materials. The CERCLA, however, does require the reporting of spills that are greater than a specified quantity for each chemical [called a *reportable quantity* (RQ)], which is given in the hazardous substance list. If a reportable quantity is not given for a hazardous substance, Section 102 of CERCLA specifies a quantity of 1 pound, unless the hazardous substance has a reportable quantity under the CWA, in which case the latter will be used.

The CERCLA requires the EPA to identify and rank facilities that require cleanup and to prioritize cleanup efforts in accordance with the National Contingency Plan, which provides the blueprints for selecting and conducting CERCLA actions. The ranked sites are placed on a list, called the *Comprehensive Environmental Response, Compensation and Liability Information System* (CERCLIS). The highest priority sites are identified in the National Priorities List (NPL), which is updated annually. Sites on the NPL qualify for the long-term remedial actions funded by the Superfund, which is partially financed by a tax on the petrochemical industry. A site that is not on the NPL may still be the subject of a short-term removal action. After a potential site is identified, the EPA researches the site and develops a list of potentially responsible parties (PRPs) that usually includes the following:

- the current owner or operator of the facility,
- the owner or operator of the facility at the time the hazardous substances were disposed at the facility,
- any party who arranged for the disposal of the hazardous substances, and
- any party who selected the disposal site and accepted the hazardous substance for transportation to the site.

The EPA is empowered to issue an administrative order under Section 106 requiring responsible parties to perform the cleanup at a site or to conduct the cleanup itself when responsible parties either refuse or are unable to obey its cleanup orders. If the EPA cleans up a site, it can recover costs from PRPs. A party that refuses to comply with an administrative order may be assessed up to $25,000 per day of the violation, and an unjustified failure or refusal to comply may also result in punitive damages equal to but not more than three times the amount of costs incurred as a result of the party's failure to take action under the order [USC Section 9607(c)(3)]. The EPA will typically issue administrative orders only to those PRPs that are the largest contributors to the site and financially viable, but under CERCLA, liability is far-reaching. Parties may be liable as a result of actions taken before CERCLA was enacted, even if they were not major contributors to contamination of the site. Liability extends to individual corporate officers or parent corporations exercising control over a corporation's hazardous waste handling and disposal activities.

The CERCLA does provide some limited opportunity for relief in those cases where there is sufficient cause not to comply with a cleanup order [42 USC Sections 9606(b)(1), 9607(c)(3)].

THE EMERGENCY PLANNING AND COMMUNITY RIGHT-TO-KNOW ACT

The Emergency Planning and Community Right-to-Know Act (EPCRA) of 1986 (Title III of the Superfund Amendments and Reauthorization Act) requires the states to establish a process for developing local emergency preparedness programs in the event of a chemical spill or accident. In addition, the states are responsible for collecting and disseminating information about the hazardous chemicals present at facilities within the local communities.

Section 301 of the Act requires any facility that produces, uses, or stores any of the substances on the EPA's list of extremely hazardous substances (EPA's *Title III List of Lists*) to participate in the local emergency planning process. This Section also establishes the State Emergency Response Commission (SERC) and local emergency planning committees (LEPCs).

Section 302 requires any facility that produces, uses, or stores any of the extremely hazardous substances in amounts equal to or greater than the threshold planning amounts to notify the SERC that the facility is subject to the emergency planning requirements. A covered facility must designate a facility representative who will participate in the emergency planning process and a facility emergency response coordinator. Further, the owner or operator must notify the SERC and LEPC within 60 days if an extremely hazardous substance (EHS) becomes present at the facility in amounts that equal or exceed the established threshold planning quantities (TPQs). Violations of this Section can result in civil penalties of up to $25,000 per day for each day that the violation or failure to comply continues.

Under Section 304, covered facilities must immediately notify the state and local authorities when there is a release of a listed hazardous substance (that is not federally permitted) that exceeds the reportable quantity established for that substance and results in exposure to persons off-site. The initial telephone notification must be followed by a written notification. Hazardous substances include those in the *Title III List of Lists* and those subject to emergency notification requirements under Section 103(a) of CERCLA. Failure to comply with Section 304 can result in administrative penalties of up to $25,000 per violation for each day the violation continues. A second or subsequent violation of Section 304 can result in civil penalties of up to $75,000 per day for each day the violation continues. A knowing or willful failure to provide notice can result in criminal penalties of up to $25,000 or 2 years of imprisonment, or both; in the case of a second or subsequent conviction, the penalties increase to a fine of up to $50,000 or imprisonment for 5 years, or both.

The Community Right-to-Know reporting imposes two requirements on the regulated facilities. Section 311 applies to facilities that must comply with OSHA's Hazard Communication Standard. This section requires the owner or operator of the facility to submit material safety data sheets (MSDSs) or a list of hazardous chemicals to the SERC, the LEPC, and the local fire department (LFD). The EPA's definition of hazardous chemicals includes hazardous chemicals present in amounts equal to or greater than 10,000 pounds and extremely hazardous substances present in amounts equal to or greater than 500 pounds or than the

TPQ established for the substance, whichever is lower. Failure to comply with Section 311 requirements can result in civil penalties of up to $10,000 per day per violation.

Section 312 requires the owner or operator of a regulated facility to submit an emergency and hazardous chemical inventory form to the SERC, LEPC, and LFD for the chemicals established under Section 311. A two-tier approach is used for reporting. The Tier One report consists of aggregate information (location, average daily quantities, and maximum annual quantities) for each health and physical category. This information must be submitted on an annual basis. The Tier Two report contains chemical-specific information (identification, location, hazard, state, average daily quantities, maximum annual quantities, length of time on-site, and storage practices). A regulated facility may elect to declare confidentiality for the location of covered materials. Failure to comply with Section 312 requirements can result in civil penalties of up to $25,000 per day per violation.

Section 313 requires owners and operators of covered facilities to complete the Toxic Release Inventory (TRI) Reporting Form R to report quantities of listed chemicals that are released to the environment or transferred off-site, waste treatment and disposal methods, and information on pollution prevention, source reduction, and recycling. Covered facilities are those in Standard Industrial Classification (SIC), Codes 20 through 39, that have 10 or more full-time employees, and exceed threshold quantities of listed toxic chemicals that are manufactured, imported, processed, or otherwise used. The threshold amount is 10,000 pounds per calendar year for listed chemicals that are otherwise used and 25,000 pounds per calendar year for manufacturing, importing, or processing of listed chemicals. In 1994, the EPA issued a final rule that established a streamlined reporting option for facilities with low annual reportable amounts of a listed chemical. Failure to comply with Section 313 requirements can result in civil penalties of up to $25,000 per day per violation.

THE CLEAN AIR ACT

The Clean Air Act of 1970, as amended in 1990, is a complex program for regulating new and existing sources of air pollution. The Act establishes primary (health-based) and secondary (welfare-based) national ambient air quality standards (NAAQS) for six outdoor air pollutants: carbon monoxide, sulfur dioxide, nitrogen oxides, particulate matter, ozone, and lead. The NAAQS are implemented through source-specific emission limitations established by the states in state implementation plans (SIPs). The SIPs for nonattainment areas (areas that are not attaining the NAAQS) must include more stringent source-specific controls than for attainment areas. Chapter 17 discusses compliance audits for air pollution sources.

New and modified sources that contribute significantly to air pollution that endangers public health or welfare are regulated under Section 111 of the CAA. To date, the EPA has identified over 60 source categories and has developed emission standards for these sources (new source performance standards). New and modified sources are also required to obtain a preconstruction review and permit before they locate in an area that either attains or does not attain the NAAQS. The prevention of significant deterioration (PSD) permit applies to sources in attainment areas that have the potential to emit over 250 tons per year of a regulated pollutant or over 100 tons per year of a regulated pollutant if the source falls within one of 28 listed source categories. These sources are required to install best available control technology (BACT) and to comply with allowable pollution increments that are designed to prevent the deterioration of air quality. Major sources (potential to emit more than 100 tons per year of a nonattainment pollutant) in nonattainment areas are required to achieve the lowest achievable emission rate (LAER).

The 1990 CAA Amendments substantially revised the regulation of toxic air pollutants under Section 112 of the CAA. An initial list of 189 substances was developed for regulation, and in 1992, the EPA published a list of all major source categories and subcategories of hazardous air pollutants (HAPs) that are to be regulated (for example, chemical plants and oil refineries). A major source of HAPs is a stationary source emitting more than 10 tons per year of any of the listed substances or 25 tons of any combination of the substances. Other sources may be regulated under an area source program. Each major source will be expected to comply with emission standards that require the installation of maximum achievable control technology (MACT). Additional controls may be required under a residual risk program that will be developed no later than 8 years after a MACT standard has been established for a source category.

The 1990 CAA Amendments also required the EPA to promulgate regulations to control and prevent accidental releases of regulated hazardous air pollutants or any other extremely hazardous substances. The Accidental Release Prevention Program requires owners and operators of facilities that have more than a threshold quantity of these substances to develop risk management plans for each substance used at the facility. Annual audits and safety inspections to prevent leaks and other episodic releases may also be required by the EPA.

Other major regulatory programs of the 1990 CAA Amendments include acid rain, visibility protection, stratospheric ozone protection, mobile sources, fuels and fuel additives, and the operating permit program.

On January 11, 1993, the EPA published final rules as required by Title IV of the CAA Amendments of 1990 for acid rain permits, sulfur dioxide emission allowance tracking and trading, emissions monitoring, excess emissions penalties and offset plans, and the administrative appeals process. Under Title IV, sulfur dioxide allowance allocations, which can be traded on a national market, are assigned to utility boilers (one allowance is an authorization to emit 1 ton of sulfur dioxide). Some other sources can opt into the program, providing they meet the applicable permitting and monitoring requirements.

Title VI of the CAA Amendments of 1990 establishes a program for the phase-out of ozone-depleting Class I (halons, carbon tetrachloride, methyl chloroform, and chlorofluorocarbons) and Class II substances (hydrochlorofluorocarbons). Beginning in 1991, the production of Class I substances is gradually restricted, and a complete ban goes into effect, beginning in the year 2000 (2002 for methyl chloroform). Phase II substances are scheduled to be phased out completely by 2030.

The 1990 CAA Amendments require manufacturers to reduce substantially the tail-pipe emissions from automobiles. Other program requirements apply to certain carbon monoxide and ozone nonattainment areas. These include the reformulated fuel program and the clean fuel program. One aspect of the clean fuel program requires the operators of centrally fueled fleets of ten or more vehicles in certain nonattainment areas to purchase and use clean fuel vehicles beginning in 1998.

Title V of the 1990 Amendments required each state to develop and implement a comprehensive operating permit system for most sources of air pollution. This program, which is similar to the National Pollutant Discharge Elimination System (NPDES) program under the CWA, consolidates into a single document all of the federal and state regulations applicable to a particular source. The Title V regulations will apply to at least the following sources:

- any major source, defined in Section 70.2 of the rules as any stationary source belonging to a major industrial class and that is 1) a major source under Section 122 of the Act, 2) a major source of air pollutants that directly emits or has the potential to emit 100 tons per year or more of any air pollutant (stack or fugitive emissions), and 3) a major source, as defined in Part D of Title I of the Act: any source subject to a standard, limitation, or other requirement under Section 111 of the Act;
- any source subject to a standard or other requirement under Section 112 of the Act;
- any affected source under Title IV of the Act; or
- any source in a source category designated by the EPA.

The Title V permit program is a state-administered program under the EPA's supervision, and the agency can veto any state permits that do not comply with the EPA's standards. The new program regulates many major and smaller sources that were not previously covered by state permits.

The 1990 CAA Amendments significantly increased the potential civil and criminal penalties that could be assessed in enforcement actions. The administrative enforcement provisions allow the administrator to impose administrative penalties up to $200,000 or more. The amendments also allow the EPA to issue a field citation and fines of up to $5,000 per violation per day for minor violations [Section 113(d)(3)]. Violators can simply pay the fine or may request a hearing. Criminal penalties include fines of up to $250,000 per day per violation and up to 5 years in prison for knowing violations against a person. The fine jumps to $500,000 per violation for a corporation. Making false statements to the EPA or failing to file or maintain records or reports required under the Act can result in fines up to $250,000 and 2 years in jail for individuals. The fine increases to up to $500,000 for corporations. This provision is important because the CAA requires an annual certification that the facility is in compliance with its permit and has reported deviations from the permit [Section 503(b)(2)].

The CAA also imposes criminal penalties for knowing or negligent release of air toxics that place another person in "imminent danger of death or serious bodily injury." A knowing release [Section 113(c)(5)] of an extremely hazardous substance that places another person in imminent danger of death or serious bodily injury is subject to fines of up to $250,000 per day and up to 15 years in prison for an individual. The fine increases to up to $1 million per day for corporations. The individual or corporation could still be liable for the release, even if they did not actually know the release posed potential problems. This Section of the CAA is particularly significant because anyone who is responsible for the management of these materials could be held accountable [Section 113(c)(4)]. A negligent release is subject to fines of up to $100,000 and up to 1 year in jail for an individual, and the fine could increase to $200,000 for a corporation.

The Clean Water Act

The goal of the CWA is to "restore and maintain the chemical, physical and biological integrity of the nation's waters." [Section 101(a), 33 USC Section 1251(a)] The Act accomplishes this goal through:

- a prohibition on discharges, except as allowed by the CWA (Section 301);
- a permit program for allowable discharges (Section 402);
- a system for determining the limitations to be imposed on regulated discharges (Sections 301, 306, 307);
- a process for cooperative federal/state implementation (Sections 401, 402);
- a system for preventing, reporting, and responding to spills (Section 311); and
- enforcement mechanisms (Sections 309, 505).

Under the CWA, sources are regulated under the NPDES program if they discharge into receiving waters or under the pretreatment program if they discharge into a public sanitary sewer system. Compliance audits for the NPDES program are discussed in Chapter 15.

NPDES PROGRAM

Sources are required to obtain an NPDES permit for every discharge of pollutants from a point source into navigable waters. The terms *pollutant*, *navigable waters*, and *point source* are broadly defined. All waters of the U.S. are navigable waters. Examples of pollutant discharges and contaminants include dredged spoil, chemical wastes, biological materials, heat, incinerator residue, and others. Point sources include discharge pipes, ditches, erosion channels, and gullies.

The NPDES permits give the permittee the right to discharge specified amounts of pollutants from specified outfalls, typically for a period of 5 years. The NPDES permit application must be completed on EPA forms and must bear the signature of a responsible corporate officer. In addition to establishing the effluent limits, the permit establishes a number of other enforceable conditions, such as monitoring and reporting requirements, a duty to operate and maintain systems properly, upset and bypass provisions, record keeping and inspection, and entry requirements. Best management practices (BMPs), which are designed to prevent or minimize the release of toxic or hazardous pollutants, may also be required. The established permit limits may be technology-based, water quality-based, or toxicity-based limitations.

TECHNOLOGY-BASED LIMITATIONS

The program for establishing technology-based effluent limits requires all dischargers to meet treatment levels that establish baseline discharges. The EPA may require more stringent treatments if it believes they are needed to maintain the quality of water into which the facility discharges. Monitoring generally consists of self-monitoring and typically takes place at the point of discharge. The permit generally contains the frequency and type of monitoring required.

The EPA has issued effluent guidelines that establish emission limits for more than 50 industrial categories. In those cases where the EPA has not established effluent guidelines, permit limits are established based on the best professional judgment of the permit writers. The required limits depend on whether the source existed at the time of the promulgation of the standard. Existing sources were required to meet the first level of control, based on the best practicable control technology available (BPT) by July 1, 1977 and the second level of control, based on the best available technology (BAT) economically achievable by July 1, 1983. The BAT standards do not apply to biochemical oxygen demand, total suspended solids, fecal coliform, pH, or oil and grease. In 1977, the CWA was amended to allow the application of the best conventional pollutant control technology (BCT) in lieu of the more restrictive BAT for conventional pollutants.

New sources are defined as those that began construction after the promulgation of an applicable new source performance standard (NSPS) or after the proposal of the NSPS if the proposed NSPS is finally promulgated within 120 days of its proposal. These sources must meet the NSPS, which may be more stringent than are standards for existing facilities. One benefit for new sources is that they are protected from more stringent technology-based (but not water quality-based or toxicity-based) limitations for a period of 10 years after the date of construction or for the Internal Revenue Service period of depreciation.

WATER QUALITY-BASED LIMITATIONS

Under the CWA, the states are required to establish water quality standards for surface waters, based on criteria developed for the intended uses of the waters in the state. Water quality criteria are quantitative limits on the physical, chemical, and biological characteristics of the water for each of the intended uses. The EPA has established water quality criteria for more than 50 pollutants, and the states typically adopt the federal criteria. The EPA regulations require water quality-based limits to be included in permits whenever a discharge may cause or have a reasonable potential to cause or contribute to any exclusion above the state's water quality standard. The limits may be based on modeling studies or on the calculation of a dilution factor, derived from the flow of the receiving stream divided by the flow from the outfall. The dilution factor, is then multiplied by the water quality standards to obtain the permit limitations.

TOXICITY-BASED LIMITATIONS

The NPDES permits may utilize toxicity-based limitations as an alternative to water quality-based limitations. Toxicity-based limitations involve exposing selected species of test organisms to an effluent in a laboratory setting to determine the effects of exposure to the effluent. Both short-term and long-term tests may be required on a regular basis (typically, quarterly or monthly).

THE PRETREATMENT PROGRAM

Under the pretreatment program, the publicly owned treatment works (POTW), rather than the state or the EPA, issues the industrial user a permit, order, or contract that specifies discharge limits. The general pretreatment regulations found in 40 CFR Part 403 regulate the discharge of pollutants that could interfere with the operation of the POTW. The pretreatment regulations prohibit discharges that have a pH less than or equal to 5.0, solid or viscous pollutants, fire and explosion hazards, and discharges with a temperature that would interfere with the biological processes in the POTW. Discharges other than pH, biochemical oxygen demand, suspended solids, and fecal coliform bacteria are regulated through categorical effluent guidelines, which are pretreatment standards for existing sources (PSES) and pretreatment standards for new sources (PSNS). The pretreatment standards are intended to require the same degree of treatment that would be needed under the NPDES program.

Dredge and Fill Permits   The discharge of dredge or fill material into any surface water is prohibited by Section 301 of the Act unless it is performed under a permit issued by the U.S. Army Corps of Engineers under Section 404 of the CWA. This section has a significant impact on development in or adjacent to designated wetland areas. In some cases, the Corps is required to conduct an Environmental Impact Statement because these permits are subject to the requirements of the National Environmental Policy Act. The Corps has developed generalized national permits for 36 activities for which the dredge and fill discharges are expected to have minimal effects (for example, some surveying activities, bank stabilization, and outfall construction projects).

Spill Prevention, Reporting, and Response   Spill prevention control and countermeasure (SPCC) plans are required of some facilities under the CWA and the Oil Pollution Act (OPA) of 1990. To date, the EPA has issued regulations only for discharges of oil. The guidelines for preparing the plans identify the types of containment structures and other measures that may be needed to control the movement of oil discharges. The SPCC plan must be developed to respond to a worst-case discharge of oil, and it must be reviewed and certified by a Registered Professional Engineer. After initial approval by the EPA, the plan must be updated periodically and resubmitted to the EPA for approval after significant changes are made.

The discharge of hazardous quantities of oil or hazardous substances to navigable waters or adjoining shorelines must be reported to the National Response Center ((800) 424-8802). Failure to report is a criminal offense, punishable by 5 years in prison. A harmful quantity of oil is defined by the EPA as any quantity that causes a film or sheen on the receiving waters or any quantity that violates an applicable water quantity standard. A harmful quantity of other hazardous substances is the EPA's RQ under CERCLA. In the event that a release of oil or of a hazardous substance in a quantity greater than the RQ occurs, the owner or operator of the facility is strictly liable for the costs of cleaning up the spill and for damages to natural resources.

Enforcement options include administrative orders (penalties may not exceed $10,000 per day for each violation, for a maximum of $125,000), civil enforcement (penalties may be assessed at up to $25,000 per violation per day), and criminal enforcement (significant fines and imprisonment). Chapter 13 covers compliance auditing of the SPCC plan and facilities.

THE POLLUTION PREVENTION ACT

The Pollution Prevention Act (PPA) of 1990 ushers in a new era of environmental management by establishing pollution prevention as a national objective. The PPA establishes pollution prevention as the most acceptable practice for managing hazardous materials and wastes, followed by recycling and treatment (see Chapter 7 for more information about pollution prevention). Disposal or other release to the environment are the least acceptable practices. The PPA also establishes the TRI by amending the reporting requirements of the EPCRA.

In response to the PPA's requirement for the EPA to establish measurable source reduction goals, the Agency developed its Pollution Prevention Strategy to incorporate pollution prevention into all of its existing programs. Data from the TRI are used by the EPA to track industrial pollution prevention efforts. Two important components of EPA's strategy to prevent pollution are the Environmental Leadership Program (ELP) and the Common Sense Initiative (CSI). The ELP encourages companies to develop and implement pollution prevention management practices and to establish environmental goals beyond those set for environmental compliance. The ELP promotes the Corporate Statement of Environmental Principles and the Model Facility Program. Companies who participate in the ELP are rewarded primarily through public recognition, but other incentives include a commitment by the EPA to accelerate the permit and product registration process, reduce monitoring, and reporting requirements and other activities to streamline the regulatory process.

The EPA's CSI will replace the pollutant-specific regulatory approach with an industry-by-industry approach. The CSI brings together representatives of the regulated community, government, environmentalists, and the public to develop a holistic approach to preventing pollution in selected industries. The initial focus of the CSI is on oil refining, auto manufacturing, printing, computers and electronics, and the metal plating and finishing industries.

THE OCCUPATIONAL SAFETY AND HEALTH ACT

The regulation of occupational safety and health involves several laws and a multitude of regulations with which an employer must comply. Some examples of applicable statutes include the Occupational Safety and Health Act (OSH Act) of 1970, the Contract Hours and Safety Standards Act of 1962 (Construction Safety Act), the Federal Mine Safety and Health Act of 1977, and the Federal Coal Mine

Health and Safety Act of 1977. There are many other industry-specific laws and regulations that may apply to a given facility.

Section 5(a)(1) of OSH Act, known as the "General Duty Clause," specifies that "each employer shall furnish to each of his employees employment and a place of employment which are free from recognized hazards that are causing or are likely to cause death or serious physical harm to his employees." The regulations that apply to the largest number of facilities are the general industry standards (29 CFR Part 1910) and the construction industry standards (29 CFR Part 1926). Figure 12–2 shows the subparts under the general industry standards.

Although the purpose of OSH Act is to protect worker health and safety and not the environment, compliance with OSHA's regulations minimizes the likelihood of discharges to the environment through the use of safe work practices. There is overlap between the EPA and OSHA, as they regulate different aspects of several issues, such as asbestos and emergency response; in addition, compliance with and utilization of some OSHA regulations is required by reference in EPA rules, such as CERCLA and EPCRA.

Important OSHA programs related to environmental management include the standards for hazard communication, emergency response, and site cleanup.

FIGURE 12–2.   OSHA's General Industry Standards:

General Industry Standards, 29 CFR Part 1910

- Subpart A—General
- Subpart B—Adoption and Extension of Established Federal Standards
- Subpart C—General Safety and Health Provisions
- Subpart D—Walking-Working Surfaces
- Subpart E—Means of Egress
- Subpart F—Powered Platforms, Manlifts, and Vehicle-Mounted Work Platforms
- Subpart G—Occupational Health and Environmental Control
- Subpart H—Hazardous Materials
- Subpart I—Personal Protective Equipment
- Subpart J—General Environmental Controls
- Subpart K—Medical and First Aid
- Subpart L—Fire Protection
- Subpart M—Compressed Gas and Compressed Air Equipment
- Subpart N—Materials Handling and Storage
- Subpart O—Machinery and Machine Guarding
- Subpart P—Hand and Portable Powered Tools and Other Hand-Held Equipment
- Subpart Q—Welding, Cutting, and Brazing
- Subpart R—Special Industries
- Subpart S—Electrical
- Subpart T—Commercial Diving Operations
- Subpart U–Y—Reserved
- Subpart Z—Toxic and Hazardous Substances

The hazard communication standard (29 CFR 1910.1200) was promulgated in 1983, and it requires chemical manufacturers, distributors, importers, and users to provide workers with information on possible exposure risks. The program requires compliance with the following provisions:

- identification of hazardous chemicals on-site and the preparation of a list of these chemicals,
- labeling of toxic chemicals,
- distribution of MSDSs on hazardous chemicals,
- use of warning signs and posters, and
- a written hazard communication program.

The Hazardous Waste Site Operations and Emergency Response (HAZWOPER) rule (29 CFR Section 1910.120) was issued under the authority of Superfund Amendments and Reauthorization Act (SARA), Title I, and is administered by OSHA. In those states without a state OSHA program, public sector personnel are covered by a similar EPA regulation (40 CFR Part 311). The emergency response portion of the rule requires covered industries to develop a Hazardous Material (Hazmat) Emergency Response Plan. Those facilities with in-house hazmat teams must develop emergency response procedures, training and medical surveillance programs, and postemergency termination procedures. Chapter 7 provides more information about emergency response.

Failure to comply with OSHA requirements can result in civil or criminal penalties. The OSHA Civil Penalty Policy allows a maximum of $70,000 penalty for each willful and knowing violation, $7,000 for each serious or other-than-serious violation, and $7,000 for each day beyond the stated abatement date for failure to correct a violation. There is also a $5,000 minimum penalty for a willful violation of the OSH Act. Penalties for record keeping and reporting violations vary from $1,000 to $3,000 per violation.

There are adjustment factors for the penalties. The potential reductions are 60% for an employer with 1–25 workers, 40% for 26–100 workers, and 20% for 101–250 workers. There is no adjustment for employers with more than 250 workers. There may be an additional 25% adjustment for evidence that the employer is making a good-faith effort to provide a safe and healthful workplace. To qualify for the 25% additional reduction, the employer must have a written and implemented safety and health program, such as the one outlined in OSHA's voluntary *Safety and Health Management Guidelines*.[2]

THE HAZARDOUS MATERIALS TRANSPORTATION ACT

The Hazardous Materials Transportation Act (HMTA) of 1974 authorized the Department of Transportation (DOT) to issue regulations governing the transport of hazardous materials (49 CFR 100–199). In 1990, the DOT published rule changes that were designed to make the rule more compatible with international shipping regulations.

These regulations potentially apply to any person who offers hazardous materials for transport. The regulations specify the requirements for bulk and nonbulk shipments of hazardous materials. Hazardous materials, under DOT, are those that are listed in the Table of Hazardous Materials (49 CFR Section 172.101), hazardous wastes, hazardous substances under CERCLA (found in Appendix A to 49 CFR Section 172.101), elevated temperature materials and marine pollutants (found in Appendix B to 49 CFR Section 172.101), or any other material determined to be hazardous by the Secretary of Transportation.

The regulations provide instructions for preparing shipping papers, chemical compatibility, labels, markings, shipping packages, placards, transportation routes, emergency response, and other aspects of transporting hazardous materials. The hazardous materials table is used by shippers to identify the proper way to package the materials and to label the packages. The regulations require immediate notice of certain hazardous materials incidents (49 CFR Section 171.15) and the submission of a detailed hazardous materials report in writing, in duplicate, to the DOT within 30 days of the date of discovery.

Shippers are responsible for completing a shipping paper with the proper name of the hazardous material, hazard class, identification number, packing group, and total amount of material. The shipper is required to certify the contents of the shipment as described in the shipping paper. The shipping paper must contain an emergency response telephone number (49 CFR Section 172.604), and the shipment must be accompanied by an MSDS or similar information about the materials (49 CFR Section 172.602). Carriers and facility operators where hazardous materials are received, stored, or handled during transportation are also required to maintain an emergency response telephone number and an MSDS or similar information about the materials. Special requirements apply to loading and unloading operations to ensure that they are accomplished safely (49 CFR Section 177.834).

The transport vehicle must have the proper placards and other identifying information, as specified by the hazardous materials classification in the Hazardous Materials Table. The regulation also requires training for individuals (called *hazmat employees*), who, in the course of employment, directly affect hazardous material transportation safety (49 CFR Section 172.704). The training requirements apply to an individual who 1) loads, unloads, or handles the hazardous materials; 2) tests, reconditions, repairs, modifies, marks, or otherwise represents containers, drums, or packagings for transporting hazardous materials; 3) prepares hazardous materials for transportation; 4) is responsible for safety of transporting hazardous materials; or 5) operates a vehicle used to transport hazardous materials. The training for these individuals must include recognition of hazards, function-specific training, and safety training that includes emergency response and personal protection. Training must be given initially within 90 days after employment or job change and at least once every 2 years thereafter. Training records must be kept for as long as the employee is employed by the employer and for 90 days thereafter.

SUMMARY

A thorough knowledge of the regulations governing the manufacture, use, storage, transport, and disposal of hazardous materials is an important aspect of managing these materials. Awareness can be achieved in a number of ways, including subscriptions to electronic databases, paper newsletters, and regulations. Regulatory compliance can avoid significant civil and criminal penalties, and it is an important first step to achieving environmental excellence.

ENDNOTES

[1] For more information on ordering the CFR and other documents published by the U.S. government, contact: Superintendent of Documents, U.S. Government Printing Office, Washington, D.C., 20402, (202) 783-3238.

[2] *Federal Register*, 54 (16): 3904-3916, January 26, 1989.

FIGURE 12–1.   Major Environmental Laws in the U.S.:

Statute	Comment
Resource Conservation and Recovery Act of 1976 (RCRA), as amended	RCRA provides for "cradle-to-grave" management of solid and hazardous wastes that includes generators, transporters, and treatment and disposal facilities.
Hazardous and Solid Waste Amendments of 1984 (HSWA)	HSWA significantly changed RCRA to include a national land disposal ban program, waste minimization programs, regulation of small-quantity generators of hazardous waste, and the regulation of underground storage tanks.
Comprehensive Environmental Response, Compensation and Liability Act of 1980 (CERCLA), as amended	CERCLA provides for the cleanup of sites where hazardous substances are released or present a threat of future release into the environment; CERCLA also provides for the Hazardous Substances Fund (Superfund) to study and clean up sites on a list of contaminated land sites.
Superfund Amendments and Reauthorization Act of 1986 (SARA)	SARA [Title III, Emergency Planning and Community Right-To-Know Act (EPCRA)] established emergency planning, reporting, and notification requirements to protect the public in the event of a release of hazardous substances.
Clean Air Act of 1970 (CAA), as amended	The CAA regulates the discharge of pollutants into the atmosphere through permits and controls on new and existing sources of air pollutants. The CAA establishes National Ambient Air Quality Standards (NAAQS) and requires state strategies to meet the standards. Hazardous air pollutants, acid rain, stratospheric ozone protection, and visibility protection are also addressed by the CAA.
The Federal Water Pollution Control Act of 1972 (FWPCA), as amended by the Clean Water Act of 1977 (CWA)	The CWA regulates the discharge of pollutants into U.S. surface waters and establishes the National Pollution Discharge Elimination System (NPDES) for sources that discharge water pollutants. Other programs established by the CWA include the dredge and permit fill program for wetlands and municipal wastewater treatment programs.
The Safe Drinking Water Act of 1974 (SDWA), as amended	SDWA regulates public drinking water supplies through national standards for levels of contaminants in public water supplies, underground injection wells, and sole source aquifers.
Toxic Substances Control Act of 1976 (TSCA), as amended	TSCA regulates the manufacture, distribution, processing, use, and disposal of certain designated "chemical substances." TSCA also requires the EPA to develop programs for asbestos in schools and in public buildings, and indoor radon levels.

FIGURE 12–1.   (*Continued*)

Statute	Comment
Federal Insecticide, Fungicide, and Rodenticide Act of 1947 (FIFRA), as amended	FIFRA requires proper labeling of pesticides and registration of pesticides every 5 years, for either general or restricted use. FIFRA also specifies tolerance levels for certain pesticides in agricultural commodities and makes recommendations for the transportation and disposal of pesticides.
National Environmental Policy Act of 1969 (NEPA)	NEPA requires federal agencies to give environmental factors the same consideration as other factors in decision making about its facilities. NEPA established the Council on Environmental Quality and the Environmental Impact Statement (EIS).
Federal Facility Compliance Act of 1992 (FFCA)	FFCA amended RCRA to ensure that federal facilities complied with all federal, state, and local solid and hazardous waste laws.
Pollution Prevention Act of 1990 (PPA)	PPA established the prevention of pollution as a national objective. It required the EPA to develop and implement a strategy to promote source reduction. The PPA amended EPCRA through the Toxic Chemical Release Inventory (TRI), which requires facilities to file an annual toxic chemical release form and to provide information on pollution prevention and recycling for each facility and for each toxic chemical.
Oil Pollution Act of 1990 (OPA)	OPA provisions establish strict liability for damages caused by oil spills in navigable waters. OPA also requires vessel and facility operators to file detailed oil spill response plans and the replacement of single-hull oil tankers and barges with double-hull vessels. OPA creates a $1 billion supplemental compensation fund for oil spills.
Marine Protection, Research, and Sanctuaries Act of 1972 (MPRSA)	MPRSA establishes a policy to regulate the dumping of all types of materials into ocean waters and to prevent or strictly limit the dumping into ocean waters of any material that would adversely affect human health, welfare, or the marine environment or systems.
Coastal Zone Management Act of 1972 (CZMA)	CZMA promotes the preservation, protection, development, and, where possible, the restoration or enhancement of coastal resources. Under CZMA, states develop management plans for their coastal resources.
Surface Mining Control and Reclamation Act of 1977 (SMCRA)	SMCRA establishes a national program for surface mining operations. Among its provisions, SMCRA provides for the reclamation of abandoned mines and controls the environmental impacts of surface mining.

FIGURE 12–1.   (*Continued*)

Statute	Comment
Endangered Species Act of 1973 (ESA)	ESA provides a program for the conservation of the endangered species, threatened species, and the ecosystems of these species. The ESA directs that all government agencies must use whatever measures are necessary to preserve these species and that no activity by a governmental agency should lead to the extinction of an endangered species.
Hazardous Materials Transportation Act of 1974 (HMTA), as amended	HMTA establishes a program of regulating the shipment of hazardous materials that may pose a threat to health, safety, property, or the environment when transported by air, water, rail, or highway.
Occupational Safety and Health Act (OSH Act) of 1970	The OSH Act regulates worker health and safety and establishes health standards for workplace hazards.
Uranium Mill Tailings Radiation Control Act (UMTRCA)	UMTRCA regulates the remedial actions for uranium mill tailings.
Low-Level Radioactive Waste Policy Act of 1980 (LRWPA) and the Low-Level Radioactive Waste Policy Act Amendments of 1985 (LRWPAA)	LRWPA controls the disposal of commercial low-level radioactive waste through performance and design standards for waste disposal facilities and interstate compacts.
Nuclear Waste Policy Act of 1982 (NWPA), as amended	NWPA controls the disposal of high-level radioactive wastes from commercial facilities through performance and design standards for these facilities. It prohibits disposal except in geological repositories.

CHAPTER 13

SPILL PREVENTION, REPORTING, AND RESPONSE

INTRODUCTION

Spill prevention programs, reporting of spills, and planned response actions are needed to prevent and minimize environmental contamination from unexpected events. The potential for spills exists whenever hazardous materials are stored, generated, used, or transported. To minimize liability and environmental damage, each facility should evaluate the potential for spills, develop plans to prevent spills, and employ countermeasures to handle spilled materials. In the United States, federal regulatory requirements that mandate spill reporting and management programs are contained under several statutes, including the Federal Water Pollution Control Act (FWPCA) for oil and hazardous substances; the Comprehensive Environmental Response, Compensation and Liability Act (CERCLA) for hazardous substances; the Toxic Substances Control Act (TSCA) for polychlorinated biphenyls (PCBs); the Hazardous and Solid Waste Amendments (HSWA) to the Resource Conservation and Recovery Act (RCRA) for leaking underground storage tanks; and the Hazardous Materials Transportation Act (HMTA) for hazardous materials. Reporting of emergency releases is also required under the Superfund Amendments and Reauthorization Act Title III, Emergency Planning and Community Right-To-Know Act (EPCRA). In addition, many states have comprehensive spill reporting regulations in conjunction with prevention and remedial action plans. More information about spill prevention reporting and response can be found through the U.S. EPA World Wide Web and other resources given in Appendix F.

www

SPILL REPORTING REQUIREMENTS

Spill reporting is required for a broad range of materials. Reporting requirements in Section 311 of the FWPCA require the submission of an immediate report to the U.S. Coast Guard National Response Center (NRC) [(800) 424-8802], followed by cleanup if a minimum quantity of hazardous substances or oil is discharged into the navigable waters of the U.S. from a vessel or facility. Failure to report is a criminal act, punishable by up to 5 years in prison. For oil spills, the Environmental Protection Agency (EPA) has determined that a "harmful quantity" is any quantity that causes a film or sheen on the receiving waters, or any quantity that violates an applicable water quality standard.

In 1980, Sections 102 and 103 of CERCLA, expanded these reporting requirements. Under CERCLA, any person in charge of a vessel or facility must notify the NRC as soon as the person has knowledge of any release from the vessel or facility of a hazardous substance[1] in an amount equal to or greater than the reportable quantity (RQ) for that substance. Under Section 102, the EPA establishes the list of RQs (amounts range from 1 pound to 5,000 pounds). If the EPA has not identified an RQ, Section 102 gives the RQ as 1 pound unless the hazardous substance has an RQ under the Clean Water Act (then the RQ from the Clean Water Act applies). Section 103(a) establishes some exemptions. It should be noted that under CERCLA's response and enforcement provisions, the release of any quantity of a hazardous substance can result in liability, even though amounts smaller than the RQ are not required to be reported. Failure to report could result in fines, civil penalties (equal to $25,000 per day), and criminal penalties (a maximum of 3 years in prison for a first conviction and 5 years for a subsequent conviction).

Under Section 304 of EPCRA, notification is required for releases of listed hazardous substances that are not federally permitted when the RQs established for the substance are exceeded and the release results in exposure to persons who are off-site. This requirement applies to the owner or operator of a facility that produces, uses, or stores a hazardous material. The notification must be made immediately via telephone, radio, or in person to the Local Emergency Planning Committee (LEPC) and the State Emergency Response Commission (SERC). As soon as possible after the release, the owner or operator must provide a written notice to the appropriate SERCs and LEPCs. If the release involves an RQ under CERCLA, the NRC must also be notified.

Polychlorinated biphenyls are regulated under TSCA Section 6, and the EPA regulations for PCBs can be found in 40 CFR Part 761. The EPA classifies PCB spills as either low- or high-concentration spills. Low-concentration spills are defined as containing a PCB concentration of less than 500 ppm and involving less than 1 pound of PCBs. High-concentration spills contain more than 500 ppm PCBs, or are low-concentration spills involving more than 1 pound of PCBs or 270 gallons or more of untested mineral oil. Any spill that involves more than 10 pounds of PCBs must be reported immediately to the appropriate Regional EPA Office (reporting to the NRC is also required under CERCLA).

Under HSWA, owners and operators of underground storage tank systems must report any suspected release to the appropriate agency within 24 hours or other reasonable time period specified by the appropriate agency (40 CFR Section 280.50).

Any carrier who transports a hazardous material[2] under the HMTA must give immediate notice at the "earliest practicable moment" after each incident in which there is death, injury, significant property damage, transportation interruptions, and special types of materials.[3] Incident reports must be given in writing, in duplicate, on Department of Transportation (DOT) Form F 5800.1 (Rev. 6/89) to the DOT within 30 days of the date of discovery of each incident that occurs during the course of transportation. Transportation includes loading, unloading, and temporary storage activities.

THE SPILL PREVENTION CONTROL AND COUNTERMEASURE PLAN

The FWPC Act requires the EPA to issue regulations requiring spill prevention control and countermeasure (SPCC) plans for oil or hazardous substance releases. To date, the EPA has issued regulations only for the discharge of oil (40 CFR Part 112) from some facilities.[4] These regulations require a facility to develop an SPCC plan if there is a potential for an oil spill from the facility to reach navigable waters and if the facility meets defined minimum capacities, which are:

- an underground storage capacity totaling over 42,000 gallons of oil; or
- an aboveground storage capacity totaling over 1,320 gallons of oil or any single container with more than 660 gallons of oil.

The regulatory definition of navigable waters is broad and includes tributaries of rivers, lakes, and ponds. The regulations do not require aboveground and underground volumes to be totaled, but they do refer to storage capacity rather than actual volumes of oil stored. In the case of aboveground storage, as few as 25 drums of stored oil trigger the need for an SPCC plan. It should be noted that transportation-related facilities are regulated by the DOT, rather than the EPA.

Facilities that existed at the time the regulations were finalized were required to prepare and implement SPCC plans by the end of 1974. Facilities constructed since 1974 are required to prepare plans within 6 months of operation and to implement them within 1 year of operation. The SPCC plan must be certified by a registered professional engineer, but this certification does not relieve the facility from responsibility. The facility still bears overall responsibility for compliance with the regulations. The SPCC plan must be available at the facility, and it must be updated and recertified every 3 years or within 6 months of any change in the plant's potential for spills.

Extensive guidelines for developing an SPCC plan are given in 40 CFR Part 112.7. These guidelines provide a framework for conducting an audit of the facility to determine whether it needs a plan or whether an existing plan is adequate and effective. One of the criteria for judging the effectiveness of the SPCC plan is usability. Is the information available and easy to find and use? Does it provide the reader with information on what constitutes a spill and how to respond in the event that one occurs? Do employees understand the plan? Is record keeping complete and updated? If the answer to each of these questions is yes, it is likely that the plan and the related management systems are working.

THE SPCC PLAN AUDIT PROCESS

OVERVIEW

During the SPCC audit, the audit team will review the facility's policies and operational procedures, will compare them to actual practices, and will determine areas of conformance or nonconformance. The audit will focus on compliance with applicable regulations and good management practices for preventing releases, response procedures during unplanned releases, and corrective actions and follow-up. Preparation for the audit includes familiarization with the design, operation, and leak detection methods for spill containment and knowledge of the regulatory requirements for spill planning.

Typically, the auditor will not perform an engineering review of containment facilities because the engineering plans were certified by a professional engineer, but the auditor will comment on general indications of adequacy of the plan. Figure 13–1 at the end of this chapter provides an example of a checklist that can be adapted to evaluate the SPCC plan.

The SPCC plan should be readily available and current. The regulations require it to be updated at 3-year intervals and to be certified by a professional engineer. The plan must document emergency contacts and procedures along with information about training, inspection and maintenance, record keeping, tank inventory, containment, emergency equipment, and other factors. The auditor may also review underground storage tank (UST) permits and contracts for services as part of the SPCC audit.

An effective training program is critical to the success of the SPCC plan. Anyone who is involved in the processing, transfer, or storage of oil must be trained. Operators should understand how to inspect tanks, should be able to locate and use spill control equipment, should perform regular maintenance, and should understand spill control procedures.

The SPCC plan requires the maintenance of certain records. Staff involved in inspections must document the inspection of tanks, initial and date the forms, and append them to the SPCC plan. Maintenance records for spill control equipment must be updated after each inspection and appended to the SPCC plan. The SPCC plan also requires documentation of spill control activities, such as inspection of impervious dikes prior to draining; the documented information must include the inspector's initials and the date.

The audit will begin with a briefing of site management and the environmental staff. The audit team will obtain information about safety requirements and will participate in an orientation tour. During the orientation tour, the audit team will become better acquainted with the physical layout of the facility and its operations. The orientation tour will be conducted with the facility representative, typically, the environmental coordinator. During the tour, the audit team will become oriented to the facility with a focus on the location of all storage containers and potential sources of unplanned releases. The tour will include the entire facility and its grounds.

After the orientation tour, the audit team will review documentation and management systems, interview site personnel, and conduct an in-depth inspection of the facility. Previously developed checklists will be used by the audit team to conduct the in-depth evaluation of the facility. At the end of each day, the audit team members will brief one another and discuss findings.

At the end of the audit, the team will debrief the management of the facility. Final activities will include drafting the audit report, responding to review comments, and submitting the final audit report. Facility or site management will follow up on audit findings, initiate corrective actions, determine the effectiveness of the actions, and report back to the audit team.

PREPARING THE INVENTORY

The first step in preparing an inventory is to determine whether the facility meets the definition of an oil storage facility under 40 CFR Part 112. In order to determine the facility's status, the auditor will inventory aboveground and underground oil storage capacities, including oil in drums, tanks, or transformers. The term *oil* means any type of oil in any form and includes:

- petroleum-based raw materials or products,
- fuel oil or gasoline,
- heated asphalt,
- lubricating, cooling, and hydraulic oil,
- hydrocarbon-based solvents,
- oil-cooled transformers that are not utility owned,
- oil-filled switches and circuit breakers,
- oil sludge, and
- waste transformer oil, oil mixed with wastes other than dredge spoil.

The inventory should include the following information: identification of the storage vessel, capacity, contents, location, and, if aboveground or underground, proximity to storm drains or surface waters and measures for containment and spill prevention. Figure 13–2 provides an example of an inventory form that could be adapted to characterize oil storage capacity. Once the inventory is compiled, it should be reviewed and evaluated to determine if the facility is regulated.

FACILITY TOUR

The auditor will inspect the facility to verify the existence of aboveground and underground storage tanks, examine spill containment measures, and identify unlisted tanks, such as drum storage areas, small-vehicle refueling points, and waste storage areas. Typical problems encountered during an audit of the SPCC program are given in Figure 13–3.

During the facility inspection, the auditor will use the data inventory form in conjunction with a plot plan of the facility to inspect and identify the location of tanks, drum storage areas, loading and unloading points, and fluid transport piping. The auditor will evaluate the potential for contaminating surface waters by identifying and examining nearby surface water, outfalls, and drainage patterns. Drainage patterns can be examined by looking for storm drains, run-off ditches, and natural swales, and by looking at the contour lines on the local quarter-section U.S. Geological Survey topographic map. The auditor will check all elements of the SPCC plan against field conditions and will note any discrepancies.

FIGURE 13–2.   Example of an Oil Storage Inventory Form:

ABC Company
Oil Storage Inventory Form

Plant: _____   Auditor: _____

Plant contact: _____   Date: _____

Oil SPCC Tank No.: _____   Capacity: _____

Contents: _____

Type: _____ Aboveground _____ Underground

Location: _____

Proximity to drains and/or surface water: _____

_____

_____

Containment: _____

_____

_____

Spill Protection: _____

_____

_____

Comments: _____

_____

_____

Rev. 1   RKB

4/10/95   Page _____of _____

The auditor will also interview the emergency coordinator and staff to verify their knowledge of the SPCC plan, the location of emergency supplies, emergency response actions, and other aspects of the plan.

UNDERGROUND TANKS

The auditor will examine underground tank locations to verify that the tank meets the definition of an underground tank and that the information in the SPCC plan about each tank is correct. The SPCC plan should provide the following information about each tank:

- location, contents, volume, construction, and age of each tank
- description of fill points, vent pipes, gauges, and heaters
- special procedures, such as safety precautions for each tank

FIGURE 13–3. Typical Problems Encountered During an SPCC Program Audit:

Problems Related to the Facility

- lack of leak testing and inspection for tanks
- inadequate capacity or condition of containment (dikes, berms)
- valves left in open position
- inadequate spill control equipment or materials
- inadequate training
- used oil not addressed
- lack of oil/water separators to handle flow prior to discharging storm water to sewer or receiving water
- accumulated rainwater in containment structures
- poor maintenance of containment, separators, drainage ditches
- inadequate security

Problems Related to Documentation

- incomplete, inconsistent, or unavailable SPCC plan
- SPCC plan not signed by professional engineer
- SPCC plan not updated in the past 3 years
- incomplete, inconsistent, or missing procedures
  - —inspection and maintenance of tanks
  - —inspection and maintenance of containment
  - —spill control
  - —release of contained rainwater
  - —spill reporting procedures
- incomplete, inconsistent, or missing inspection logs
- incomplete, inconsistent, or missing maintenance logs
- incomplete or missing training program and records
- incomplete or missing spill reports

- leak testing procedures
- records of leak tests

ABOVEGROUND TANKS

The auditor will examine aboveground tank locations to verify that the tank is an aboveground tank and that the information in the SPCC plan about each tank is correct. The SPCC plan should provide the following information about each tank:

- location, contents, volume, construction, and age of each tank
- description of fill points; floats, sensors and alarms; vent pipes; gauges; heaters; fume collection systems
- special procedures, such as draining condensate and safety precautions for each tank
- records of inspections and maintenance

The auditor will review the log of the tank inspection program. An annual inspection of the tanks is typical, and it consists of examining the tank for signs of damage, rust and corrosion, weeping seams, deformation of the tank walls, or other stresses. The auditor will visually examine each of the tanks and note evidence of deterioration. The base of each tank will be checked for cracks or breaks in the curbing. The auditor will verify the presence of level-sensing equipment and will observe that it is operational.

The auditor will examine spill control measures for aboveground tanks (including watertight concrete dikes, earthen dikes, flow diversion barriers, and drainage ditches) to determine whether they are present, appear to have adequate capacity, and are in good condition. The SPCC regulations require dikes to have enough capacity to hold the entire contents of the largest tank plus precipitation, typically a 5-year storm for the region. The condition of concrete and earth dikes will be examined to be sure the structures are sound. Dikes will be examined for standing water, rust, or oil. The drainage procedure and log will be examined. Concrete dikes will be examined for proper drainage controls; valves will be checked to be sure they are closed and tagged if the dike is not being drained. Earthen dikes will be examined for the presence of oil-soaked dirt, which, over time, can build up and reduce the ability of water to flow out of the dike. This increases the possibility of standing water in the dike, which reduces the space available in the event of a spill. The presence of high grass, undergrowth, erosion, and trash are other signs of poor maintenance.

The auditor will look at curbing and drainage ditches to determine whether they correspond to the SPCC plan and will evaluate their capacity and general condition. The auditor will also check for the presence of a flow diversion system, which might be a pond, impoundment, settling pit, or separator. These structures will be examined for general condition and evidence that they work.

LOADING/UNLOADING AND TRANSFER OPERATIONS

The auditor will review the SPCC plan requirements, personnel training for loading/unloading and transfer operations, and written procedures for conducting these operations, including safety precautions and emergency procedures. The most likely problems at transfer stations are minor spills, which could be due to incomplete sealing between the fill hose and truck, and draining of the fill hose between fills. The auditor will check for procedures such as using drip pans or drain buckets to handle these problems. The auditor will also check for the presence of protection against overfilling, such as tank-vent liquid alarms, metered fill volumes, or sensors on top-fill nozzles.

RUNOFF CONTROLS

The auditor will examine the runoff patterns at the facility with respect to potential sources of spills and will check for the presence of run-off controls. Storm drains can be protected from contamination by installing water seals into the drains. Positive oil spill control for run-off from the entire facility can be accomplished by an oil/water separator or catch basin designed to be pumped out whenever oil is present.

SECURITY CONTROLS

The auditor will review the facility's security measures. These should include adequate lighting, fencing around oil storage areas or the entire facility, a locked or guarded entrance, and locked or inaccessible tank valves and pump switches.

SUMMARY

The storage of petroleum and hazardous materials poses significant dangers to the environment and to human health in the event of unplanned releases. The focus of planning efforts should be to create a complete inventory of storage containers and to provide for proper controls for loading/unloading and transfer operations, run-off, and plant security. These measures should be in place for both petroleum products and hazardous materials stored at the facility. A well-designed SPCC plan can minimize the facility's financial liability by reducing the likelihood that serious releases will occur.

ENDNOTES

[1] Under CERCLA, a hazardous substance includes:
- hazardous wastes defined in Section 101(14) of RCRA
- characteristic hazardous wastes under Section 3001 of RCRA
- hazardous substances defined in Sections 311(b)(2)(A) and 307(a) of the FWPCA
- toxic pollutants defined in Section 307 of the CWA
- hazardous air pollutants listed under Section 112 of the Clean Air Act
- imminently hazardous chemical substances or mixtures under Section 7 of TSCA
- substances that may present substantial danger to health or the environment under Section 102 of CERCLA

[2] Hazardous materials under HWTA include:
- hazardous materials in the Hazardous Materials Table
- reportable quantities of hazardous substances listed in CERCLA
- hazardous waste
- marine pollutant
- elevated-temperature materials

[3] Incident reporting is required (49 CFR Section 171.15) when:
1. a person is killed; or
2. a person receives injuries requiring his or her hospitalization; or
3. estimated carrier or other property damage exceeds $50,000; or
4. an evacuation of the general public occurs, lasting 1 or more hours; or
5. one or more transportation arteries or facilities are closed or shut down for 1 hour or more; or
6. the operational flight pattern or routine of an aircraft is altered; or
7. radioactive materials are involved; or

8. etiological agents are involved; or

9. certain quantities of marine pollutants are released; or

10. the carrier believes a hazardous situation exists.

[4] The Oil Pollution Act (OPA) of 1990 also requires SPCC plans for some facilities that handle, transport, or store oil and if, because of its location, it could reasonably be expected to cause substantial harm to the environment. Regulated facilities include:

- facilities that transfer oil over water to or from vessels that have a total oil storage capacity greater than or equal to 42,000 gallons; and

- facilities with a total oil storage capacity of at least 1 million gallons, where one or more of the following is true:

  a. the facility does not have secondary containment for each aboveground storage area sufficiently large to contain the capacity of the largest tank plus sufficient freeboard for precipitation;

  b. the facility is located at a distance from fish and wildlife or sensitive environments such that a discharge could cause injury to them;

  c. the facility is located such that a discharge could shut down operations at a public drinking water intake; or

  d. the facility had a reportable spill greater than or equal to 10,000 gallons within the past 5 years [40 CFR Section 112.20(f)]

FIGURE 13–1.   SPCC Plan Checklist:

Responses			Questions to Ask
YES	**NO**	**N/A**	*General Requirements*
————	————	————	1. Does the facility have an SPCC Plan?
————	————	————	2. Has the plan been prepared or updated within the past 3 years?
————	————	————	3. Is the plan certified by a professional engineer registered in the same state?
————	————	————	4. Is there a designated spill coordinator and at least one alternate?
————	————	————	a. Are current home telephone numbers listed?
————	————	————	5. Is the phone number of the National Response Center [800-424-8802 or 202 267-2675] listed?
————	————	————	6. Are the phone numbers of the regional EPA office, nearest state agency, county health agency listed?
————	————	————	7. If previous spills have occurred in the past, are the phone numbers of neighbors who might have been affected listed?
————	————	————	8. Is documentation controlled?
YES	**NO**	**N/A**	*Training*
————	————	————	9. Does the facility have a spill training program for operators involved in oil processing, transfer, or storage?
————	————	————	10. Does the training program include a statement of management policy on spills?
————	————	————	a. Does the policy prohibit intentional oil pollution and commit resources to the cleanup of the spill and impacts caused by the spill?
————	————	————	11. Does the training program provide for a description of the SPCC regulations?
————	————	————	12. Does the training program provide for a description of the facility's spill control and spill contingency plans?
————	————	————	13. Does the training program provide for familiarization with the spill control equipment and how to use it?
————	————	————	14. Does the training program provide for routine maintenance of spill control equipment?

FIGURE 13–1. (Continued)

YES	NO	N/A	*Training*
_____	_____	_____	a. Does it include regular checks of all valves, pumps, and fittings for malfunction or leakage?
_____	_____	_____	b. Does it include preventive maintenance of oil skimmers, overfill prevention devices, and other mechanical spill control equipment?
_____	_____	_____	15. Does the training program provide for record keeping?
_____	_____	_____	a. Does it include tank inspection records?
_____	_____	_____	b. Does it include large equipment maintenance records?
_____	_____	_____	c. Does it include spill control records?
_____	_____	_____	16. Are all the tanks listed and characterized by type, location, construction, and content?
_____	_____	_____	17. Are all loading or unloading facilities listed with the location, material, and frequency of use?
_____	_____	_____	18. Are containment or diversionary structures such as dikes, berms, and culverts listed?
_____	_____	_____	19. Is emergency equipment such as sorbents and booms specified by type and location?
_____	_____	_____	20. Does the tank inspection program include type and frequency (for example, yearly inspection of aboveground tanks or periodic pressure tests of underground tanks)?
_____	_____	_____	a. Are the test results appended to the plan?
_____	_____	_____	b. Are the test results initialed by the inspector?
_____	_____	_____	21. If the plan includes impervious dikes, are records included in the plan of each time accumulated precipitation was drained?
_____	_____	_____	a. Does the plan address the locking of drain valves and sump pumps, and checking for oil contamination by the responsible person prior to opening and draining?
_____	_____	_____	b. Is each opening recorded on a form and appended to the plan?
_____	_____	_____	c. Are the forms initialed by a responsible person?

FIGURE 13–1.   (*Continued*)

YES	NO	N/A	*Maintenance*
————	————	————	22. Does the facility have a routine maintenance program for spill control equipment?
————	————	————	23. Does it include regular checks of all valves, pumps, and fittings for malfunction or leaks?
————	————	————	24. Does it include preventive maintenance of mechanical spill control equipment such as oil skimmers and overfill prevention devices?
————	————	————	25. Does it include maintenance records for larger equipment?
————	————	————	26. Are maintenance records appended to the plan along with tank inspection records?

YES	NO	N/A	*Security*
————	————	————	27. Are plant security measures such as gate control, perimeter control, and lighting specified?

YES	NO	N/A	*Previous Spills*
————	————	————	28. Are previous reportable spills included?
————	————	————	29. Does the report of previous spills include quantity and direction of flow?
————	————	————	30. Does the report include countermeasures taken and measures to prevent recurrence?

CHAPTER 14

UNDERGROUND STORAGE TANKS (USTS)

INTRODUCTION

A widespread environmental problem that is responsible for significant soil and water contamination is leaking underground storage tanks (LUSTs). The major reasons for LUST systems are corrosion, faulty installation, piping failures, and overfills. The extent of the potential contamination is staggering. A single gallon of gasoline can contaminate 1 million gallons of water. Over 1 million USTs in the U.S. contain petroleum or hazardous substances regulated by the EPA. As of April 1995, more than 287,000 releases had been confirmed, and the EPA estimates that about half of these releases reached groundwater.[1]

Underground storage tanks are regulated by the EPA under the authority of Subtitle I (Regulation of Underground Storage Tanks) of the Hazardous and Solid Waste Amendments (HSWA) of 1984, which amended the Resource Conservation and Recovery Act (RCRA) of 1976. The final regulations, which were adopted in 1988, have been amended several times. The states have the authority to manage the LUST program in lieu of the federal government, providing that state regulations are no less stringent than federal requirements and that they provide for adequate enforcement. The regulatory requirements apply to owners and operators of new and existing tanks.

The regulations are intended to minimize the possibility of leaks and filling problems with existing tanks that contain petroleum products or hazardous substances by requiring closure if the tanks cannot meet certain design and operating standards. Notification requirements apply to the installation, transfer, and closure of all tanks. Existing tanks (brought into use prior to December 22, 1988) were required to have leak detection systems by December 1993 and spill, overfill, and corrosion protection by December 1998. New tanks (brought into use after December 22, 1988) must meet strict design and operating standards that include corrosion protection and spill and overflow prevention equipment. If a leak occurs, the owner or operator must take corrective actions and follow closure requirements if a tank is temporarily or permanently closed.

Finally, the owner or operator must be able to demonstrate financial responsibility for the costs involved with correcting and mitigating a spill. Required coverage is either $500,000 or $1 million for each occurrence and either $1 million or $2 million in aggregate coverage, depending on the ownership category (the

www

EPA has developed 5 major groups of owners and operators that depend on size of the operation and type of governmental unit).

This chapter provides an overview of UST regulation and management. The U.S. EPA's World Wide Web site (http://www.epa.gov) provides access to other useful resources.

NOTIFICATION AND REPORTING REQUIREMENTS

Notification and reporting are required for installation, closure, suspected releases, and corrective actions. Owners and operators of USTs must register the USTs with the responsible state agency within 30 days when UST systems are:

- brought into service,
- acquired by a new owner,
- upgraded (tank relining, piping replacement, leak detection system or equipment installation, installation of spill/overfill prevention equipment or corrosion protection),
- repaired (restoration of a tank or UST system component that has caused or could potentially cause a release of product from the UST system),
- temporarily closed,
- undergoing a change in service,
- closed permanently.

INSTALLATION

Notification requires new owners to certify that they have complied with requirements for installation of tanks and piping, including industry codes of practice, cathodic protection, release detection, and financial responsibility. Additional notifications are required prior to relining, repairing, or permanently closing USTs.

CLOSURE

There are three types of UST closures: removal, in-place closure, and change-in-service closure. Closures may be temporary or permanent. When closing the UST system, all federal, state, and local regulations must be met.

Permanent closure requires the tanks to be emptied, cleaned, and either removed from the ground or filled with an inert solid material. To close the UST system permanently, owners and operators must notify the responsible agency within 30 days before beginning the closure. The agency will respond in writing with a closure approval date, which will expire 90 days after the date given. The closure approval letter must be kept on site at all times. At closure, a certified contractor must be used to perform the work. Before the completion of the permanent closure, owners and operators must conduct a UST system closure site assessment to determine whether contamination is present. The closure site assessment must be maintained for at least 3 years after completion of permanent closure or change in service by the owners and operators who took the UST out of service, the current owners and operators of the UST system, or the implementing agency if the records cannot be maintained at the closed facility.

Temporary closures do not exempt owners and operators from regulatory requirements. Owners and operators must continue to comply with the operation and maintenance of corrosion protection and release detection systems (not required if tank is empty). They must comply with release reporting, investigation, confirmation, and corrective action requirements in the event of suspected or confirmed releases. If the UST closure is for 3 months or more, owners and operators must leave vent lines open and functioning, and must cap and secure all other lines, pumps, manways, and ancillary equipment. In the event that the closure is for more than 12 months, the owner or operator may be required to permanently close the UST.

LEAKING USTs

Suspected or leaking USTs must be reported and investigated. Signs of suspected releases include:

- erratic behavior of product-dispensing equipment,
- sudden loss of product through inventory control checks,
- tank tightness test failure,
- water present in the UST,
- free product present at the site,
- vapors in basements and/or nearby utility lines, or
- results from leak detection monitoring and testing that indicate a leak exists

Suspected releases should be reported to the responsible agency and investigated within the specified time limits to confirm the release. Leaks are investigated by first conducting a tightness test of the entire UST system and then checking the site for additional information on the presence and source of the contamination. Two failed tank tightness tests are considered a confirmed release. Confirmed releases must be reported within specified time limits and investigated to define the extent of the contamination. Once the investigation is completed, corrective action plans must be developed for all sites that have soil and/or groundwater contamination.

Confirmed releases are handled in two stages: short-term actions and long-term actions. The regulatory agency must be contacted for specific requirements for handling UST releases, but a brief summary of needed actions follows. Short-term actions seek to contain the problem, make necessary notifications, and perform an initial investigation. Long-term actions are needed when short-term actions do not result in the cleanup of a site.

Specific short-term actions include:

- Take immediate action to stop and contain the release.
- Report the release to the appropriate regulatory agency within 24 hours. Petroleum releases and overfills of less than 25 gallons do not have to be reported if they are immediately contained and cleaned up.

- Remove explosive vapors and fire hazards. Make sure the release poses no immediate hazard to human health and safety.

- Remove petroleum from the UST system to prevent further releases into the environment. Report progress to the regulatory agency no later than 20 days after confirming a release.

- Investigate to determine whether environmental damage exists or may occur in the future as a result of the release. The site assessment must determine the extent of soil and groundwater contamination. The results of the assessment and a corrective action plan must be submitted to the regulatory agency within the appropriate time frame.

Long-term actions are needed when the short-term assessment and cleanup are not sufficient to correct the environmental impacts. Long-term actions include developing and submitting a corrective action plan that details how the cleanup will be performed. After the regulatory agency approves part or all of the plan, the organization is responsible for the final cleanup at the site and for submitting a final report on the cleanup.

Leaking tanks can be repaired, providing that industry standards are followed for performing the repair work. Within 30 days of the repair, the tank must be tested using an internal inspection or tightness testing, a monthly leak detection monitoring method, or another approved method. Within 6 months of repair, USTs with cathodic protection must be tested to confirm that the cathodic protection is working properly. Records of repair and follow-up testing must be maintained for as long as the UST remains in service.

Leaking metal piping must be replaced because it cannot be repaired. Fiberglass-reinforced plastic piping can be repaired, providing that the manufacturer's instructions or national codes are followed. The piping must be tested within 30 days of the repair.

GENERAL OPERATING REQUIREMENTS FOR PETROLEUM USTS

Owners and operators of UST systems must ensure that the UST system is made of or lined with materials that are compatible with the regulated substances stored in the UST system. New and existing tanks should be equipped with the proper protection for spills, overfills, and corrosion. Figure 14–1 shows leak detection methods for tanks and piping. Personnel should be properly trained to perform their assigned tasks, and records must be maintained for the proper time intervals. Following the regulatory requirements for new and existing petroleum tanks will minimize the likelihood that product will enter the environment and will reduce costs associated with cleanup operations.

NEW PETROLEUM USTS

Owners and operators must meet four requirements when a new UST that contains petroleum is installed:

FIGURE 14–1.   Leak Detection Methods for Tanks and Piping:

1. Groundwater Monitoring
2. Vapor Monitoring
3. Secondary Containment with Interstitial Monitoring
4. Automatic Tank Gauging Systems
5. Tank Tightness Testing and Inventory Control
6. Manual Tank Gauging
7. Leak Detection for Underground Suction Pump
8. Leak Detection for Pressurized Underground Piping

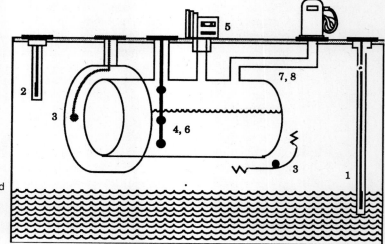

- The owner or operator must certify that the tank and piping are installed according to industry codes.
- The UST must have leak detection.
- The UST must have spill and overflow protection, and correct filling practices must be observed.
- The UST must have corrosion protection.

INSTALLATION

The steps for installing the UST system include excavation to the proper area and depth, laying out and assembling the tank system in the excavated site, backfilling around the tank system, and grading the surface. Improper installation can result in the failure of the UST or associated piping. To prevent problems such as damage to the tank or the cathodic protection, the installation of the UST must be done by qualified installers who follow industry codes.

LEAK DETECTION

All regulated USTs must have leak detection systems that are installed, calibrated, operated, and maintained according to the manufacturer's instructions.[2] The leak detection system must be capable of detecting a leak from any portion of the tank or its piping that routinely contains petroleum, and leak detection systems must comply with performance standards found in 40 CFR Sections 280.43 and 280.44.

Permissible leak detection systems for unpressurized tanks include:

- **Groundwater monitoring.** Groundwater monitoring involves installing monitoring wells at strategic locations in the ground near the tank and along the piping runs. A site assessment is needed prior to installation to ensure that the

balance the volume recorded from deliveries and sales. Figure 14–2 shows an example of a daily reconciliation form.

Manual tank gauging is another leak detection method that can be used by itself for small tanks up to 1,000 gallons. It can be used in combination with tank tightness testing for tanks between 1,001- and 2,000-gallon capacity, but only for the first 10 years of tank installation. This method requires the tank to be undisturbed for 36 hours each week and its contents to be measured twice at the beginning and twice at the end of the test period.

Leak detection for piping is especially important as most leaks come from piping. Pressurized piping must have a device that automatically shuts off or restricts flow, or it must have an alarm that indicates a leak. An annual tightness test or one of the methods noted above for testing tanks is required for pressurized piping systems.

If the UST has suction piping, leak detection may or may not be required. There are two configurations that do not require leak detection. Leak detection is not required if the piping is below grade at less than atmospheric pressure and sloped so that the piping's contents drain back into the tank when the suction is released. Also, leak detection is not required if each suction line has only one check valve that is located directly below the suction pump. Any other configurations require monthly monitoring using one of the monthly methods noted above or tightness testing of the piping once every 3 years.

Regardless of the leak detection method selected, it must be compatible with the stored product and be used in an appropriate environment. Monitoring equipment must be calibrated according to the manufacturer's instructions or at the frequency required by the regulations. Equipment must be maintained on a regular basis.

SPILL PROTECTION

Spills often occur at the fill pipe when the delivery truck's hose is disconnected.[3] Although these spills are typically small, they can add up over time and result in significant environmental damage. To minimize the damage from spills, new USTs must have catchment basins to contain spills, and the owner or operator must ensure that industry standards are followed for correct filling practices. A catchment basin is simply a bucket sealed around the fill pipe to prevent spills from entering the environment. The basin, which can range in capacity from only a few gallons to much larger sizes, must be large enough to contain the spill when the delivery hose is uncoupled from the fill pipe. The catchment basin must be equipped with a drain or pump to remove spilled product and water that may have collected in the basin.

OVERFILL PROTECTION

Overfills typically result in the release of much larger quantities of product than do spills. Owners and operators are required to take measures to prevent overfills. This is accomplished by installing overfill protection for new USTs (automatic shutoff devices, overfill alarms, and ball float valves) and by following industry standards for correct filling practices. If the UST never receives more than

25 gallons of product at a time, overfill protection is not required. Many small used oil tanks fall into this category. Overfill prevention equipment includes:

- **Automatic Shutoff Devices.** These devices are installed in the fill pipe to slow down and then stop the delivery of product when it has reached a certain level in the tank. These devices automatically shut off flow into the tank when the tank is no more than 95% full.
- **Overfill Alarms.** Overfill alarms use probes installed in the tank to activate an alarm when the tank is either 90% full or within 1 minute of being overfilled. The alarm alerts the driver to shut off the fill valve.
- **Ball Float Valves.** Ball float valves are floats placed at the bottom of the vent line several inches below the top of the UST. The ball float rises as product fills the tank and eventually restricts vapor flowing out of the vent line before the tank is full. If the fittings are tight, the product flow into the tank will be restricted, which alerts the driver to close the shutoff valve. However, if the fittings are loose, an overfill could result if the product flow is not restricted.

CORROSION PROTECTION

Tanks and piping corrode when, over time, bare metal in contact with the soil and moisture produces an underground electric current that destroys the metal. New tanks must meet performance standards to prevent corrosion. These include:

- Installing a tank and piping made of noncorrodible material such as fiberglass.
- Protecting the tank and piping by enclosing them in a noncorrodible material.
- Installing a tank made of steel clad with a thick layer of noncorrodible material.
- Installing a tank and piping made of steel coated with corrosion-resistant materials and having cathodic protection. The corrosion-resistant coating protects the system from corrosion by electrically isolating the metal from the surrounding environment. Asphaltic coatings do not qualify as corrosion-resistant coating.

Cathodic protection can be provided through the sacrificial anode system or the impressed current system. The sacrificial anode system relies on pieces of metal (anodes) that are attached to the UST. Because the anodes are more electrically active than the UST's steel, the corrosive current passes through the anode rather than the UST. Over time, the anode corrodes, but the tank does not. The impressed current system uses a rectifier to convert alternating current to direct current, which is sent through an insulated wire to metal bars (anodes) buried in the soil near the UST. The current then flows through the soil to the UST system and returns to the rectifier through an insulated wire attached to the UST. The UST is protected because the current flows toward the UST rather than away from it.

EXISTING PETROLEUM USTs

Existing petroleum USTs must comply with the following requirements:

- have leak detection no later than December 1993,
- have spill and overfill protection by December 1998, and
- have corrosion protection by December 1998.

LEAK DETECTION

Existing USTs should have leak detection, and any of the monthly monitoring leak detection methods discussed for new USTs are acceptable. In addition, existing USTs can use the combined method of inventory control and tank tightness testing, but the combined method can be used for only 10 years after upgrading the UST with spill, overfill, and corrosion protection. After 10 years, one of the monthly monitoring methods must be used. Manual gauging can be used for tanks that have less than 2,000-gallon capacity, providing that tanks between 1,001 and 2,000 gallons in capacity use the combination of manual tank gauging and tank tightness testing for only 10 years after upgrading with spill, overflow, and corrosion protection. After 10 years, one of the monthly monitoring methods must be used. If the UST has not been upgraded, combined manual tank gauging and annual tank testing can be used, but only until December 1998. Other methods of leak detection approved by the regulating agency can be used.

SPILL AND OVERFILL PROTECTION

Existing UST systems must be upgraded with spill and overfill protection by December 1998. If this deadline is not met, the UST must be replaced with a new UST system or permanently closed. The required spill and overfill protection is identical to methods used for new tanks.

CORROSION PROTECTION

Any existing UST that does not have corrosion protection by December 1998 must be replaced with a new UST or permanently closed. An existing tank meets the corrosion protection requirements if it already has one of the corrosion protection methods for new tanks. If the tank does not have corrosion protection, the owner or operator has three options for retrofitting existing tanks:

- **Add cathodic protection.** Before cathodic protection can be added, the tank must be evaluated by an approved method to be sure that it is structurally sound and does not leak. If the tank is sound, a qualified corrosion expert must design the cathodic protection system that will be installed at the UST site. Once installed, the system must be tested within 6 months of installation and at least every 3 years thereafter. Results of the last two tests must be maintained. Impressed current systems work best with existing USTs, and these must be inspected every 60 days to verify that the system is operating. The last three inspection results should be retained to prove that the impressed current system is operating properly.
- **Add interior lining to the tank.** The interior of the tank can be lined with a thick layer of noncorrodible material. If this is the only method of protection, the lining must pass an internal reinspection in 10 years and every 5 years thereafter to make sure that the lining is sound. Records of the inspection results should be retained.
- **Combined cathodic protection and interior lining.** The combined method provides greater protection than does either method alone, and the interior lining does not have to be inspected periodically. The cathodic system must be inspected as described above.

Existing piping must have cathodic protection that is designed by a qualified corrosion expert and tested and inspected periodically, and records of inspections must be maintained. Cathodic protection is not needed if the piping is made of a noncorrodible material (fiberglass).

GENERAL OPERATING REQUIREMENTS FOR HAZARDOUS SUBSTANCE USTS

Owners and operators of UST systems must ensure that the UST system is made of or lined with materials that are compatible with the hazardous substances stored in the UST system. New and existing tanks should be equipped with the proper protection for spills, overfills, and corrosion. Personnel should be properly trained to perform assigned tasks, and records must be maintained for the proper time intervals. Following the regulatory requirements for new and existing hazardous substance tanks will minimize the likelihood that product will enter the environment and will reduce the costs associated with cleanup operations.

In the event of a release, the procedures followed for hazardous substance USTs are similar to those for petroleum tanks. The notification procedure is slightly different. Hazardous substance spills and overfills that meet or exceed the reportable quantity must be reported immediately to the National Response Center (800-424-8802 or 202-267-2675). These spills and overfills must also be reported to the appropriate regulatory agency within 24 hours. However, reporting is not required if the spill or overfill is less than the reportable quantity and is immediately contained and cleaned up.

NEW HAZARDOUS SUBSTANCE USTS

New hazardous substance USTs are those installed after December 22, 1988. These USTs must meet the same requirements as the new petroleum USTs: correct installation; spill, overfill, and cathodic protection; corrective action; and closure. In addition, new hazardous substance USTs must have secondary containment and interstitial monitoring for leak detection. The secondary containment can be accomplished by placing one tank inside another tank or one pipe inside another pipe. Other methods include placing the UST system inside a concrete vault or lining the excavation zone around the UST system with a liner that is impermeable to the hazardous substance. A variance may be granted from the requirement for secondary containment and interstitial monitoring if an alternative leak detection method can be shown to be effective.

EXISTING HAZARDOUS SUBSTANCE USTS

Existing hazardous substance USTs are those installed before December 22, 1988. At that time, owners/operators were required to start using filling procedures that prevented spills and overfills immediately. In addition, these tanks were required to have leak detection systems by December 22, 1993. Leak detection requirements can be met in one of three ways. First, any leak detection method for existing petroleum USTs can be used until December 1998 if the selected method can effectively detect releases of the stored hazardous materials. Second, after December 22, 1998, existing hazardous substance USTs must meet the same requirements for secondary containment and interstitial monitoring that apply to new hazardous substance USTs. Third, an existing hazardous substance UST may be granted a variance from the secondary containment and interstitial monitoring requirement if an alternative method can be shown to be effective.

Existing hazardous substance USTs must also be fitted with spill and overfill protection devices and corrosion protection devices by December 22, 1998.

THE AUDIT PROCESS

During the UST audit, the audit team will review the facility's policies and operational procedures and compare them to actual practices and determine areas of conformance or nonconformance. The audit will determine whether the facility complies with applicable regulations and whether it utilizes good management practices for preventing releases, for response procedures in the event leaks occur, and in other areas shown in Figure 14–3 at the end of this chapter.

Preparation for the audit includes familiarization with the design, operation, and leak detection methods of USTs, and the auditor should be familiar with Subtitle I of RCRA, 40 CFR Part 280, and state regulations. The RCRA/Comprehensive Environmental Response, Compensation and Liability Act (CERCLA) hotline or the state UST office can provide many useful sources of information (see Endnotes 2 and 3). It is important for the facility to have permits and other documentation available in an organized, easy-to-use format.

The audit will begin with a briefing of site management and the environmental staff. The audit team will obtain information about required safety requirements and will participate in an orientation tour. During the orientation tour, the audit team will become better acquainted with the physical layout of the facility and its operations. The orientation tour will be conducted with the facility representative, typically, the environmental coordinator. During the tour, the audit team will become oriented to the facility, with a focus on the location of all storage tanks—aboveground and underground. The tour should include the entire facility and its grounds.

After the orientation tour, the audit team will review records and management systems, interview site personnel, and conduct an in-depth inspection of the facility. Previously developed checklists will be used by the audit team to conduct the in-depth evaluation of the facility. At the end of each day, the audit team members will brief one another and discuss findings.

The review of documents and records should include the facility's UST permits; contracts for services; operational procedures; procedures for training, maintenance, and record keeping; and the plan and procedures for reporting and responding to leaks and emergency conditions. The auditor will examine the UST permit file to complete the inventory forms and will identify citations and notices of violation (NOVs), site assessments, corrective action reports, agency inspection reports, and other data that have been submitted for regulatory purposes. For leak detection equipment, required records include maintenance, repair, and calibration records of on-site leak-detection equipment, yearly monitoring results, recent tightness test results, and copies of performance claims by manufacturers of leak-detection equipment. Records of the required inspections and tests are needed for the corrosion protection system. If UST systems are upgraded or repaired, records demonstrating that the work was done properly must be maintained. If USTs are permanently closed, records of the site assessment must be maintained for at least 3 years after closing the USTs. Records are also required to document financial responsibility.

One of the first tasks during the audit process is to compile an inventory of existing USTs. The regulatory definition of a UST is a tank and connected underground pipe that is 10% or more by volume beneath the surface of the ground and which is used to contain an accumulation of petroleum or other regulated substances (40 CFR Section 280.12). Other regulated substances are hazardous substances as defined in Section 101(14) of the CERCLA of 1980. Figure 14–4 shows the types of tank systems that are excluded, exempted, or deferred in Subtitle I.

The inventory should include the following information: tank identification number; operational date; contents; construction; method of leak detection; spill, overfill, and corrosion protection; release history and closure status. Figure 14–5 shows an example of a form that could be used to create an inventory of each tank. The inventory can be prepared by examining existing records. Each UST can be numerically coded on the plot plan of the facility to facilitate the plant tour and final report preparation. The plot plan also gives the auditor an easy way to identify any USTs that were omitted from the initial inventory. Omitted sources can be added to inventory forms during the audit.

The auditor uses the prepared inventory of USTs to examine the entire facility to locate all USTs, inspect them, and note apparent problems. If possible, the auditor examines practices to determine whether procedures are conducted properly. The auditor also takes photographs to provide documentation for unanswered questions or problems, and the auditor may ask for facility staff for additional information. In addition to examining documented USTs, the auditor will look for evidence (such as pipes protruding from the ground) of undocumented USTs and evidence of potential leaks (evidence of product, recently disturbed ground, areas with no vegetation, and so forth).

At the end of the audit, the audit team will debrief the management of the facility. Final activities will include drafting the audit report, responding to review comments, and submitting the final audit report. Facility or site management will follow up on audit findings, initiate corrective actions, determine the effectiveness of the actions, and report back to the audit team. Figure 14–6 gives examples of typical problems that might be noted during an audit of USTs.

FIGURE 14–4. Tank Systems that Are Not Covered by UST Regulations (40 CFR Part 280):

Tank Systems Excluded from the Definition of UST

a. farm or residential tanks of 1,100 gallons or less capacity that are used noncommercially for the storage of motor fuel;

b. heating oil storage tanks that are used on the premises where the tank is stored;

c. septic tanks;

d. surface impoundments, pits, ponds, or lagoons;

e. storm water or wastewater collection systems;

f. pipeline facilities (including gathering lines) that are regulated under:
   i. the Natural Gas Pipeline Safety Act of 1968, or
   ii. the Hazardous Liquid Pipeline Safety Act of 1979, or
   iii. state laws comparable to the provisions of i and ii;

g. flow-through process tanks;

h. liquid traps or associated gathering lines directly related to oil or gas production and gathering operations;

i. storage tanks that are situated in an underground area (for example, basement) if the tank is situated upon or above the surface of the floor in that area; or

j. pipes connected to any of the tanks described in items a. through i.

Tank Systems Specifically Exempted from Regulation

a. UST systems holding hazardous wastes listed or identified under Subtitle C of RCRA or a mixture of such hazardous wastes and other regulated substances. These are regulated under RCRA, Subtitle C;

b. wastewater treatment tank systems that are part of a wastewater treatment facility regulated under Section 402 or 307(b) of the Clean Water Act;

c. equipment or machinery that contains regulated substances for operational purposes (for example, hydraulic lift tanks);

d. UST systems with capacities of 110 gallons or less;

e. UST systems that contain a de minimis concentration of regulated substances; or

f. emergency spill or overflow containment UST systems that are expeditiously emptied after use.

Deferred Tank Systems

EPA has deferred five other tanks systems from some of its regulations. These include wastewater treatment tank systems, USTs that contain radioactive materials regulated under the Atomic Energy Act of 1954, airport hydrant fuel distribution systems, UST systems with field-constructed tanks, and UST systems that are part of an emergency generator system at nuclear power generation facilities regulated by the Nuclear Regulatory Commission under 10 CFR, Part 50, Appendix A. The installation of these deferred systems must meet some construction standards.

FIGURE 14–5.   Example Inventory Form for USTs:

ABC Company
UST Inventory Form

Plant: _____   Auditor: _____

Plant Contact: _____   Date: _____

UST Tank No.: _____   Contents: _____

Operational Date: _____   _____

Tank Status:

Currently in use?   _____ yes   _____ no

Temporarily out of use?   _____ yes   _____ no

Date of last use: _____

Permanently out of use?   _____ yes   _____ no

  Date removed from ground: _____

  Date filled in place: _____

Requesting closure?   _____ yes   _____ no

Tank Construction/Protection:

Construction:

_____ steel   _____ fiberglass/plastic

_____ concrete   _____ unknown

_____ other, specify _____

Internal protection:

_____ cathodic   _____ inner lining

_____ none   _____ unknown

_____ other, specify _____

External protection:

_____ cathodic   _____ inner lining

_____ none   _____ unknown

_____ other, specify _____

Piping:

Type: _____ bare steel   _____ galvanized steel

         _____ cathodic protection   _____ fiberglass reinforced plastic

         _____ unknown   _____ other, specify _____

Method:

         _____ pressurized   _____ European suction

         _____ American suction

Release Detection:

         _____ manual tank gauging

         _____ tank tightness testing with inventory control

         _____ automatic tank gauging

         _____ vapor monitoring

FIGURE 14–5.   (*Continued*)

_____ groundwater monitoring

_____ interstitial monitoring within a secondary barrier

_____ interstitial monitoring within secondary containment

_____ automatic line leak detectors

_____ line tightness testing

_____ statistical inventory reconciliation

_____ other, specify _____

Cathodic Protection:

For coated steel tanks with cathodic protection—

_____ impressed current   _____ sacrificial anodes

For coated steel pipes with cathodic protection—

_____ impressed current   _____sacrificial anodes

Other, specify _____

Spill Control:

_____ catchment basins   _____ overfill alarms

_____ automatic shutoff devices   _____ ball float valves

_____ other, specify _____

Proper installation, upgrade or closure (check all that apply):

_____ The installer has been certified by the tank and piping manufacturers.

_____ The installer, upgrade, or closure contractor has been certified by the Office of the State Fire Marshall.

_____ The installation or upgrade has been inspected and certified by a registered professional engineer.

_____ The installation, upgrade, or closure has been inspected by the Office of the State Fire Marshall.

_____ All work listed on the manufacturer's installation lists has been completed.

_____ Another method of compliance was used, specify _____

Previous Release History:

_____

_____

_____

_____

Comments:_____

_____

_____

_____

Rev. 1

7/15/94

RKB

Page _____ of _____

FIGURE 14–6.   Typical Problems Encountered During an Audit of USTs:

Problems Related to Tanks

- missing or incomplete inventory
- USTs not identified
- evidence of leaks (staining of ground, spilled product)
- leak prevention measures
    — monthly inventory control not conducted
    — groundwater monitoring wells not checked on a monthly basis
    — monitoring equipment not working properly or not calibrated as required by manufacturer
    — monitoring wells not marked and locked
    — tank tightness testing not conducted every 5 years
    — suction piping not tested as required (monthly monitoring or tightness testing once every 3 years)
    — pressurized piping not tightness tested
    — pressurized piping not outfitted with automatic flow restrictor, automatic shut-off device, or continuous alarm system
    — incorrect filling practices
    — missing catchment basin (or pump or drain) or improperly graded surface
- overfill protection
    — missing or improperly functioning automatic shutoff devices, overfill alarms, or ball float valves
    — loose fittings
    — incorrect filling practices
- missing or improper installations of corrosion protection systems
- improperly closed tank (temporary or permanent)

Problems Related to Documentation

- notification and record keeping incomplete or unavailable for installation or closure
- incomplete or unavailable leak detection performance and maintenance records
    — tightness test records
    — performance claims for leak detection equipment
    — maintenance, repair, and calibration records of on-site leak detection equipment
- missing or improper documentation for action in the event of an emergency or spill
- missing or incomplete reports of suspected releases, site assessments, and corrective actions

SUMMARY

The regulatory framework (and good management practices) for USTs is directed toward identifying tanks that contain petroleum or other hazardous materials, preventing leaks, ensuring that leaking tanks will be detected, and assessing and correcting environmental hazards that result from unplanned releases of spilled product. Tanks should have adequate leak detection equipment, spill and overflow protection, and cathodic protection. Record keeping should be up to date and complete. Good management practice dictates that any facility that maintains tanks should have a spill prevention and countermeasures contingency plan, regardless of the amount of product that is maintained at the facility. Following these measures can minimize environmental impacts and can reduce financial liability in the event of unplanned releases.

ENDNOTES

[1] U.S. Environmental Protection Agency. *Musts for USTs, A Summary of Federal Regulations for Underground Storage Systems, EPA 510-K-95-002.* Washington D.C.: U.S. Environmental Protection Agency, Solid Waste and Emergency Response, 1995.

[2] U.S. Environmental Protection Agency. *Musts for USTs, A Summary of Federal Regulations for Underground Storage Systems, EPA 510-K-95-002.* Washington D.C.: U.S. Environmental Protection Agency, Solid Waste and Emergency Response, 1995.

[3] EPA has a series of videos that can be ordered at nominal cost for the Environmental Media Center, Box 30212, Bethesda, MD 20814. One of these, *Keeping it Clean,* shows how deliveries can be made safely with no spills.

FIGURE 14–3.   UST Checklist:

Responses			Questions to Ask
YES	**NO**	**N/A**	***General Requirements***

			1. Does the facility have regulated tanks?
			a. how many? _____
			2. Does the facility have nonregulated tanks?
			a. how many? _____
			3. Does the facility have plans for emergency response and spills? (Cross-reference emergency response and spill response checklist)

YES	**NO**	**N/A**	***Record Keeping***
			4. Was notification given for regulated tanks?
			a. new USTs (installed after 12/22/88)?
			b. existing UST (installed before 12/22/88)?
			c. change of owner or operator?
			d. a system upgrade?
			e. an address change?
			f. a temporary closure?
			g. a permanent closure and site assessment review?
			h. a request for closure?
			i. other, specify _____
			5. Does the facility maintain leak-detection records?
			a. previous year's monitoring results?
			b. most recent tightness test?
			c. manufacturer's performance claims?
			d. maintenance records for monitoring equipment?
			e. repair records for monitoring equipment?
			f. calibration records of monitoring equipment?
			6. Does the facility maintain records for corrosion protection system?
			a. inspections?
			b. repair records?
			c. maintenance records?
			7. Does the facility maintain records showing proper repairs of USTs?
			8. Does the facility maintain records showing proper upgrading of USTs?
			9. Does the facility have records of site assessment for permanent closure of UST (for at least 3 years after closure)?

FIGURE 14–3. (*Continued*)

YES	NO	N/A	*Record Keeping*
―――	―――	―――	10. Does the facility have documentation of financial responsibility?
―――	―――	―――	a. is coverage adequate?
―――	―――	―――	b. specify amount _____
―――	―――	―――	11. Does the facility have documentation of spills and countermeasures taken, and measures to prevent recurrence?
―――	―――	―――	a. Does the facility have proper documentation for operating procedures and work instructions related to equipment and transfers of materials?
			Comments: _____

YES	NO	N/A	*Operation*
―――	―――	―――	12. Are the tanks correctly installed?
―――	―――	―――	a. new tanks?
―――	―――	―――	b. existing tanks?
			Comments: _____
―――	―――	―――	13. Do the tanks have proper leak detection?
―――	―――	―――	a. new tanks?
―――	―――	―――	b. existing tanks?
			Comments: _____
―――	―――	―――	14. Do the tanks have proper corrosion control?
―――	―――	―――	a. new tanks?
―――	―――	―――	b. existing tanks?
			Comments: _____
―――	―――	―――	15. Are the tanks properly closed?
―――	―――	―――	a. permanent closure?
―――	―――	―――	b. temporary closure?
			Comments: _____

Source: Adapted from U.S. Environmental Protection Agency, Solid Waste and Emergency Response, *RCRA Inspection Manual 1993 Edition*, OSWER Directive 9938.02B (Washington, D.C.: U.S. Environmental Protection Agency, 1993).

CHAPTER 15

WATER POLLUTION

INTRODUCTION

The regulatory framework for preventing pollution of the nation's waterways is a complex system of regulations that dates to the Federal Water Pollution Control Act (FWPCA) of 1972 [now known as the Clean Water Act (CWA)]. The objective of the FWPCA was "to restore and maintain the chemical, physical, and biological integrity of the Nation's waters" (33 USCA Sections 1251–1387). The goals of Section 1251 included eliminating the discharge of pollutants into navigable waters by 1985, prohibiting the discharge of toxic pollutants in toxic amounts, and controlling nonpoint sources. These goals are accomplished through water pollution control regulations that apply to existing sources, new or modified sources, sources that discharge to Publicly Owned Treatment Works (POTW), and sources of toxic pollutants. Individual programs include the construction of treatment works, effluent limitations, water quality standards, best management practices, the National Pollution Discharge Elimination System (NPDES), ocean discharge criteria, dredged or fill material permits, and other permits and licenses. The federal government has the overall responsibility for water quality programs but some programs such as the construction program and implementation of the NPDES program are delegated to the states.

Because of the complexity of the regulatory framework, this chapter will be limited to a brief introduction to industrial sources of water pollution and commonly used treatment methods, followed by an overview of compliance auditing of the NPDES program. The U.S. EPA World Wide Web site at the Office of Water (*http://www.epa.gov/ow*) and other sites in Appendix F provide access to other useful resources.

INDUSTRIAL SOURCES OF WATER POLLUTION

Water can become contaminated by microbes, organic and inorganic chemicals, radioactive materials, and thermal pollution from point sources (sanitary wastewater, industrial process water, noncontact cooling water, boiler and cooling tower blowdown) and nonpoint sources (urban storm-water runoff, spills, agricultural runoff). Figure 15–1 lists examples of specific areas in industrial facilities that can contribute to water pollution.

Sanitary wastewater includes human wastes from bathrooms and food-related wastes from cafeterias and break rooms. In some cases, laboratory wastewater can be included in the sanitary system, but this practice should be evalu-

FIGURE 15–1. Potential Sources of Wastewater in Industrial Facilities:

- Sanitary wastewater
  —rest rooms
  —lunch rooms
  —locker rooms
- Process water
  —spray booths
  —wash water
  —valve leaks
- Cooling water
  —blowdown
  —noncontact
- Boiler blowdown
- Air pollution control equipment
- Tank farm drainage
- Containment area drainage

- Laboratories
  —industrial
  —municipal
  —schools and universities
- Fire water
- Seal water
- Quench tanks
- Roof drains
- Floor drains (process areas, warehouses)
- Waste storage piles
- Raw material storage piles
- Shipping and receiving areas
- Condensate water
- Waste storage areas

ated carefully because of potential damage to the POTW and the possibility that wastes may not be treated by the system. Sanitary wastewater can be managed by discharging it to the public sewer system followed by treatment at the local POTW or by discharging it to on-site treatment systems. On-site options include septic tanks and tile fields or package treatment plants that, in turn, discharge to local streams.

Industrial process water is the most diverse of the point sources because it comes into contact with a variety of contaminants through many different types of manufacturing operations, including heating and cooling, washing, the application of coatings and paints, and the use of binders and adhesives. Process water is also generated by pollution control equipment such as the venturi scrubber or packed tower. Depending on the contaminants, process water may be discharged with or without pretreatment to a POTW, a package treatment plant, or a water source.

Noncontact cooling water is water that is circulated in closed systems and does not contact the product or raw material. It may be drawn from a lake, river, or the public water supply system. Cooling water may be discharged into a POTW or into the water source from which it was drawn.

Blowdown is water that is used to help control corrosion and the buildup of scale from sources such as cooling towers and boilers. The water used in these sources is typically treated with chemicals such as anticorrosives and zinc or sodium chromate. Blowdown may be discharged with or without pretreatment to the POTW or to the water source.

Uncontaminated stormwater can be handled by surface drains, ditches, and impoundment basins. Stormwater that has the potential to be contaminated by chemicals from raw material storage areas, shipping and receiving areas, or process areas, and it may need to be collected and treated prior to discharge into receiving waters or on-site treatment plants. Surge basins provide controlled releases of the runoff to prevent washout of the wastewater treatment facility.

WASTEWATER TREATMENT METHODS

Wastewater can be treated with physical, chemical, and biological processes. Regardless of the type of treatment method, the treated water may require testing for physical, chemical, and biological contaminants to evaluate the effectiveness of the treatment method. The collection, analysis, and disposal of samples must be conducted according to prescribed or standard methods.

Wastewater may require pretreatment to condition or control the flow of the effluent. Pretreatment methods include screening, pH control, flow equalization, and mixing. Large debris that can damage pumps and other equipment can be removed from the waste stream by passing the water through screens. Screens must be cleaned by hand or mechanical means on a periodic basis to prevent plugging, and the screened material requires disposal. Acids and alkalis are added to wastewater to control the pH. If the effluent has the potential for large changes in pH, the facility should have an emergency bypass basin. Mechanical mixers and other equipment are used to ensure that the effluent has uniform quality.

PHYSICAL TREATMENT

Physical treatment methods, including flocculation, separation, filtration, and adsorption are used to remove particulates. Flocculation involves adding a chemical to the effluent to provide a binding site for small particles so they can increase in size and settle. The chemical addition must occur at a controlled rate accompanied by gentle mixing in a tank or basin to ensure contact and promote growth of the floc.

Separation involves removing particulates and oil or other organics from the wastewater. Separators include pits, chambers, and basins that hold the wastewater and allow the contaminants to float to the surface or settle to the bottom. Flow-through separators may contain a baffling system to reduce flow and increase retention time. After removal by skimming or pumping, the collected contaminants require disposal.

Filtration is used to remove solids that cannot be removed by flocculation or separation alone. Filtration involves passing the wastewater through a bed of granular material (such as sand, gravel, or anthracite) to remove particles, followed by periodic cleaning of the filter bed. The cleaning is accomplished by backwashing (reversing the flow of water) to remove the accumulated solids. The used water can be returned to the settling device to remove the solids for disposal.

Adsorption involves passing the wastewater through a suitable material (activated carbon, activated alumina, or molecular sieves) that attracts and retains soluble contaminants. Carbon is the most commonly used adsorbent, and it can be regenerated by thermally oxidizing the organics. Typically, the activated carbon is contained in a fixed bed, and the wastewater is passed through the bed, but powdered activated carbon can also be added directly to the wastewater, followed by settling and treatment for regeneration.

Typical problems with physical treatment units include improper chemical addition rates, exceeding design flow rates, exceeding design particulate or oil concentrations, and inadequate cleaning. Improper disposal of collected solids and oil is another concern.

CHEMICAL TREATMENT

There are a number of chemical treatment methods that can control individual organic and inorganic contaminants. Precipitation involves adding a chemical (for example, alum, lime, ferrous sulfate, ferric chloride) at a controlled rate to the wastewater, followed by mixing and sedimentation to remove the precipitate that must be dewatered prior to disposal. Other chemical treatment methods include using hydrogen peroxide to oxidize phenol prior to biological treatment and chlorine to destroy pathogens.

Typical problems with chemical treatment methods include improper chemical addition rates, improper handling and storage of hazardous materials, and improper disposal of wastes that may contain heavy metals, organics, and other hazardous constituents.

BIOLOGICAL TREATMENT

Biological treatment uses microorganisms to convert organic and inorganic contaminants into less hazardous products. The sludge that is produced from the treatment process is typically dewatered, and it may be incinerated prior to final disposal. Activated sludge is an aerobic process in which microorganisms oxidize the organic components of a wastewater into carbon dioxide, water, and new biomass. The process takes place inside a tank or basin that has been inoculated with the microorganisms. The tank is aerated with oxygen or air, and after an appropriate retention time the digested wastewater is sent to a settling tank, where the solids separate from the effluent.

Anaerobic digestion is a process in which anaerobic and facultative microorganisms decompose biological sludges in oxygen-deficient atmospheres. The digestion process can take place in a single tank or in a two-stage process. Temperature, pH, and ammonia can affect the efficiency of digestion, and these parameters must be monitored closely.

THE NPDES AUDIT PROCESS

OVERVIEW

Prior to conducting the audit, the audit team will gain an understanding of the facility processes, the types of wastewater generated at the facility, and the types of equipment it utilizes to control water pollution. In addition, the auditor will review the NPDES regulations applicable to the facility, which include industry effluent standards, best management practices, pretreatment standards, and POTW standards. Other documents that will be reviewed by the auditor include the last audit, if any, protocols, and preaudit questionnaires. The auditor may also review NPDES permits and POTW contracts.

Depending on the scope of the audit, the following areas might be considered:

- NPDES receiving water body discharge
- Discharge to POTW
- Runoff discharges
- Unpermitted discharges
- Pretreatment discharge
- Flow measurement program
- Biomonitoring program
- Training

- Sampling program
- Laboratory quality assurance
- On-site treatment facility
- Air releases
- Hazardous waste storage and disposal
- Sludge management
- Records and record keeping

It is important for the facility to have permits and other documentation available in an organized, easy-to-use format. The auditor will examine the NPDES permit file to complete the inventory forms; identify citations and notices of violation (NOVs); and review site assessments, corrective action reports, agency inspection reports, and other data that have been submitted for regulatory purposes.

The audit will begin with a briefing of site management and the environmental staff. The audit team will obtain information about safety requirements and participate in an orientation tour. During the orientation tour, the audit team will become better acquainted with the physical layout of the facility and its operations. The orientation tour is conducted with the facility representative, typically, the environmental coordinator. During the tour, the audit team will become oriented to the facility and acquainted with the processes that generate wastewater and with the locations of discharge points and run-off flow patterns. The tour should include the entire facility, its grounds, and any on-site wastewater treatment works.

After the orientation tour, the audit team will review records and management systems, interview site personnel, and conduct an in-depth inspection of the facility. Figure 15–2 contains a list of records that might be reviewed, depending on the nature of the facility. Previously developed checklists will be used by the audit team to conduct the in-depth evaluation of the facility. Figure 15–3 at the end of this chapter is an example of a checklist that can be used to audit record keeping and reporting requirements. At the end of each day, the audit team members will brief one another and discuss findings.

During the inspection, the audit team will identify and evaluate the sources of wastewater generation, discharge points directly to receiving waters or indirectly to POTWs, and runoff flow patterns on a plot plan of the facility. The audit team will look for undocumented discharges and consider operational factors such as the condition of equipment, availability of safety controls and equipment, and characteristics of the effluent and receiving stream. The audit team will also note discharges with noxious odors, visible solids in discharges, and the presence of fish or vegetation kills.

FIGURE 15–2.   Typical Documentation Elements of an NPDES Audit:

- Permits
  —current copy on site
  —correctness of information
- Sampling and analysis data
  —dates, times, locations of sampling
  —analytical methods and results
  —results of analyses
  —dates and times of analyses
  —name(s) of analysis and sampling personnel
  —instantaneous flow at grab sample stations
- Monitoring Records
  —submission of discharge monitoring reports with information required by permits
  —maintenance of original charts from continuous monitoring instrumentation
  —maintenance of flow monitoring calibration records
  —maintenance of records for required time periods (typically, 3 years)
- Laboratory Records
  —calibration and maintenance of equipment
  —calculations
  —quality assurance/quality control data
- Facility Operating Records
  —daily operating log
  —summary of all laboratory tests and other required measurements
  —treatment chemicals used
  —weather conditions
  —equipment maintenance completed and scheduled
- Treatment Plant Records (required by Federal Construction Grants program)
  —plant Operations and Maintenance Manual
  —percent removal records
  —"as-built" engineering drawings
  —copy of construction specifications
  —equipment supplier manual
  —data cards on all equipment
- Management Records
  —average monthly operating records
  —annual reports
  —emergency conditions (power failures, bypass reports, and so forth)
- Pretreatment Records
  —POTW and industrial monitoring and reporting requirements
  —industrial user discharge data
  —compliance status records
  —POTW enforcement initiatives
- SPCC Plan
- Best Management Practices

The auditor will compile the discharge information into an inventory and determine whether the facility is in compliance with permitted discharge limits and policies developed by the facility. The characterization of discharges will begin with those point sources and run-off sources that are permitted through the NPDES permit and POTW contract or limits. Finally, the auditor will develop an inventory of any other discharges of wastewater. For each source of wastewater, the auditor will develop an inventory of the source, water flow, treatment methods, and discharge quality and quantity. The auditor will also determine whether unregulated discharges are in compliance with regulations and other policies developed by the facility.

At the end of the audit, the team will debrief the management of the facility. Final activities will include drafting the audit report, responding to review comments, and submitting the final audit report. Facility or site management will follow up on audit findings, initiate corrective actions, determine the effectiveness of the actions, and report back to the audit team.

DISCHARGES TO THE RECEIVING WATER BODY

The auditor will obtain copies of all current NPDES permits for discharges to receiving water bodies and will determine whether the permit conditions are being met. The permit will identify effluent limits for specific constituents, flow, and temperature. The permits may also specifiy best management practices (BMP) to minimize or eliminate the discharge of toxic or hazardous materials to public waters. Examples of general BMPs are operations and maintenance practices or good housekeeping procedures. The permit could also include site-specific BMPs such as monitoring requirements or the construction of berms or dikes. The auditor will compare the permit requirements to the facility's discharge data to evaluate compliance. If the permit contains a compliance schedule, the auditor will determine whether the facility is conforming to the schedule. Figures 15–4 and 15–5 show examples of data forms that can be adapted to create an inventory of NPDES sources and to evaluate compliance with the NPDES permit. Figure 15–6 shows an example of a noncompliance form for NPDES discharges.

DISCHARGES TO THE POTW

Facilities that discharge effluent to a POTW are required to comply with the National Pretreatment Program (NPP) regulations, which were developed to protect the POTWs and the environment from damage that could result from industrial discharges to the sanitary sewer system. Industrial dischargers must comply with prohibited discharge standards, appropriate pretreatment standards, and reporting requirements.

Under the NPP, facilities are prohibited from discharging to the POTW those contaminants that have a potential to cause:

- a fire or explosion hazard,
- a corrosive hazard (no pH <5),
- an obstruction to flow in the sewage collection system to the POTW,

FIGURE 15–4.   Example of an NPDES Receiving Water Body Inventory Form:

ABC Company
NPDES Receiving Water Body Inventory Form

Plant: _____   Auditor: _____

Plant Contact: _____   Date: _____

Discharge No: _____

Discharge Description: _____

_____

_____

Location:_____

_____

Receiving water: _____

Is discharge covered by NDPES permit?   _____ Yes   _____ No
  If yes,
    Permit No.: _____   Expiration Date: _____
If permit has expired, is permission notice on file?   _____ Yes _____ No
  If no, why not? _____
Is water treated prior to discharge?   _____ Yes   _____ No
  If yes, describe: _____

_____

_____

Description of best management practices, if applicable: _____

_____

_____

_____

Agency reporting requirements:_____

_____

_____

Comments: _____

_____

_____

Rev. 2   RKB
4/01/94   Page ___ of ___

- temperature elevations at the POTW that would inhibit biological activity at the POTW, or
- interference with the POTW or sludge process, which, in turn, could cause the POTW to violate its NPDES permit or prevent the POTW from using its chosen sludge or disposal practice

The facilities must also meet the appropriate categorical pretreatment standards (65 toxic pollutants) for 34 industrial categories based on the characteristic wastes produced by the manufacturing process at each type of industry, the

FIGURE 15–5.   Example of an NPDES Receiving Water Body Flow and Monitoring
Data Inventory Form:

NPDES Receiving Water Body Inventory Form Flow and Monitoring Data Summary				

Plant: _____   Auditor: _____

Plant Contact: _____   Date: _____

Discharge No.: _____

Flow data (daily):

  min _____ ave _____ max _____

Monitoring Data:

Parameter		Permit Requirements		Latest Test Data
(units)	Discharge Limit	Test Frequency	Results	Laboratory ID
____	_____	_____	____	_____
____	_____	_____	____	_____
____	_____	_____	____	_____
____	_____	_____	____	_____
____	_____	_____	____	_____
____	_____	_____	____	_____

Rev. 2   RKB

4/01/94   Page ___ of ___

wastewater control technologies available to the industry, and economical considerations. These standards apply to the point of discharge from the pretreatment unit for the regulated process or to the point of discharge for untreated process wastewater. In addition, the facility may be required to meet other local requirements developed by the POTWs that operate approved pretreatment programs as part of the contract between the facility and the POTW. For facilities that discharge to POTWs not operating approved pretreatment programs, facilities are issued Industrial Waste Pretreatment (IWP) permits by the state environmental agency.

Even if it is not required by regulation, all facilities should develop an inventory of their discharges to the POTW. These data will help protect the facility's interests in the event that it becomes suspected of causing a problem at the POTW. Figure 15–7 and Figure 15–8 provide examples of inventory forms that could be adapted to evaluate discharges to a POTW.

STORM-WATER RUNOFF

During the plant tour and review of the facility's permits, the auditor will identify potential areas where surface runoff could come into contact with storage piles, process areas, or loading areas, and will evaluate the facility's practices to minimize contamination problems. Potential problem areas include parking lots, shipping and receiving areas, storage piles, and roof drains. During the plant in-

FIGURE 15–6. Example of an NPDES Receiving Water Body Noncompliance Form:

ABC Company
NPDES Receiving Water Body Noncompliance Form

Plant: _____   Auditor: _____

Plant Contact: _____   Date: _____

Discharge No.: _____   Test Frequency: _____
Receiving water: _____

Parameter (units)	Measured Discharge Limit	Date Last Exceeded	No. of Noncompliance Events During the Past 12 Months
BOD	_____	_____	_____
COD	_____	_____	_____
TOC	_____	_____	_____
TSS	_____	_____	_____
Oil and grease	_____	_____	_____
Ammonia	_____	_____	_____
pH	_____	_____	_____
Sulfide	_____	_____	_____
Sulfate	_____	_____	_____
Hexavalent chromium	_____	_____	_____
Total chromium	_____	_____	_____
Lead	_____	_____	_____
Mercury	_____	_____	_____
Zinc	_____	_____	_____
Copper	_____	_____	_____
Iron	_____	_____	_____
Phenol(s)	_____	_____	_____
Other	_____	_____	_____
	_____	_____	_____
	_____	_____	_____

Comments: _____
_____
_____
_____
_____

Rev. 2   RKB
4/01/94   Page ___ of ___

BOD—biochemical oxygen demand
COD—chemical oxygen demand
TOC—total organic carbon
TSS—total suspended solids

FIGURE 15–7.   Example of an Inventory Form for Evaluating Discharges to the POTW:

ABC Company
NPDES Discharge to POTW Audit Form

Plant: _____

Auditor:_____

Plant Contact: _____   Date: _____

Name of POTW: _____

POTW Contact: _____

Discharge Description: _____

_____

_____

Is discharge covered by POTW contract?   _____ Yes   _____ No

  If no, why not? _____

  If no, has the facility characterized its effluent?   _____ Yes   _____ No

Is treatment required prior to discharge?   _____ Yes   _____No

  If yes, describe: _____

_____

_____

Comments: _____

_____

_____

Rev. 2   RKB
4/01/94   Page ___ of ___

---

spection, the auditor will examine the grounds for potential signs of contamination. The slope of the land will be checked for obvious problems. The auditor will look for areas of dead or dying vegetation and discolored soil or pavement. Containment areas will be examined for overflow potential.

The facility may be required to use best management practices by the NPDES permit, or the facility may have its own policies for managing runoff. The auditor will determine whether systems and practices are in place to comply with runoff requirements. Figure 15–9 shows an example of a form that can be adapted to evaluate a facility's runoff sources.

ON-SITE WASTEWATER TREATMENT PLANT

The auditor will tour the on-site treatment facility and make general observations about the condition, operation, and maintenance of the facility. The auditor's evaluation is not an in-depth inspection of the treatment plant, but the auditor will look for physical conditions that indicate existing or potential problems. Signs of problems might include treatment units out of service; excessive noise

FIGURE 15–8. Example of an Inventory Form for Evaluating Discharges to the POTW, Flow, and Monitoring Data Inventory:

POTW Discharge
Flow and Monitoring Data Inventory

Plant: _____   Auditor: _____

Plant Contact: _____   Date: _____

Name of POTW: _____

POTW Contact: _____

Flow Data (daily):

   min _____   ave _____   max _____

Monitoring Data:

Parameter	*Contract Limits/POTW Limits*		*Latest Test Data*	
	Discharge Limit	*Test Frequency*	*Results*	*Laboratory ID*
_____	_____	_____	_____	_____
_____	_____	_____	_____	_____
_____	_____	_____	_____	_____
_____	_____	_____	_____	_____
_____	_____	_____	_____	_____
_____	_____	_____	_____	_____
_____	_____	_____	_____	_____

Comments: _____
_____
_____

Rev. 2   RKB
4/01/94   Page ___ of ___

from process or treatment areas; evidence of corrosion at the plant; flow-through bypass channels; open-ended pipes; erosion of stabilization pond banks or dikes; pH problems in digestors, ammonia stripping units, or denitrification units; poor sludge distribution on drying beds; and a lengthy list of other potential warning signals.

The auditor will also evaluate the facility's programs for operations and maintenance (see Figure 7–1 for a checklist that can be used to evaluate operations and maintenance); sampling, analyses, and quality assurance (see Figure 8–6 for a checklist for laboratory evaluation); biomonitoring; and flow measurement (see below).

BIOMONITORING

The NPDES permit may require acute and chronic toxicity testing procedures. A number of EPA documents are available on detailed procedures for biomonitor-

FIGURE 15–9.   Example of an Inventory Form for Facility Storm-Water Runoff:

ABC Company
Storm-Water Runoff Inventory Form

Plant: _____   Auditor: _____

Plant Contact: _____   Date: _____

Runoff Discharge No.: _____
Discharge Description: _____
_____
_____

Is discharge covered by NPDES permit?   _____ Yes   _____ No
  If yes, describe best management practices and implementation status: _____
_____
_____
_____

If no, does facility have a best management practice policy?   _____ Yes   _____ No
  If yes, describe best management practices and implementation status:_____
_____
_____
_____

Potential sources of runoff:

Source	Contamination Potential	Receiving Water	Treatment Method
Process areas	_____	_____	_____
Shipping areas	_____	_____	_____
Receiving areas	_____	_____	_____
Parking lots	_____	_____	_____
Roof drains	_____	_____	_____
Other	_____	_____	_____
	_____	_____	_____
	_____	_____	_____

Comments: _____
_____
_____
_____
_____

Rev. 2   RKB
4/01/94   Page ___ of ___

ing, including *Methods for Measuring the Acute Toxicity of Effluents to Freshwater and Marine Organisms* and *Methods for Estimating the Chronic Toxicity of Effluents and Receiving Waters to Freshwater Organisms.* The auditor will evaluate a self-biomonitoring program to determine:

- compliance with the discharger's NPDES permit limitations and requirements;
- whether the records and reports required by the NPDES permit are being maintained
- the adequacy of the permittee's reports;
- whether representative samples are being collected and analyzed properly;
- whether toxicity tests have been conducted properly; or
- the need for toxicity limits.

The auditor will evaluate the sampling of effluent, facilities and equipment, conditions for rearing or holding test organisms, acceptability of dilution water, adequacy of test procedures, validity of test results, and proper record-keeping and data reporting.

FLOW MEASUREMENT

The measurement of discharge flows is an integral part of the NPDES program. Discharge flows are used to compute mass loadings from measured concentrations of pollutants. Flows in closed channels are usually measured by a metering device (Pitot tube, venturi meter, electromagnetic flow meter) inserted into the conduit. Closed channel flow is normally encountered between treatment units in a wastewater treatment plant where liquids and/or sludges are pumped under pressure.

Open channel flow is the most common type of flow at NPDES discharge points. It is flow that occurs in conduits that are not full of liquid. It can be measured by primary and secondary devices. Primary devices are calibrated hydraulic structures, such as flumes and weirs, that are inserted in the open channels. Accurate flow measurements can be obtained by measuring the depth of liquid (head) at the specific point in the primary device. Secondary devices are used in conjunction with primary devices to automate the flow measuring process. Examples of secondary devices are floats, ultrasonic transducers, and bubblers. The output of the secondary device is generally transmitted to a recorder and/or totalizer to provide flow data to the operator.

Commonly encountered problems with primary and secondary devices include faulty fabrication or construction and improper installation. For example, weirs that are not level, plumb, and perpendicular to the flow direction will not provide accurate flow measurements. Corrosion, scale formation, or accumulation of solids on primary devices can bias the flow measurement. The auditor will evaluate whether the flow measuring system conforms to the conditions of the NPDES permit. This evaluation will include an inspection of records (strip charts, logs, and so forth) to document discontinuities in flow measurement and to look for bypasses. The auditor will also review and evaluate calibration and maintenance programs to determine whether flow measurements are accurate and precise. Figure 15–10 at the end of this chapter shows an example of a checklist for evaluating flow measurements.

SUMMARY

The NPDES permit system regulates facilities that discharge wastewater directly to a receiving water body. Discharges to POTWs are regulated by local or state rules. These discharges must meet specific limits on chemical, physical, or biological contaminants in the wastewater stream. The NPDES compliance audit determines whether the facility is meeting its requirements as specified in its permit. Attaining and maintaining compliance requires a management system that allows sources and contaminants to be identified, ensures good record keeping and document control, provides adequate operational controls, and ensures competent laboratory and measurement support.

FIGURE 15–3.   NPDES Verification, Record Keeping, and Reporting Evaluation Checklist:

Responses			**Questions to Ask**
YES	**NO**	**N/A**	***General***
_____	_____	_____	1. Is there a current copy of the permit on site?
_____	_____	_____	2. Is the identifying information correct?
_____	_____	_____	3. Was notification given to the EPA/state of new, different, or increased discharges?
_____	_____	_____	4. Are accurate records of influent volume maintained, when appropriate?
_____	_____	_____	5. Are the number and location of discharge points as described in the permit?
_____	_____	_____	6. Are the names and locations of receiving waters correct?
_____	_____	_____	7. Are all discharges permitted?
YES	**NO**	**N/A**	***Record Keeping and Reporting***
_____	_____	_____	8. Are records and reports maintained as required by the permit?
_____	_____	_____	9. Is all required information available, complete, and current?
_____	_____	_____	10. Is information maintained for 3 years?
_____	_____	_____	11. Are analytical results consistent with data reported on daily monitoring reports?
_____	_____	_____	12. Are sampling and analyses data adequate, and do they include:
_____	_____	_____	a. dates, times, and location of sampling?
_____	_____	_____	b. name of individual performing sampling?
_____	_____	_____	c. analytical methods and techniques?
_____	_____	_____	d. results of analyses and calibration?
_____	_____	_____	e. dates of analyses?
_____	_____	_____	f. name of person performing analyses?
_____	_____	_____	g. instantaneous flow at grab sample stations?
_____	_____	_____	13. Are monitoring records adequate, and do they include:
_____	_____	_____	a. flow, pH, dissolved oxygen, and so forth, as required by permit?
_____	_____	_____	b. monitoring charts maintained for 3 years?
_____	_____	_____	c. flow meter calibration records?
_____	_____	_____	14. Are laboratory equipment, calibration, and maintenance records adequate?

FIGURE 15–3. *(Continued)*

YES	NO	N/A	*Record Keeping and Reporting*
_____	_____	_____	15. Are plant records (required only for facilities built with federal construction grant loans) adequate, and do they include:
_____	_____	_____	a. an operation and maintenance manual?
_____	_____	_____	b. an "as-built" engineering drawing?
_____	_____	_____	c. schedules and dates of equipment maintenance repairs?
_____	_____	_____	d. an equipment supplies manual?
_____	_____	_____	e. equipment data cards?
_____	_____	_____	16. Are pretreatment records adequate, and do they include an inventory of industrial waste contributors?
_____	_____	_____	17. Do the records include:
_____	_____	_____	a. monitoring data?
_____	_____	_____	b. inspection reports?
_____	_____	_____	c. compliance status records?
_____	_____	_____	d. enforcement actions?
YES	**NO**	**N/A**	*Compliance Schedule Status Review*
_____	_____	_____	18. Is the compliance schedule being met?
_____	_____	_____	19. Have the necessary approvals been obtained to begin construction?
_____	_____	_____	20. Are financing arrangements complete?
_____	_____	_____	21. Have contracts for engineering services been executed?
_____	_____	_____	22. Have design plans and specifications been completed?
_____	_____	_____	23. Has construction started?
_____	_____	_____	24. Is construction on schedule?
_____	_____	_____	25. Has startup begun?
			26. Has an extension of time been requested?

Source: Adapted from U.S. Environmental Protection Agency, Office of Water, Office of Water Enforcement and Permits, *NPDES Compliance Inspection Manual* (Washington, D.C.: U.S. Environmental Protection Agency, 1988).

FIGURE 15–10.   Flow Measurement Checklist:

Responses			**Questions to Ask**
YES	**NO**	**N/A**	*General*
_____	_____	_____	1. a. Are primary flow measuring devices properly installed and maintained?
_____	_____	_____	b. Is flow measured at each outfall? Number of outfalls _____.
_____	_____	_____	c. Is there a straight length of pipe or channel before and after the flow meter of at least 5–20 diameters?
_____	_____	_____	d. If a magnetic flow meter is used, are there sources of electric noise in the near vicinity?
_____	_____	_____	e. Is the magnetic flow meter properly grounded?
_____	_____	_____	f. Is the full pipe requirement met?
_____	_____	_____	2. a. Are flow records properly kept?
_____	_____	_____	b. Are all charts maintained in a file?
_____	_____	_____	c. Are all calibration data entered into a log book?
_____	_____	_____	3. Is actual discharged flow measured?
_____	_____	_____	4. Is effluent flow measured after all return lines?
_____	_____	_____	5. Are secondary instruments (recorders, totalizers) properly operated and maintained?
_____	_____	_____	6. Are spare parts stocked?
_____	_____	_____	7. Are effluent loadings calculated using effluent flow?
YES	**NO**	**N/A**	*Flumes*
_____	_____	_____	8. Is flow entering the flume reasonably well distributed across the channel and free of turbulence, boils, or other disturbances?
_____	_____	_____	9. Are cross-sectional velocities at the entrance relatively uniform?
_____	_____	_____	10. Is the flume clean and free of debris or deposits?
_____	_____	_____	11. Are all dimensions of the flume accurate and level?
_____	_____	_____	12. Are the sides of the flume vertical and smooth?
_____	_____	_____	13. Is the flume head being measured at the proper location?
_____	_____	_____	14. Is the measurement of the flume head zeroed to the flume crest?
_____	_____	_____	15. Is the flume properly sized to measure the range of existing flow?
_____	_____	_____	16. Is the flume operating under free-flow conditions over the existing range of flows?

FIGURE 15–10.   (*Continued*)

YES	NO	N/A	*Flumes*
_____	_____	_____	17. Is the flume submerged under certain flow conditions?
_____	_____	_____	18. Is the flume operation invariably free-flow?

YES	NO	N/A	*Weirs*
_____	_____	_____	19. Type of weir being used: _____
_____	_____	_____	20. Is the weir exactly level?
_____	_____	_____	21. Is the weir plate plumb, and are its top and edges sharp and clean?
_____	_____	_____	22. Is the downstream edge of the weir champered at 45°?
_____	_____	_____	23. Is there free access for air below the nappe of the weir?
_____	_____	_____	24. Is the upstream channel of the weir straight for at least four times the depth of water level, and is it free from disturbances?
_____	_____	_____	25. Is the distance from the sides of the weir to the side of the channel at least 2 H (H = head in feet)?
_____	_____	_____	26. Is the area of approach channel at least (8 × nappe area) for an upstream distance of 15 H? a. If not, is the velocity of approach too high?
_____	_____	_____	27. Are head measurements properly made by facility personnel?
_____	_____	_____	28. Does leakage occur around the weir?
_____	_____	_____	29. Are proper flow tables used by facility personnel?

YES	NO	N/A	*Other Flow Devices*
_____	_____	_____	30. Type of flow meter used: _____ _____
_____	_____	_____	31. What are the most common problems that the operator has had with the flow meter? _____ _____ _____ _____
_____	_____	_____	32. Measured wastewater flow: _____ million gallons/day (mgd) Record flow: _____mgd; Error _____%

FIGURE 15–10.   (*Continued*)

YES	NO	N/A	*Calibration and Maintenance*
_____	_____	_____	33. Is the flow totalizer properly calibrated?
_____	_____	_____	34. Is the flow routinely checked by the proper operator? Frequency/day _____
_____	_____	_____	35. Are maintenance inspections routinely done by plant personnel? Frequency/year _____
_____	_____	_____	36. Are flow meter calibration records kept? Frequency of flow meter calibration: ____ /month
_____	_____	_____	37. Is the flow measurement equipment adequate to handle the expected ranges of flow rates?
_____	_____	_____	a. Is the calibration frequency adequate?

Source: U.S. Environmental Protection Agency, Office of Water, Office of Water Enforcement and Permits, *NPDES Compliance Inspection Manual* (Washington, D.C.: U.S. Environmental Protection Agency, 1988).

CHAPTER 16

HAZARDOUS WASTE MANAGEMENT

INTRODUCTION

The goals of the Resource Conservation and Recovery Act (RCRA) include reducing the amount of waste generated, protecting health and the environment from the potential hazards of waste disposal, ensuring that wastes are managed in an environmentally sound manner, and conserving energy and natural resources. These goals apply to hazardous and nonhazardous wastes. Subtitle C of RCRA establishes the national hazardous waste management program, and the federal RCRA hazardous waste regulations are given in 40 CFR Parts 260–272. The states may be authorized to administer specific aspects of the RCRA program.

In 1984, the Hazardous and Solid Waste Amendments (HSWA) significantly amended the hazardous waste components of RCRA by including small generators and imposing new requirements on treatment, storage, and disposal facilities (TSDFs). Hazardous wastes are also regulated under the authority of the Comprehensive Environmental Response, Compensation and Liability Act (CERCLA), but CERCLA applies to inactive or abandoned sites that have been contaminated with hazardous wastes; RCRA applies primarily to facilities that actively generate and manage hazardous wastes.

The core of RCRA's regulatory program is the "cradle-to-grave" management of hazardous wastes through seven major sets of regulations:

- identification and listing of hazardous waste (Part 261)
- standards for generators of hazardous waste (Part 262)
- standards for transporters of hazardous waste (Part 263)
- standards for owners/operators of hazardous waste TSDFs (Parts 264, 265, 267)
- standards for the management of specific hazardous wastes and specific types of hazardous waste management facilities (Part 266 and 267)
- land disposal restriction standards (Part 268)
- requirements for the issuance of permits to hazardous waste facilities (Part 270)
- standards and procedures for authorizing state hazardous waste programs to be operated in lieu of the federal program and approved state hazardous waste management programs (Parts 271 and 272)

Under these regulations, generators of hazardous waste must comply with standards for handling wastes properly, packaging and labeling materials for shipment, and preparing manifests to track the waste shipment. Transporters are responsible for complying with regulations for manifests, making sure that materials are labeled, and delivering the wastes. Transporters must also comply with Department of Transportation (DOT) regulations relating to transport vehicles, placarding, and spill response. TSDFs must comply with performance standards, groundwater monitoring requirements, and a prohibition on the land disposal of untreated hazardous wastes. Because of the complexity of the regulations, this section provides an overview of selected RCRA compliance issues for generators, transporters, and TSDFs. Access to other useful resources can be found through the U.S. EPA World Wide Web site (http://www.epa.gov) (see Appendix F).

www

WHAT IS REGULATED?

HAZARDOUS WASTE DETERMINATION

One of the first tasks toward achieving regulatory compliance and properly managing waste materials is determining whether or not a waste is hazardous. To make this determination (40 CFR Section 262.11), a generator must first classify a waste as a solid waste, as defined by 40 CFR Section 261.2,[1] then evaluate it as a hazardous waste by determining whether the waste:

- is excluded under Section 261.4[2] or
- is listed as hazardous in Part 261 Subpart D or
- meets a characteristic(s) in Part 261, Subpart C, either by testing or applying knowledge and, for purposes of part 268, determining whether the listed waste exhibits a characteristic(s).

If the waste is determined to be hazardous, the generator must refer to Parts 264 (for permitted facilities), 265, and 268 for possible exclusions or restrictions pertaining to the waste.

Figure 16–1 provides a flowchart that shows the steps required for making a hazardous waste determination.

LISTED WASTES

Listed wastes are those that have been assigned an identifying number that can be used to identify the waste for reporting and land ban requirements. Facilities can easily use the "F" and "K" lists to determine whether a waste is hazardous. The F list (40 CFR Section 261.31) contains hazardous wastes from nonspecific sources (for example, F001 identifies spent solvents used in degreasing, such as 1,1,1-trichloroethane). The K list (40 CFR Section 261.32) includes hazardous wastes from specific sources (for example, K103 wastes are process residues from aniline extraction from the production of aniline). Two additional lists, "P" and "U," identify commercial chemical products, including off-specification species, container residues, and spill residues that must be treated as hazardous when discarded, mixed with waste oil, used oil, or other material, and applied to the

FIGURE 16–1.   Hazardous Waste Determination Flowchart:

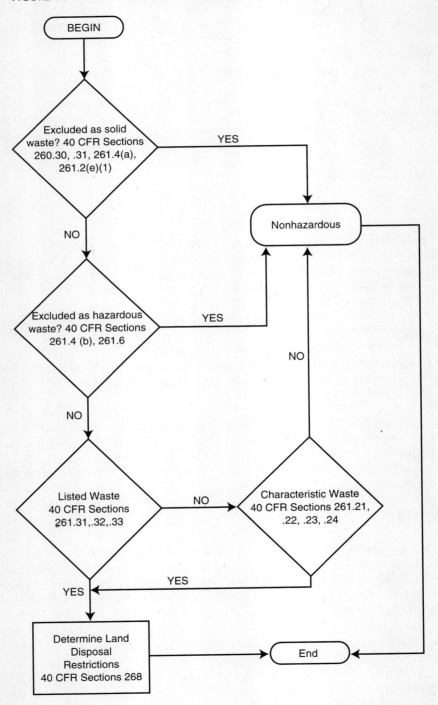

land, or when produced for use as a fuel (or as a component of fuel) (40 CFR Section 261.33). The "P" list identifies chemicals that are acutely hazardous and subject to more rigorous management requirements (for example, P063 is hydrogen cyanide). The "U" list identifies toxic chemicals that are not as hazardous as the P list (for example, U002 is acetone).

Companies may file a "delisting petition" with the EPA to remove wastes generated at their facilities from the EPA's lists. If a company wishes to delist a waste, it must demonstrate that the waste does not contain the hazardous constituents that motivated the EPA to list the waste or any other constituents that would cause the waste to be hazardous.

CHARACTERISTIC WASTES

Wastes that are not listed may still be hazardous if the waste exhibits one of four characteristics: ignitability (D001 wastes), corrosivity (D002 wastes), reactivity (D003 wastes), or toxicity (D004–D043 wastes). Appendix I of 40 CFR Part 261 contains representative determination methods, but alternate approved methods may be used.

A waste exhibits the ignitable characteristic if a representative sample is 1) a liquid having a flashpoint <140°F (60°C); 2) a nonliquid that causes fire through friction, absorption of moisture, or spontaneous chemical changes and, when ignited, burns so vigorously and persistently as to create a hazard; 3) an ignitable compressed gas; or 4) an oxidizer.

A waste exhibits the corrosivity characteristic if a representative sample is 1) aqueous, with a pH <2 or ≥12.5; or 2) a liquid and corrodes steel at a rate >6.5 mm/hr when applying a National Association of Corrosion Engineers Standard Test Method.

A waste exhibits the reactivity characteristic if it 1) is normally unstable and readily undergoes violent change; 2) reacts violently with water; 3) forms potentially explosive mixtures with water; 4) generates toxic gases, vapors, or fumes when mixed with water; 5) is a cyanide or sulfide-bearing waste which, when exposed to pH conditions between 2 and 12.5 can generate toxic gases, vapors, or fumes; 6) is capable of detonation or explosion if subjected to a strong initiating source or if heated under confinement; 7) is readily capable of detonation or explosive decomposition or reaction at standard temperature and pressure; or 8) is a forbidden explosive as defined by DOT.

A waste exhibits the toxicity characteristic if the extract of a representative sample of the waste contains any of the contaminants listed in Table 1 in 40 CFR Section 261.24, at or above the specified regulatory levels. The extract should be obtained using the Toxicity Characteristic Leaching Procedure (TCLP) or an equivalent approved method.

MIXTURE RULE, DERIVED-FROM RULE, AND CONTAINED-IN PRINCIPLE

Wastes may also be hazardous if they fall under the EPA's "mixture rule," "derived-from rule," or "contained-in principle." These aspects of hazardous waste regulation are important because they broaden the definition of hazardous waste and impose restrictions on the management of qualified materials. It is easy for facilities to overlook these special materials.

The mixture rule says that whenever listed hazardous wastes are mixed with solid wastes, the EPA considers the mixture to be a hazardous waste unless the mixture qualifies for an exemption. The EPA considers the mixing of some wastes to be a treatment that may require a RCRA permit. However, if a waste is listed in 40 CFR Part 261, Subpart D solely because it exhibits a characteristic, and a mixture of the waste no longer exhibits that characteristic, it is not considered to be a hazardous waste. Under the derived-from rule, a waste that is generated from the treatment, storage, or disposal of a hazardous waste (for example, the ash from incineration) is also a hazardous waste unless it is exempted. The contained-in principle says that soil, groundwater, surface water, and debris that is contaminated with hazardous waste must be managed as hazardous waste.

USED, REUSED, RECYCLED, OR RECLAIMED HAZARDOUS WASTES

The regulatory requirements that apply to used, reused, recycled, or reclaimed hazardous waste can be complex. In general, hazardous wastes that fall into these categories must be managed according to the regulations for generators, transporters, and TSDFs.

WHO IS REGULATED AND WHAT ARE THEIR RESPONSIBILITIES?

Anyone who manages a hazardous waste must notify the EPA within 90 days after the promulgation of regulations about that waste. The notification requirement applies to generators, transporters, owners and operators of TSDFs, and persons who produce, market, or burn hazardous waste-derived fuels. The notification must be made on Form 8700-12 for each site at which hazardous waste is managed. The notification form requires the name of the company, its location, and the EPA identification numbers for the listed and characteristic wastes it manages. In addition, the type of hazardous waste activity or used oil activity must be identified. If the facility burns waste fuel, the type of combustion device must be identified. Transporters must indicate the mode of transportation.

GENERATORS

The identification of hazardous wastes begins with the generator. A *generator* is "any person, by site, whose act or process produces hazardous waste identified or listed in Part 261 of this chapter or whose act first causes hazardous waste to become subject to regulation." (40 CFR Section 260.10) Generators are required to retain records related to their determinations that a waste is hazardous for at least 3 years from the date the waste is sent to a TSD facility. In practice, most generators retain these records indefinitely.

Once the stream has been characterized, the generator must determine the quantity of hazardous waste present at the facility in order to identify the applicable regulations. Hazardous waste generators fall into three categories: Conditionally Exempt Small-Quantity Generator (CESQG), Small-Quantity Generator (SQG), and Large-Quantity Generator (LQG). These categories are determined by the amount of waste generated by monthly calendar count, as shown in Figure 16–2.

FIGURE 16–2.   Categories of Hazardous Waste Generators:

You are a Large-Quantity Generator (LQG) if . . .

In one calendar month, you . . .

- generate 1,000 kilograms (2,200 pounds) or more of hazardous waste *or*
- generate 1,000 kilograms (2,200 pounds) or more of spill cleanup debris containing hazardous waste *or*
- generate more than 1.0 kilograms (2.2 pounds) of acutely hazardous waste *or*
- generate more than 100 kilograms (220 pounds) of spill cleanup debris containing an acutely hazardous waste *or*

At any time you . . .

- accumulate more than 1.0 kilograms (2.2 pounds) of acutely hazardous waste on site.

You are a Small-Quantity Generator (SQG) if . . .

In one calendar month, you . . .

- generate more than 100 kilograms (220 pounds) and less than 1,000 kilograms (2,200 pounds) of hazardous waste *or*
- generate more than 100 kilograms (220 pounds) and less than 1,000 kilograms (2,200 pounds) of spill cleanup debris containing hazardous waste *or*

At any time you . . .

- accumulate more than 1,000 kilograms (2,200 pounds) of hazardous waste on site.

You are a Conditionally Exempt Small-Quantity Generator (CESQG) if . . .

In one calendar month, you . . .

- generate 1.0 kilogram (2.2 pounds) or less of acutely hazardous waste *or*
- generate 100 kilograms (220 pounds) or less of hazardous waste *or*
- generate 100 kilograms (220 pounds) or less of spill cleanup debris containing hazardous waste *or*

At any time you . . .

- accumulate up to 1,000 kilograms (2,200 pounds) of hazardous waste on site.

Note: 1 barrel is approximately equal to 55 gallons or 440 pounds of hazardous waste.
Source: U.S. Environmental Protection Agency, Solid Waste and Emergency Response, *RCRA Inspection Manual 1993 Edition,* OSWER Directive 9938.02B (Washington, D.C.: U.S. Environmental Protection Agency, 1993).

LQG RESPONSIBILITIES

After determining its waste streams, the LQG must notify EPA of its hazardous waste generation activity. The LQG must obtain an EPA identification number before transporting, treating, storing, or disposing of hazardous wastes, and it must use transporters and TSDFs that have obtained their EPA identification numbers. The generator must prepare the Uniform Hazardous Waste Manifest (EPA Form 8700-22) for any transport of hazardous waste (Figure 16–3). The manifest provides a chain-of-custody record for tracking the shipment of waste. The generator is required to identify each authorized transporter and designated TSD facility by name and EPA identification number. The generator must describe the waste according to the DOT regulations, which specify the proper shipping name, hazard class, and identification number. The generator must also certify that the manifest "fully and accurately" describes the shipment and that the waste has been properly packaged, labeled, and marked. LQGs are also required to certify that they have a program in place to reduce the volume and toxicity of waste generated to the degree determined to be economically practicable and that they have selected the TSD method that minimizes threats to human health and the environment.

The generator must sign the manifest certifications by hand, and a copy of the final signed manifest must be maintained for at least 3 years. If the manifest is not returned to the generator in a timely or properly executed manner (within 45 days from the shipping date), the generator must file an exception report with the EPA or the state. The exception report consists of a copy of the manifest in question and a cover letter that describes the efforts taken to locate the waste or manifest and the result of those efforts.

The generator is responsible for complying with the DOT regulations for properly packaging, labeling, and marking the hazardous wastes for shipment. The EPA also requires any container of 110 gallons or less to be marked with the generator's name, address, manifest document number, and the words: *Hazardous Waste: Federal law prohibits improper disposal. If found contact the nearest police or public safety authority or the United States Environmental Protection Agency.*

LQGs are allowed to accumulate hazardous wastes without obtaining an RCRA storage permit in two situations. The generator is allowed to accumulate hazardous wastes on site without a permit for 90 days or less, providing that the generator meets certain requirements (40 CFR Section 262.34). These conditions include placing the wastes in specified containers or tanks, managing waste on drip pads (applies to wood-treating facilities), providing documentation of procedures to remove the waste at least once every 90 days, marking the start date for accumulation on each container, and complying with emergency preparedness and response requirements.

A generator may accumulate up to 55 gallons of hazardous waste or 1 quart of acutely hazardous waste in containers at a place near the point of generation without obtaining a permit, providing that the containers are marked with the words HAZARDOUS WASTE or other words identifying the contents of the container and that the generator complies with regulations governing the condition (40 CFR Section 265.171), compatibility (40 CFR Section 265.172), and management (40 CFR Section 265.171) of containers. A generator who accumulates haz-

FIGURE 16–3.   Uniform Hazardous Waste Manifest:

Please print or type.   *(Form designed for use on elite (12-pitch) typewriter.)*   *Form Approved OMB No. 2050-0039. Expires 9-30-91*

UNIFORM HAZARDOUS WASTE MANIFEST	1. Generator's US EPA ID No.		Manifest Document No.	2. Page 1 of	Information in the shaded areas is not required by Federal law.

GENERATOR

3. Generator's Name and Mailing Address

A. State Manifest Document Number

B. State Generator's ID

4. Generator's Phone

5. Transporter 1 Company Name	6.	US EPA ID Number

C. State Transporter's ID

D. Transporter's Phone

7. Transporter 2 Company Name	8.	US EPA ID Number

E. State Transporter's ID

F. Transporter's Phone

9. Designated Facility Name and Site Address	10.	US EPA ID Number

G. State Facility's ID

H. Facility's Phone

11. US DOT Description *(Including Proper Shipping Name, Hazard Class, and ID Number)*	12. Containers		13. Total Quantity	14. Unit Wt/Vol	I. Waste No.
	No.	Type			
a.					
b.					
c.					
d.					

J. Additional Descriptions for Materials Listed Above

K. Handling Codes for Wastes Listed Above

15. Special Handing Instructions and Additional Information

16. GENERATOR'S CERTIFICATION: I hereby declare that the contents of this consignment are fully and accurately described above by proper shipping name and are classified, packed, marked, and labeled, and are in all respects in proper condition for transport by highway according to applicable international and national government regulations

If I am a large quantity generator, I certify that I have a program in place to reduce the volume and toxicity of waste generated to the degree I have determined to be economically practicable and that I have selected the practicable method of treatment, storage, or disposal currently available to me which minimizes the present and future threat to human health and the environment; OR, if I am a small quantity generator, I have made a good faith effort to minimize my waste generation and select the best waste management method that is available to me and that I can afford.

Printed/Typed Name	Signature	Month	Day	Year

TRANSPORTER

17. Transporter 1 Acknowledgment of Receipt of Materials

Printed/Typed Name	Signature	Month	Day	Year

18. Transporter 2 Acknowledgment of Receipt of Materials

Printed/Typed Name	Signature	Month	Day	Year

FACILITY

19. Discrepancy Indication Space

20. Facility Owner or Operator. Certification of receipt of hazardous materials covered by this manifest except as noted in Item 19.

Printed/Typed Name	Signature	Month	Day	Year

EPA Form 8700-22 (Rev. 9-88) Previous edition is obsolete.

FIGURE 16–3.   *(Continued)*

Please print or type   *(Form designed for use on elite (12 pitch) typewriter.)*   Form approved QMB No. 2050 0039 Expires 9-30-91

UNIFORM HAZARDOUS WASTE MANIFEST (Continuation Sheet)	21. Generator's US EPA ID No.	Manifest Document No.	22. Page	Information in the shaded areas is not required by Federal law

23. Generator's Name	L. State Manifest Document Number
	M. State Generator's ID

24. Transporter _____ Company Name	25. US EPA ID Number	N. State Transporter's ID
		O. Transporter's Phone
26. Transporter _____ Company Name	27. US EPA ID Number	P. State Transporter's ID
		Q. Transporter's Phone

28. US DOT Description *(Including Proper Shipping Name, Hazard Class, and ID Number)*	29 Containers No	Type	30 Total Quantity	31 Unit Wt/Vol	R Waste No.
a.					
b.					
c.					
d.					
e.					
f.					
g.					
h.					
i.					

G E N E R A T O R

S. Additional Descriptions for Materials Listed Above	T. Handling Codes for Wastes Listed Above

32. Special Handling Instructions and Additional Information

T R A N S P O R T E R

33. Transporter _____ Acknowledgment of Receipt of Materials		Date
Printed/Typed Name	Signature	Month Day Year
34. Transporter _____ Acknowledgment of Receipt of Materials		Date
Printed/Typed Name	Signature	Month Day Year

F A C I L I T Y

35. Discrepancy Indication Space

EPA Form 8700-22A (Rev. 9-88) Previous edition is obsolete.

ardous wastes for more than 90 days is an operator of a storage facility and must apply for a permit.

Generators must manage all containers according to Subpart I of Part 265. Each container of hazardous waste stored on site must be clearly marked with the words HAZARDOUS WASTE and with the date that the generator began to collect waste in that container. The generator must ensure that containers are in good condition, compatible with the wastes stored in them, stored closed, handled properly, and inspected to ensure that they remain in good condition. Generators who manage tanks must comply with all of Subpart J (except Sections 265.197(c) and 265.200). (SQGs who manage tanks must comply with requirements in Section 265.201 only.)

Generators must comply with the preparedness and prevention requirements in Subpart C of Part 265 and must comply with the contingency plan and emergency procedures contained in Subpart D of Part 265. Meeting these requirements means that generators must develop a contingency plan designed to minimize hazards to human health or the environment from fires, explosions, or other unexpected hazards related to the hazardous waste or constituents at the facility. An existing Spill Prevention, Control and Countermeasures (SPCC) Plan can be used, providing that it is amended to incorporate hazardous waste management provisions. A copy of the contingency plan must be maintained at the facility and submitted to all local emergency response teams. When an emergency occurs, the emergency coordinator must activate the plan, report the events, and clean up the hazards and contaminated materials.

The generator must provide emergency equipment (phone, alarm, fire extinguisher, spill control equipment) and must test and maintain this equipment, provide adequate aisle space, make appropriate arrangements with authorities in the event of emergencies, and provide training for personnel according to 40 CFR Section 265.16 to ensure their safety in the event of an emergency.

A generator (except SQG) who ships any hazardous waste off site to a TSD facility must prepare and submit to the EPA a biennial report that is filed on March 1 of even-numbered years for the preceding calendar year. This report is filed on Form 8700-13A, and it provides a record of all the transporters and TSDFs who handled the waste and information about the type and quantity of wastes managed. In addition, the reports must include information about the waste minimization efforts undertaken to reduce the toxicity and volume of hazardous wastes and the results actually achieved in comparison with previous years (see Chapter 7 for information about waste minimization and pollution prevention). The biennial report must be maintained for at least 3 years. Some states require annual reports to be filed. In practice, generators typically maintain all reports and supporting records for periods longer than required to minimize their liability in the event of problems at later times.

SQG AND CESQG RESPONSIBILITIES

SQGs must meet substantially the same requirements as LQGs, but there are some differences. SQGs may accumulate up to 6,000 kg (13,200 lb) of hazardous waste on site for up to 180 days without a permit. The allowable storage time increases to 270 days if the waste will be shipped over 200 miles. The certification statement for waste minimization is modified for the SQGs.

An exception to manifest requirements applies to SQGs if the waste is reclaimed under a contractual agreement that specifies the type of waste and frequency of shipments and if the vehicle used to transport the waste and return the regenerated material is owned by the reclaimer. The generator must maintain a copy of the agreement for at least 3 years after the termination of the agreement [40 CFR Section 262.20(e)].

CESQGs must meet certain minimum standards that include determining whether or not their wastes are hazardous wastes, accumulating not more than 1,000 kg of hazardous waste at any time, and ensuring that hazardous wastes are treated on site, disposed of on site, or sent to an allowable facility. When determining the quantity of hazardous waste generated, the CESQG does not need to include hazardous waste under the following conditions: 1) when it is moved from on-site storage; 2) when it is produced by on-site treatment (including reclamation) of hazardous waste, so long as the hazardous waste that is treated was counted once; or 3) when materials are generated, reclaimed, and subsequently reused on site, so long as the materials have been counted once (40 CFR Section 261.5).

An SQG that does not receive the return copy of the manifest with the handwritten signature of the owner/operator of the designated facility within 60 days of the date the waste was accepted by the initial transporter must submit a copy of the manifest with an indication that the SQG has not received confirmation of delivery to the regional EPA administrator.

TRANSPORTERS

A transporter is any person engaged in the off-site movement of hazardous waste by air, rail, highway, or water (40 CFR Section 260.10). Any off-site transport of hazardous wastes is covered under RCRA—even if the transport is only a short distance along a right-of-way. On-site transport of hazardous wastes by generators or operators of TSDFs is not covered under RCRA. Transporter requirements are given in 40 CFR Sections 263.20, 263.21, and 264.22(a).

Transporters must obtain an EPA identification number before transporting hazardous wastes. The EPA requirements for physically managing the wastes mirror the DOT requirements for the transport of hazardous materials, and transporters are subject to enforcement by both the EPA and DOT.

Transporters are allowed to accept only those wastes that are accompanied by a manifest (see Figure 16–3), signed, and dated by the generator [40 CFR Section 263.20(a)]. Before the waste can be transported, the transporter must sign and date the manifest to acknowledge the receipt of the materials and must note any discrepancies. One copy of the executed manifest must be left with the generator before the transporter leaves the generator's property. The manifest must accompany the waste at all times, and when the transporter delivers the waste to another transporter or to the TSD facility, additional chain-of-custody requirements apply. The person receiving the waste must date and sign the manifest and must note any discrepancies. The transporter retains one copy of the signed manifest and gives the remaining copies to the person receiving the wastes. The transporter must maintain the executed copy of the manifest for a period of 3 years, but the holding period is typically longer because of liability concerns. If the wastes cannot be delivered, the transporter is responsible for contacting the generator for instructions and for revising the manifest as instructed.

Transporters are permitted to hold a properly contained hazardous waste for up to 10 days during the normal course of transportation at a transfer facility without being required to obtain a RCRA storage permit. Additional requirements apply to transporters who import hazardous wastes, mix hazardous wastes, or allow hazardous wastes to accumulate beyond the transfer allowance.

TSD FACILITIES

Because the regulatory standards and best management practices that apply to TSDFs are too numerous to detail, this section provides a cursory review of general requirements. A *facility* is defined as all contiguous land and structures, appurtenances, and improvements on the land used for treating, storing, or disposal operations units (that is, landfills, surface impoundments, waste piles, or combinations of them) (40 CFR Section 260.10). A *treatment* is broadly defined to include any method, technique, or process, including neutralization, designed to change the physical, chemical, or biological character or composition of any hazardous waste so as to 1) neutralize the waste; 2) recover energy or material resources from the waste; 3) make it nonhazardous or less hazardous; 4) make it safer to transport, store, or dispose of; or 5) make the waste amenable for recovery, amendable for storage, or reduced in volume (40 CFR Section 260.10). A *disposal facility* is defined as a facility or a part of a facility at which hazardous waste is intentionally placed into or on any land or water, and at which waste will remain after closure (40 CFR Section 260.10). Some facilities and activities excluded from regulation are:

1. The disposal of hazardous waste by means of ocean disposal that is regulated by a permit issued under the Marine Protection, Research and Sanctuaries Act (except as provided in an RCRA permit-by-rule).

2. The disposal of hazardous waste by underground injection that is regulated by a permit issued under the Safe Drinking Water Act (except as provided in an RCRA permit-by-rule).

3. The treatment, storage, or disposal of hazardous waste by a publicly owned treatment works (except as provided in an RCRA permit-by-rule).

4. The treatment, storage, or disposal of hazardous waste in states with approved programs, which are therefore subject to regulation under the state program.

5. Facilities authorized by a state to manage industrial or municipal waste, provided that any hazardous wastes received are excluded under 40 CFR Section 261.5.

6. Facilities that recycle or recover hazardous waste, except as provided in 40 CFR Part 266.

7. Temporary on-site accumulation of hazardous wastes by generators pursuant to 40 CFR Section 262.34.

8. Farmers who dispose of their own hazardous wastes in compliance with 40 CFR Section 262.34.

9. Totally enclosed treatment facilities as defined in 40 CFR Section 260.10.

10. Elementary neutralization units and wastewater treatment units as defined in 40 CFR Section 260.10.

11. Actions taken in immediate response to a spill.

12. Transporters storing waste for 10 days or less in approved containers at a transfer facility.

13. The addition of absorbent material to hazardous waste in a container to reduce the amount of free liquid in the container, providing that the absorbent material is added when the waste is first placed into the container.

The facilities that are regulated can be categorized as interim status facilities and permitted facilities. An *interim status facility* is one that is operating without a final RCRA permit until one can be issued. In 1984, the HSWA specified timetables for the issuance of final permits to all interim status facilities. Interim status was awarded to those facilities that:

- existed on November 19, 1980 or on the effective date of statutory or regulatory changes that make the facility subject to an RCRA permit,
- notified EPA pursuant to RCRA Section 3010(a) of its hazardous waste management activities, and
- filed a preliminary permit application (Part A Permit, requirements found in 40 CFR Part 265).

All other TSDFs must obtain an individual RCRA permit (Part B Permit, requirements found in 40 CFR Part 264) before the facility can be built. The topic areas covered in Parts 264 and 265 include administrative and nontechnical requirements and technical requirements (see Figure 16–4).

Every facility owner/operator must have an EPA identification number. Before a hazardous waste can be managed, the owner/operator must obtain a detailed chemical and physical analysis of a representative sample of the waste (40 CFR Section 265.13). A written waste analysis plan must be developed that specifies the parameters to be analyzed, test methods, sampling methods, frequency of analysis, and additional items. Off-site facilities must have procedures to verify that the waste matches the identity of the waste on the manifest. Unless exempt, the facility must have adequate security (40 CFR Section 265.14) that includes a 24-hour surveillance system or an artificial or natural barrier that completely surrounds the active portion of the facility, and the facility must have a way to control entry.

The owner/operator is responsible for conducting general inspections of the facility (40 CFR Section 265.15). This requirement includes inspections for malfunctions and deteriorations, operator errors, and discharges that may lead to any release of hazards. The owner/operator must also develop and follow a written schedule for inspecting all monitoring equipment, safety and emergency equipment, security devices, and operating and structural equipment. The sched-

FIGURE 16–4.   Interim Status Facility and Permitted Facility Regulatory Requirements:

The standards in 40 CFR Parts 264 and 265 include administrative and nontechnical requirements, and technical requirements:

Subpart A	Who is regulated
Subpart B	General facility standards
Subpart C	Preparedness and prevention
Subpart D	Contingency plan and emergency procedures
Subpart E	Manifest system, record keeping, and reporting
Subpart F	Groundwater monitoring requirements
Subpart G	Closure and postclosure requirements
Subpart H	Financial assurance

Technical requirements include general standards applying to several types of facilities and specific standards applying to each waste management method. Specific standards of 40 CFR Part 265 are given in Subparts I–R:

Subpart I	Containers
Subpart J	Tanks
Subpart K	Surface impoundments
Subpart L	Waste piles
Subpart M	Land treatment units
Subpart N	Landfills
Subpart O	Incinerators
Subpart P	Other thermal treatment units
Subpart Q	Chemical, physical, and biological treatment units
Subpart R	Underground injection wells.

Specific standards for Part B permits are given in subparts I–O and X of 40 CFR Part 264:

Subpart I	Containers
Subpart J	Tanks
Subpart K	Surface impoundments
Subpart L	Waste piles
Subpart M	Land treatment units
Subpart N	Landfills
Subpart O	Incinerators
Subpart X	Miscellaneous units

ule must be kept at the facility, identify the types of problems to be looked for during the inspection, and include the frequency of inspection for each item. Observed problems must be corrected. The inspections must be recorded in an inspection log, and the records must be kept for at least 3 years.

Personnel training provisions (40 CFR Section 265.16) require the facility to provide training to employees. The training program must be directed by a person trained in hazardous waste management procedures and must include instruction in hazardous waste management procedures relevant to each employee's position. The training must, at a minimum, ensure that personnel are able to respond effectively in an emergency. The training program must be completed within 6 months after the effective date of employment or assignment to the facility or to a new position at the facility. Annual refresher training is required. The facility owner/operator must maintain personnel documents on the job title for each employee and the name of the person holding the position, a written job description for each position, a written training program, and documentation of training, including training records on each employee until closure. Training records of former employees must be kept for at least 3 years from the date the employee last worked at the facility.

The requirements for Subpart C, *Preparedness and Prevention,* and Subpart D, *Contingency Plan and Emergency Procedures,* are reviewed in the above section, *LQG Responsibilities.*

The manifest system requires the owner/operator of a facility to sign and date each copy of the manifest that accompanies the hazardous waste shipment and to note any discrepancies. The transporter must be given one copy and, within 30 days, the owner/operator must send a signed copy of the manifest to the generator. A copy of the manifest should be retained at the facility for at least 3 years from the date of delivery. A significant manifest discrepancy is a difference >10% in weight for bulk waste between the manifested amount and the amount received, and any variation in piece count for batch waste. At the time the discrepancy is discovered, the owner/operator must attempt to reconcile the discrepancy with the waste generator or transporter. If it cannot be resolved within 15 days, the owner/operator must immediately submit a letter describing the discrepancy and attempts to reconcile it, and a copy of the manifest to the EPA's Regional Administrator.

A written operating record must be kept at each facility (40 CFR Section 265.73). The operating record must include a description and quantity of each hazardous waste received and the method(s) and date(s) of its treatment, storage, or disposal. The location of the waste within the facility must be noted, along with cross-references to manifest document numbers. Additional information includes records and results of waste analyses and trial tests; summary reports and details of all incidents that require implementing the contingency plan; records and results of inspections; monitoring and analytical data; all closure cost estimates and postclosure cost estimates; and notices and certification for land disposal ban. All records must be available upon request, and a copy of records of waste disposal locations and quantities must be submitted upon closure of the facility.

Additional record keeping includes the submission of the biennial report (40 CFR Section 265.75) and the unmanifested waste report (40 CFR Section 265.76), which must be submitted within 15 days of receiving a shipment of waste that is not accompanied by a manifest. The owner/operator must also report to the Regional Administrator releases, fires, and explosions; groundwater contamination and monitoring data; facility closure; and reports required under 40 CFR Parts 265 AA and BB.

THE HAZARDOUS WASTE AUDIT PROCESS

OVERVIEW

During the hazardous waste management audit, the audit team will review the facility's policies and operational procedures and will compare them to actual practices to determine areas of conformance and nonconformance. The audit should give a complete picture of the waste streams, amounts generated, management techniques, waste minimization program, and disposal methods. The audit should include waste generation and accumulation areas, waste shipping and receiving areas, all on-site TSDFs, and any remaining plant areas. If possible, visits to off-site properties or transporters should also be arranged to ensure that these contractors are providing high-quality services that comply with existing regulations. If site visits cannot be arranged, documentation should be requested and reviewed.

The audit team should be knowledgeable about the applicable regulations and good management practices for preventing releases, response procedures in the event that leaks occur, and other areas shown in Figure 16–4. Prior to the audit, the auditor will review the federal and state hazardous waste regulations and EPA background documents on specific industries and waste streams. The auditor will want to obtain and review copies of hazardous waste permit applications filed by the facility, sample manifests, and any prior evaluations. Specific aspects of the facility and its RCRA program that will be of interest to the auditor are:

- lists of responsible persons involved in the hazardous waste program
- flowchart or other illustration showing processes used and design features of present and planned units and processes at the facility
- for TSDFs, a list of the wastes that are treated, stored, or disposed of by type of waste and related management practices
- for generators, a list of the wastes generated, including their origin
- previous inspection or audit reports
- biennial, annual, and other reports submitted by the facility to the state or region, including the most recent monitoring reports, where applicable
- previous EPA studies, consultants' reports, and laboratory reports
- a detailed map or plot plan showing the facility layout and location of waste management units
- facility's RCRA Notification Form
- facility's RCRA Part A Permit application (for TSDFs)
- facility's RCRA Part B Permit application (for certain TSDFs)
- facility's final RCRA permit (for certain TSDFs)

The RCRA hazardous waste permit may be issued by the state or the EPA, or in some instances, the facility may have a combination of state and the EPA permits, depending on the state's authority to administer the RCRA program. Even though a permit does not reflect new regulations promulgated since the permit's original issuance, the facility still needs to comply with these requirements. The EPA has determined that many regulations are self-implementing and do not need to be incorporated into permits with specific language.

The audit will begin with a briefing of site management and the environmental staff. The audit team will obtain information about safety requirements and will participate in an orientation tour. During the orientation tour, the audit team will become better acquainted with the physical layout of the facility and its operations. The orientation tour will be conducted with the facility representative, typically, the environmental coordinator. During the tour, the audit team will become oriented to the facility, with a focus on the areas involving hazardous wastes. The tour should include the entire facility and its grounds.

After the orientation tour, the audit team will review records and management systems, interview site personnel, and conduct an in-depth inspection of the facility. Previously developed checklists will be used by the audit team to conduct the in-depth evaluation of the facility. At the end of each day, the audit team members will brief one another and discuss findings.

At the end of the audit, the audit team will debrief the management of the facility. Final activities will include drafting the audit report, responding to review comments, and submitting the final audit report. Facility or site management will follow up on audit findings, initiate corrective actions, determine the effectiveness of the actions, and report back to the audit team.

PREPARING THE INVENTORY

The inventory should include process chemicals and materials, scrap and recycled materials, and cleaners. The input inventory should also include the containers used to receive and store the chemicals and materials because the containers could be considered to be hazardous unless they are properly cleaned. All of this information can help identify the waste streams and indicate opportunities for reducing the volume or toxicity of wastes. Sources of information for the input inventory include the environmental coordinator, health and safety coordinator, purchasing agent, and other key personnel, such as supervisors and staff.

WASTE IDENTIFICATION AND DETERMINATION

The auditor will prepare a detailed map or plot plan showing the facility layout and location of waste management units and groundwater monitoring wells, if any. The auditor will become familiar with the facility, its processes, and any waste management units (permitted and unpermitted). The status of each permitted unit (that is, under construction, operational, under corrective action, being closed, or closed) will be determined. The auditor will identify the types and characteristics of wastes being stored, treated, or disposed of at the facility. The inventory should include all waste streams, large and small. Attention should be given to waste streams generated as a result of cleaning equipment. Using all of this information, the auditor will prepare an inventory of wastes generated at the facility (see Figure 16–5 for an example form).

The inventory will be used to determine whether all hazardous waste streams have been identified and whether the facility's waste determinations are correct. The auditor will also evaluate the facility's procedures for making determinations on new waste streams. Although RCRA regulations allow facilities to make self-determinations of hazard based on knowledge of the processes used by the facility, analytical data are preferred in most cases because of the liability potential if the determinations are incorrect.

Many facilities have identified and classified their waste streams and have computerized databases available for review. Even in these cases, the auditor will

FIGURE 16–5.   Example Inventory Form for Identifying and Classifying
Hazardous Wastes:

ABC Company
Hazardous Waste Inventory Form

Plant: _____   Auditor: _____

Process Area: _____   Date: _____

Facility Contact: _____

Waste Stream No.: _____   State (solid, liquid, gas): _____

Description: _____

_____

Location:_____

Type of Waste: _____ hazardous _____ nonhazardous _____ undetermined

U.S. EPA/State Hazardous Waste ID No.: _____

Determination: _____ based on testing _____ based on knowledge

Lab Data Available: _____ yes _____ no

Certified Lab: _____ yes _____ no   ID No. _____

Date Last Analysis: _____   Data Attached: _____ yes _____ no

Waste Generation Rate: _____ T/yr _____ gal/yr

Accumulation Area: _____

Type of Containers: _____

_____

Condition of Containers: _____

_____

Disposal Procedure:

 Current: _____

 Past: _____

Is waste subject to land disposal restrictions under 40 CFR Section 268? _____ yes _____ no

Is waste subject to exclusions under 40 CFR Section 261.4? _____ yes _____ no

Have manifests been used for off-site shipment? _____ yes _____ no

Comments: _____

_____

_____

_____

Rev. 2   05/01/96

---

still evaluate this information to ensure that all waste streams have been classified, that classifications are correct, and that analytical data are sound and available.

FACILITY TOUR

During the facility inspection, the auditor will examine the entire facility and will conduct interviews to determine whether the facility complies with required regulations and with its own management policies and practices. The focus of the audit depends on the type of facility. For instance, during the inspection of generators, the auditor will look for hazardous wastes that may not currently be reported as hazardous, noncomplying procedures or management practices, steps in the management process during which wastes may be mishandled or misidentified, opportunities for spills or releases, and any deviations from the facility's stated normal operating procedures.

The entire facility and its property should be examined to determine whether dump sites exist or whether there is evidence of soil or water contamination. Unmanaged scrap or salvage materials should be documented. These must be properly managed to minimize potential soil and water contamination. Figure 16–6 gives examples of common RCRA compliance problems. Figure 16–7 at the end of this chapter is an EPA facility checklist for evaluating the general requirements that apply to facilities.

DOCUMENTATION AND RECORD-KEEPING REVIEW

The auditor will review the facility's operational procedures; procedures for training, maintenance, testing and analysis, inspections, record keeping; and procedures for reporting and responding to leaks and emergency conditions. There are numerous documentation and record-keeping requirements under RCRA that differ for generators, transporters, and TSDFs. Figure 16–8 identifies selected record keeping that is required by RCRA regulations. The auditor will determine whether required records and plans are present, organized, accurate, kept up to date, controlled, and maintained for required periods of time.

WASTE GENERATION AND ACCUMULATION AREAS

The auditor will examine all locations used by the facility to manage the wastes identified in the inventory. The locations of all bulk waste containers, drum storage areas, waste tanks, and other waste-related areas will be identified on the facility plot plan. The waste containers and locations will be examined to determine whether they accurately reflect the inventory and stated management procedures. The auditor will look for evidence of spills, leaks or improper storage, and for the potential for these events. The auditor will examine waste mixing areas to determine whether mixing practices, materials segregation, and containment are appropriate.

Typically, an auditor will not perform sampling of the materials to verify the inventory, but during some types of audits, the auditor may request addi-

FIGURE 16–6.   Common RCRA Compliance Problems:

Generators

- Complete lack of or improper waste determination
- Failure to notify EPA of waste generation activity and to obtain an EPA ID number
- Failure to manage waste containers properly (for example, labeling, marking accumulation start dates, storing containers closed, providing adequate aisle space, inspections)
- Storage of waste longer than allowable time frames
- Accepting waste from off site (for example, accepting waste from a parent or sister company)
- Failure to provide annual personnel training
- Failure to update contingency plan when necessary
- Failure to provide tanks with required equipment
- Failure to maintain and operate the facility to minimize the possibility of a release (that is, sloppy housekeeping as it relates to waste management practices)

Transporters

- Failure to notify EPA of the waste transportation activity and to obtain an EPA ID number
- Storage of waste for more than 10 days
- Failure to unload entire amount of waste, resulting in formation of "heels"

TSDFs

- Failure to provide and/or maintain emergency equipment
- Failure to document inspections adequately and/or to remedy problems identified during those inspections
- Failure to maintain adequate personnel training records
- Sloppy housekeeping, as described for generators
- Failure to amend contingency plan when necessary
- Failure to manage waste containers or tanks properly (for example, labeling, marking, and so forth as listed above for generators)
- Failure to maintain adequate financial assurance to closure
- Failure to maintain groundwater monitoring system properly
- Failure to conduct groundwater sampling properly
- Failure to provide tanks with required equipment

FIGURE 16–8. Selected Record-Keeping Requirements under RCRA:

Generators (40 CFR citation)

262.34	Job titles and personnel records, agreements with local authorities, and contingency plan
262.40, .42	Manifests, exception reports, waste analyses and test results (or other bases for determining the hazardous nature of a waste and its classification), shipping papers for bulk shipments by rail or water
262.52–.55	Exporter notification, manifest, exception reports, annual reports
262.41	Biennial report
268.7	Land disposal notification and certification

Transporters

263.20–.22	Manifests, shipping papers for bulk shipments by rail or water, and manifests for foreign shipments
279.46	Tracking records for shipments of used oil

Interim Status Facilities

General facility standards, including the following:

265.13	Waste analysis plan
265.15	Inspection schedule
265.16	Job titles and personnel records
265.51–.53	Contingency plan
265.71–.77	Manifest system (records of manifests)
265.73	Operating record
265.75	Biennial report
265.76	Unmanifested waste report
265.77	Additional reports for releases, fires, explosions, and so forth
265.92	Groundwater monitoring plan
265.94	Groundwater monitoring record
265.112	Closure plan
265.118	Postclosure plan
265.119	Postclosure notice
265.120	Certification of completion of postclosure
265.142	Cost estimate for closure
265.147	Proof of liability coverage
268.7	Land disposal notification and certification
268.19(d)	Special notification for characteristic wastes

FIGURE 16–8. *(Continued)*

Facility-specific standards, including the following:

265.193	Annual assessment for tanks
265.195	Tank inspection records
265.196	Certification of major repairs to tanks
265.226	Surface impoundment inspection records
265.279	Land treatment, requirements for operating record
265.309	Landfills, requirements for operating record, contents, and organizations of cells
265.347	Incineration monitoring and inspection records
265.377	Thermal treatment monitoring and inspection records
265.402	Chemical, physical, and biological treatment waste analyses records and trial tests
265.403	Chemical, physical, and biological treatment waste inspection records
265.440	Drip pad contingency plan
265.441	Drip pad assessment and plan for upgrade, repair, or modification
265.444	Drip pad inspection report
265.1035	Air emissions from process vents testing and operational records
265.1064	Air emissions from equipment leaks testing and operational records

Permitted Facilities

264	Requirements are similar to Part 265 for general facility standards; preparedness and prevention; contingency plan and emergency procedures; manifest system, record keeping and reporting; and financial requirements. Some additional requirements apply for specific facilities.

Specific Hazardous Wastes and Specific Types of Facilities

266.42	Used oil analysis
266.44	Used oil fuel analysis
266.100	Boiler and industrial furnace exemption for metals recovery units
266.103	Boiler and industrial furnace operating record
266.108	Small quantity boiler and industrial furnace burner exemption waste quantity records
266.111	Direct transfer equipment inspection records for boilers and industrial furnaces

FIGURE 16–8.   (*Continued*)

266.112   Boiler and industrial furnace waste residue data

Used Oil Processors and Rerefiners

279.55   Used oil analysis plan
279.56   Tracking records
279.57   Operating records

Off-Specification Used Oil Burners

279.65   Tracking records
279.66   Off-specification used oil certification

Used Oil Fuel Marketers

279.72   Analysis of used oil fuel
279.74   Tracking records
279.75   Off-specification used oil certification

tional sampling data. Empty drum accumulation areas will be examined to determine whether drums comply with EPA's empty container rule (40 CFR Section 261.7). Compliance with this rule is important because residues in an "empty container" are exempted from hazardous waste regulation under RCRA. Under RCRA, the following definitions apply to empty containers.

- A container or an inner liner is considered empty if:
  — all wastes have been removed using commonly employed practices, and
  — no more than 1 inch of residues remains on the container bottom or liner, or no more than 3% by weight of the total capacity of the container remains in the container or inner liner if the container is greater than 110 gallons in size.
- A compressed gas container is empty when the pressure approaches atmospheric.
- A container or liner that held an acute hazardous waste is empty if:
  — it has been triple-rinsed using a solvent or cleaned by a method that is equivalent, or
  — the inner liner was prevented from coming into contact with the commercial chemical product, and it has been removed.

The auditor will examine containers for proper markings, labels, and condition. Some of the warning signs for a release or potential release and hazardous situations from various units include:

- bungs and lids that are not secure
- rusty or deformed containers
- poorly stacked drums

- puddles around units, stained soil, stained concrete
- leaking valves on tanks
- strong odors from encapsulated units
- dead vegetation
- erosion
- dusty conditions around piles
- any other indication that the unit is not containing the waste or is poorly managed, such as lack of segregation

Discrepancies between field conditions and regulatory requirements and policies will be noted. Figure 16–9 at the end of this chapter shows an EPA form for evaluating containers.

GENERATORS

The auditor will examine the record-keeping documentation for generators to validate the determinations, conformance with requirements, and completeness of record keeping. Containers will be examined for proper markings (DOT label, manifest number, and generator's name and address) and for signs of damage, leaks, and corrosion. Motor vehicles, freight containers, and rail cars will be checked for appropriate placards and markings. Containers and tanks will be checked for "hazardous waste" marking, accumulation periods, and accumulation amounts. Waste minimization practices will be reviewed. Figure 16–10, at the end of this chapter, shows an EPA form for evaluating generators, and Figure 16–11 at the end of this chapter shows an RCRA waste minimization checklist.

TRANSPORTERS

The auditor will examine the record-keeping documentation for transporters to validate conformance with manifest and other requirements, and for completeness of record keeping. Transport vehicles will be examined to check for signs of leaking containers, appropriate placards and markings, appropriate locations of manifests and/or shipping papers, and vehicle condition. Figure 16–12, at the end of this chapter, shows an EPA form for evaluating transporters. This form does not include all of the DOT requirements.

TSD FACILITIES

The auditor will examine each TSD facility to ensure that each meets the regulatory and management requirements specified for the facility. The auditor will examine site security, communication equipment, and protocols. The auditor will check all loading and unloading areas and drum accumulation areas for proper containment, labeling of containers, stacking, and segregation of containers. Recycling and reuse operations that are not permitted should be evaluated to ensure that storage and treatment permits are not required.

The auditor will examine tanks and surface impoundments for integrity, availability, and construction of containment areas; presence of alarms and waste cutoff systems; and evidence of leaks and spills. Freeboard will also be checked for surface impoundments.

Waste piles will be examined to be sure they are properly segregated, compatible with the waste pile base, protected from the wind, and adequately contained. Additionally, they must have adequate evaluation and disposal methods for runoff, and replacement units and lateral expansions must have the required liners and a leachate collection system.

Land treatment areas will be examined for the adequacy of the runoff management system, adequacy of the evaluation and disposal method for runoff, type of cover crop, and compatibility of wastes. The auditor will examine landfills for technological requirements, adequacy of the cell/container map, adequacy of the evaluation and disposal method for runoff, integrity of the final cover, and maintenance of the groundwater monitoring system.

Incinerators will be checked for waste types, feed rates, operating conditions, ash handling, and ultimate disposal. Thermal treatment units will be checked for waste types, operating conditions, residue handling, and ultimate disposal.

The auditor will check a chemical, physical, or biological treatment unit for waste type being treated, operating conditions, residue handling and ultimate disposal, treatment effectiveness records, and whether emissions of gases, vapors, mists, or odors are present.

Air emissions from process vents and equipment leaks will be evaluated in terms of required record keeping. Additionally, the auditor will look for visible emissions from flares, leaks from pumps, and barrier fluid systems with sensors on pumps exempted from monthly monitoring, and will check equipment that has previously leaked and was repaired.

SUMMARY

The regulatory framework for managing hazardous waste management is complex and far-reaching. The regulated facilities are responsible for classifying themselves and for correctly identifying their waste streams. These two determinations affect the remaining management programs required of the facilities. Because of the potential liability associated with hazardous waste disposal, all facilities should maximize prevention of pollution and waste minimization strategies.

FIGURE 16–7.   RCRA General Facility Checklist:

Responses			**Questions to Ask**
YES	**NO**	**N/A**	***General***
_____	_____	_____	1. Does facility have an EPA identification no.? (Sections 264/5.11)
_____	_____	_____	a. If yes, specify EPA ID _____
_____	_____	_____	b. If no, explain _____
_____	_____	_____	_____
_____	_____	_____	2. Has facility received hazardous waste from a foreign source? (Sections 264/5./12)
_____	_____	_____	a. If yes, has it filed a notice with the Regional Administrator?
YES	**NO**	**N/A**	***Waste Analysis***
_____	_____	_____	3. Does facility maintain a copy of the waste analysis plan on site? (Sections 264/5.13)
			a. If yes, does it include:
_____	_____	_____	1) Parameters for which each waste will be analyzed? [Sections 264/5.13(b)(1)]
_____	_____	_____	2) Test methods used to test for these parameters? [Sections 264/5.13(b)(2)]
_____	_____	_____	3) Sampling method used to obtain sample? [Sections 264/5.13(b)(3)]
_____	_____	_____	4) Frequency with which the initial analyses will be reviewed or repeated? [Sections 264/5.13(b)(4)]
_____	_____	_____	5) (For off-site facilities) Waste analyses that generators have agreed to supply? [Sections 264/5.13(b)(5)]
			6) (For off-site facilities) Procedures used to inspect and analyze each movement of hazardous waste, including: [Sections 264/5.13(c)]
_____	_____	_____	a) Procedures to be used to determine the identity of each movement of waste?
_____	_____	_____	b) Sampling method to be used to obtain representative sample of the waste to be identified?
			4. Does the facility provide adequate security through: (Sections 264/5.14)
_____	_____	_____	a. 24-Hour surveillance system (for example, television monitoring or guards)?
			or
_____	_____	_____	b. 1) Artificial or natural confining barrier around facility (for example, fence or fence and cliff)?

FIGURE 16–7. *(Continued)*

YES	NO	N/A	***Waste Analysis***
			[Sections 264/5.14(b)] Describe:

_____

_____

_____

and

_____	_____	_____

2) Means to control entry through entrances (for example, attendant, television monitors, locked entrance, controlled roadway access)? [Sections 264/5.14(b)(2)(ii)]

Describe: _____

_____

_____

YES	NO	N/A	***General Inspection Requirements***

5. Does the owner/operator maintain a written schedule at the facility for inspecting: (Sections 264/5.15)

YES	NO	N/A	
_____	_____	_____	a. Monitoring equipment?
_____	_____	_____	b. Safety and emergency equipment? [Sections 264/5.15(b)]
_____	_____	_____	c. Security devices?
_____	_____	_____	d. Operating and structural equipment?
			e. Types of problems with equipment:
_____	_____	_____	1) Malfunction? [Sections 264/5.15(a)]
_____	_____	_____	2) Operator error?
_____	_____	_____	3) Discharges?
_____	_____	_____	6. Does owner/operator maintain an inspection log? [Sections 264/5.15(d)]
			a. If yes, does it include:
_____	_____		1) Date and time of inspection?
_____	_____		2) Name of inspector?
_____	_____		3) Notation of observations?
_____	_____		4) Date and nature of repairs or remedial action?
_____	_____		b. Have problems been corrected? (Use narrative explanation sheet.) Are there any malfunctions or other deficiencies not corrected? [Sections 264/5.15(c)]

YES	NO	N/A	***Personnel Training***
_____	_____	_____	7. Does the owner/operator maintain personnel training records at the facility? (Sections 246/5.16)
			How long are they kept? _____
_____	_____	_____	a. If yes, do they include:

FIGURE 16–7. (*Continued*)

YES	NO	N/A	*Personnel Training*
——	——	——	1) Job title and written job description of each position? [Sections 264/5.16(d)]
——	——	——	2) Description of type and amount of training?
——	——	——	3) Records of training given to facility personnel?

YES	NO	N/A	*Requirements for Ignitable, Reactive, or Incompatible Waste*
——	——	——	8. Does facility handle ignitable or reactive wastes? (Sections 264/5.17)
——	——	——	a. If yes, is waste separated and confined from sources of ignition or reaction (open flames, smoking, cutting and welding, hot surfaces, frictional heat), sparks (static, electrical, or mechanical), spontaneous ignition (for example, from heat-producing chemical reactions and radiant heat)?
			1) If yes, use narrative explanation sheet to describe separation and confinement procedures.
			2) If no, use narrative explanation sheet to describe sources of ignition or reaction.
——	——	——	b. Are smoking and open flames confined to specifically designated locations?
——	——	——	c. Are "No Smoking" signs posted in hazardous areas?
——	——	——	d. Are precautions documented (Part 264 only)? [Sections 264.17(c)]
——	——	——	9. Are containers leaking or corroding? (Sections 264/5.171)
——	——	——	a. If yes, use narrative explanation sheet to explain.
——	——	——	10. Is there evidence of heat generation from incompatible wastes?

YES	NO	N/A	*Preparedness and Prevention (40 CFR Sections 264/5 Subpart C)*
——	——	——	11. Is there evidence of fire, explosion, or contamination of the environment?
			a. If yes, use narrative explanation sheet to explain.
			12. Is the facility equipped with: (Sections 264/5.32)
——	——	——	a. Internal communication or alarm system?
——	——	——	1) Is it easily accessible in case of emergency? (Sections 264/5.34)
			b. Telephone or two-way radio to all emergency response personnel? [Sections 264/5.32(b)]
——	——	——	c. Portable fire extinguishers, fire control equipment, spill control equipment, and decontamination equipment? [Sections 264/5.32(c)]

FIGURE 16–7. *(Continued)*

YES	NO	N/A	**Preparedness and Prevention (40 CFR Sections 264/5 Subpart C)**
_____	_____	_____	d. Water of adequate volume for hoses, sprinklers, or water spray system? [Sections 264/5.32(d)] 1) Describe source of water: _____ _____
_____	_____	_____	13. Is there sufficient aisle space to allow unobstructed movement of personnel and equipment? (Sections 264/5.35)
_____	_____	_____	14. Has the owner/operator made arrangements with the local authorities to familiarize them with characteristics of the facility? (Layout of facility, properties of hazardous waste handled and associated hazards, places where facility personnel would normally be working, entrances to roads inside facility, possible evacuation routes.) (Sections 264/5.37)
_____	_____	_____	15. In the case that more than one police or fire department might respond, is there a designated primary authority [Sections 264/5.37(a)(2)]
_____	_____	_____	a. If yes, name primary authority: _____
_____	_____	_____	16. Does the owner/operator have phone numbers of and agreements with state emergency response teams, emergency response contractors, and equipment suppliers? [Sections 264/5.37 (a)(3)]
_____	_____	_____	a. Are they readily available to all personnel?
_____	_____	_____	17. Has the owner/operator arranged to familiarize local hospitals with the properties of hazardous waste handled and types of injuries that could result from fires, explosion, or releases at the facility? [Sections 264/5.37(a)(4)]
_____	_____	_____	18. If state or local authorities decline to enter into the arrangements called for under Sections 264/5.37, is this entered in the operating record? [Sections 264/5.37(b)]
YES	**NO**	**N/A**	**Contingency Plan and Emergency Procedures (40 CFR Section 264/5 Subpart D)**
_____	_____	_____	19. Is a contingency plan maintained at the facility? (Sections 264/5.51)
_____	_____	_____	a. If yes, is it a revised SPCC Plan? [Sections 264/5.52(b)]
			b. Does contingency plan include:
_____	_____	_____	1) Arrangements with local emergency response organizations? [Sections 264/5.52(c)]
_____	_____	_____	2) Emergency coordinators names, phone numbers, and addresses? [Sections 264/5.52(d)]

FIGURE 16–7. *(Continued)*

YES	NO	N/A	**Contingency Plan and Emergency Procedures (40 CFR Section 264/5 Subpart D)**
_____	_____	_____	3) List of all emergency equipment at facility and description of equipment? [Sections 264/5.52(e)]
_____	_____	_____	4) Evacuation plan for facility personnel? [Sections 264/5.52(f)]
_____	_____	_____	20. Is there an emergency coordinator on site or on call at all times? (Sections 264/5.55)

YES	NO	N/A	**Manifest System, Record Keeping, and Reporting (40 CFR Section 264/5 Subpart E)**
_____	_____	_____	21. Does facility receive waste from off site? [Sections 264/5.71(a)]
_____	_____	_____	a. If yes, does the owner/operator retain copies of all manifests?
_____	_____	_____	1) Are the manifests signed, dated, and returned to the generator?
_____	_____	_____	2) Is a signed copy given to the transporter?
_____	_____	_____	22. Does the facility receive any waste from a rail or water (bulk shipment) transporter? [Sections 264/5.71(b)]
_____	_____	_____	a. If yes, is it accompanied by a shipping paper?
_____	_____	_____	1) Does the owner/operator sign and date the shipping paper and return a copy to the generator?
_____	_____	_____	2) Is a signed copy given to the transporter?
_____	_____	_____	23. Has the owner/operator received any shipments of waste that were inconsistent with the manifest (manifest discrepancies)? (Sections 264/5.72)
_____	_____	_____	a. If yes, has an attempt been made to reconcile the discrepancy with the generator and transporter?
_____	_____	_____	1) If no, has Regional Administrator been notified?
_____	_____	_____	24. Does the owner/operator keep a written operating record at the facility? [Sections 264/5.73(a)]
			a. If yes, does it include: [Sections 264/5.73(b)]
_____	_____	_____	1) Description and quantity of each hazardous waste received?
_____	_____	_____	2) Methods and dates of treatments, storage, and disposal?
_____	_____	_____	3) Location and quantity of each hazardous waste at each location?
_____	_____	_____	4) Cross-references to manifests/shipping papers?
_____	_____	_____	5) Records and results of waste analyses?
_____	_____	_____	6) Reports of incidents involving implementation of the contingency plan?

FIGURE 16–7.   (Continued)

YES	NO	N/A	**Manifest System, Record Keeping, and Reporting (40 CFR Section 264/5 Subpart E)**
___	___	___	7) Records and results of required inspections?
___	___	___	8) Monitoring or testing analytical data? (Part 264)
___	___	___	9) Closure cost estimates and, for disposal facilities, postclosure cost estimates? (Part 264)
___	___	___	10) Notices of generators as specified? [Section 264.12(b)]
___	___	___	11) Certifications of permittee waste minimization program? [Section 264.73 (b)(9)]
___	___	___	12) Land disposal restriction records required by Sections 268.5, 268.6, 268.7(a), and 268.8, as applicable? [Sections 264.73(b)(10)–(16)]
___	___	___	25. Does the facility submit a biennial report by March 1 every even-numbered year? [Sections 264/55.75]
			a. If yes, do reports contain the following information:
___	___	___	1) EPA ID number? [Sections 264/5.75(a)]
___	___	___	2) Date and year covered by report? [Sections 264/5.75(b)]
___	___	___	3) Description/quantity of hazardous waste? [Sections 264/5.75(d)]
___	___	___	4) Treatment, storage, and disposal methods? [Sections 264/5.75(e)]
___	___	___	5) Monitoring data under Sections 265.94(a)(2) and (b)(2)? [Sections 265.75(f)]
___	___	___	6) Most recent closure and postclosure cost estimates? [Sections 264/5.75(g)]
___	___	___	7) For TSD generators, description of efforts to reduce volume/toxicity of waste generated and actual comparisons with previous year? [Sections 264/5.75(h)]
___	___	___	8) Certification signed by owner/operator? [Sections 264/5.75(j)]
___	___	___	26. Has the facility received any waste (that does not come under the small generator exclusion) not accompanied by a manifest? (Sections 264/5.76)
___	___	___	a. If yes, has an unmanifested waste report been submitted to the Regional Administrator?
___	___	___	27. Does the facility submit to the Regional Administrator reports on releases, fires, and explosions; contamination and monitoring data; and facility closure? (Sections 264/5.77)

Source: Adapted from U.S. Environmental Protection Agency, Solid Waste and Emergency Response, *RCRA Inspection Manual 1993 Edition*, OSWER Directive 9938.02B (Washington, D.C.: U.S. Environmental Protection Agency, 1993).

FIGURE 16–9.   RCRA Containers Checklist:

Responses			Questions to Ask
YES	**NO**	**N/A**	*Use and Management (Sections 264/5.171)*
_____	_____	_____	1. Are containers in good condition?
YES	**NO**	**N/A**	*Compatibility of Waste with Containers (Sections 264/5.172)*
_____	_____	_____	2. Is container made of a material that will not react with the waste that it stores?
YES	**NO**	**N/A**	*Management of Containers (Sections 264/5.173)*
_____	_____	_____	3. Are containers always closed while holding hazardous waste?
_____	_____	_____	4. Are containers opened, handled, or stored in a manner that will prevent ruptures and leaks?
YES	**NO**	**N/A**	*Inspections (Sections 264/5.175)*
_____	_____	_____	5. Does owner/operator inspect containers at least weekly for leaks and deterioration?
YES	**NO**	**N/A**	*Containment (Section 264.175)*
_____	_____	_____	6. Do container storage areas have a containment system?
YES	**NO**	**N/A**	*Ignitable and Reactive Waste (Sections 264/5.176)*
_____	_____	_____	7. Are containers holding ignitable and reactive waste located at least 15 m (50 ft) from facility property lines?
YES	**NO**	**N/A**	*Incompatible Waste (Sections 264/5.177)*
_____	_____	_____	8. Are incompatible wastes or materials placed in the same containers?
_____	_____	_____	9. Are hazardous wastes placed in washed, clean containers when they previously held incompatible waste?
_____	_____	_____	10. Are incompatible hazardous wastes separated from each other by a berm, dike, wall, or other device?
YES	**NO**	**N/A**	*Closure (Section 264.178)*
_____	_____	_____	11. At closure, were all hazardous wastes and associated residues removed from the containment system?

Source: Adapted from U.S. Environmental Protection Agency, Solid Waste and Emergency Response, *RCRA Inspection Manual 1993 Edition*, OSWER Directive 9938.02B (Washington, D.C.: U.S. Environmental Protection Agency, 1993).

FIGURE 16–10.   RCRA Generator Checklist:

Responses			**Questions to Ask**
YES	**NO**	**N/A**	***General***
____	____	____	1. Does generator have EPA ID No.? (Section 262.12) If yes, specify EPA ID No. _____
YES	**NO**	**N/A**	***Manifest***
____	____	____	2. Does generator ship waste off site? (Section 262.20) If no, do not fill out manifest and pretransport sections. If yes, identify primary off-site facility(s). Use narrative explanation sheet.
____	____	____	3. Does generator use manifest? (Section 262.20)
____	____	____	a. If no, is generator a small-quantity generator (generating between 100 and 1,000 kg/month?

Note: SQCs are exempt only if wastes are reclaimed. [See Section 262.20(e).]

____	____	____	b. If yes, does generator indicate this when sending waste to a TSD facility?
			c. If yes, does manifest include the following information? (Part 262 Appendix)
____	____	____	1) Manifest document no.?
____	____	____	2) Generator's name, mailing address, telephone no.?
____	____	____	3) Generator EPA ID no.?
____	____	____	4) Transporter name(s) and EPA ID no.(s)?
____	____	____	5) a) Facility name, address, and EPA ID no.?
____	____	____	b) Alternate facility name, address, and EPA ID no.?
____	____	____	c) Instructions to return to generator if undeliverable?
____	____	____	6) Waste information required by DOE—shipping name, quantity (weight or vol.), containers (type and number)?
____	____	____	7) Emergency information (optional)? (Special handling instructions, telephone number)
____	____	____	8) Is the following certification on each manifest form? *This is to certify that the above-named materials are properly classified, described, packaged, marked, and labeled and are in proper condition for transportation according to the applicable national and international regulations.*
____	____	____	4. Does generator retain copies of manifests? (Section 262.40) If yes, complete a–e. (Section 262.23)
____	____	____	a. 1) Did generator sign and date all manifests? 2) Who signed for generator?

FIGURE 16–10 *(Continued)*

YES	NO	N/A	*Manifest*

Name _____Title _____

| | | | b. Did generator obtain handwritten signature and date of acceptance from initial transporter? |

Who signed and dated for transporter? (Section 262.23)

Name _____ Title _____

			c. Does generator retain one copy of manifest signed by generator and initial transporter? (Section 262.40)
			d. Do returned copies of manifest include facility owner/operator signature and date of acceptance? (Section 262.40)
			e. Does generator retain copies for 3 years? (Section 262.40)

YES	NO	N/A	*Hazardous Waste Determination*

| | | | 5. Does generator generate solid waste(s) listed in Subpart D (List of Hazardous Waste)? |

a. If yes, list wastes and quantities (including EPA Hazardous Waste No.) _____

_____

_____

| | | | 6. Does generator generate solid waste(s) listed in Subpart C that exhibit hazardous characteristics (corrosivity, ignitability, reactivity, EP toxicity)? |

a. If yes, list waste and quantities (include EPA Hazardous Waste No.) _____

_____

_____

b. Does generator determine characteristics by testing or by applying knowledge of processes? _____

| | | | 1) If determined by testing, did generator use test methods in Part 261, Subpart C (or equivalent)? |

2) If equivalent test methods used, attach copy of equivalent methods used.

| | | | 7. Are there any other solid wastes generated by generator? |
| | | | a. If yes, did generator test all wastes to determine whether or not they were hazardous? |

1) If no, list wastes and quantities deemed nonhazardous or processes from which nonhazardous waste was produced (use additional sheets if necessary) _____

_____

_____

FIGURE 16–10.   *(Continued)*

YES	NO	N/A	*Pretransport Requirements*
____	____	____	8. Does generator package waste in accordance with 49 CFR Sections 173, 178, and 179 (DOT requirements)? (Section 262.30)
____	____	____	9. a. Are containers to be shipped leaking or corroding? Use additional sheet to describe containers and condition.
____	____	____	b. Is there evidence of heat generation from incompatible waste in the containers?
____	____	____	10. Does generator follow DOT labeling requirements in accordance with 49 CFR Section 172? (Section 262.31)
____	____	____	11. Does generator mark each package in accordance with 49 CFR Section 172? (Section 262.32)
____	____	____	12. Is each container of 110 gallons or less marked with the following label? (Section 262.32)

Label saying: *HAZARDOUS WASTE—Federal Law Prohibits Improper Disposal. If found, contact the nearest police or public safety authority or the U.S. Environmental Protection Agency.*

Generator name(s) and address(es) _____

_____

Manifest document no. _____

YES	NO	N/A	
____	____	____	13. Does generator have placards to offer to transporters? (Section 262.33)
			14. Accumulation time (Section 262.34)
____	____	____	a. Are containers used to temporarily store waste before transport?
____	____	____	1) If yes, is each container clearly dated? [Section 262.34(a)(2)]
____	____	____	b. 1) Does generator inspect containers for leakage or corrosions? (Section 265.174—*Inspections*)
			2) If yes, with what frequency? _____
____	____	____	c. Does generator locate containers holding ignitable or reactive waste at least 15 m (50 ft) from the facility's property line? (Section 265.176—*Special Requirements for Ignitable or Reactive Waste*)

Note: If tanks are used, fill out checklist for tanks. (*See RCRA Hazardous Waste Tank Systems Inspection Manual,* OSWER Dir. No. 9938.4)

YES	NO	N/A	
____	____	____	d. Are the containers labeled and marked in accordance with questions 10, 11, and 12 of this form?

Note: If generator accumulates waste on site, fill out checklist for *General Facilities, Sections on Contingency Plan and Emergency Procedures and Manifest Systems.*

FIGURE 16–10. *(Continued)*

YES	NO	N/A	**Pretransport Requirements**
___	___	___	e. Does generator comply with requirements for personnel training? (Attach checklist for Section 265.16—*Personnel Training*.)
			15. Describe storage area. Use photos and narrative explanation sheet.

YES	NO	N/A	**Record Keeping and Records (40 CFR Section 262.40)**
			16. Does generator keep the following reports for 3 years?
___	___	___	a. Manifest or signed copies from designated facilities?
___	___	___	b. Biennial reports?
___	___	___	c. Exception reports?
___	___	___	d. Test results?
			17. Where are the records kept (at facility or elsewhere)?
			_____
			18. Who is in charge of keeping the records?
			Name _____ Title _____

YES	NO	N/A	**Special Conditions**
___	___	___	19. Has the primary exporter received from or transported to a foreign source any hazardous waste?
___	___	___	a. If yes, has he/she filed a notice with the Regional Administrator? (Section 262.53)
___	___	___	b. Is this waste manifested and signed by a foreign consignee? (Section 263.54)
___	___	___	c. If generator transported waste out of the country, has he/she received confirmation of delivered shipment? (Section 262.54)

Source: Adapted from U.S. Environmental Protection Agency, Solid Waste and Emergency Response, *RCRA Inspection Manual 1993 Edition,* OSWER Directive 9938.02B (Washington D.C.: U.S. Environmental Protection Agency, 1993)

FIGURE 16–11.   RCRA Waste Minimization Checklist:

Responses			Questions to Ask
YES	**NO**	**N/A**	***Statutory/Regulatory Requirements***

_____	_____	_____	1. Has the manifest been certified by an authorized representative? [Section 262.20(a)]
_____	_____	_____	2. Has the waste minimization statement on the manifest been altered or deleted? [Section 262.20(a)]
_____	_____	_____	3. Does the facility have a written description of their waste minimization program? [Sections 262.20(a), 264/5.75]
_____	_____	_____	If a written description is not provided, can the facility personnel provide a verbal description of the waste minimization program?

Description of program:

_____

_____

_____

_____

_____	_____	_____	4. Is there any visual evidence of the facility's waste minimization efforts?
			If yes, describe the activities/program observed.

_____

_____

_____

5. Does the description in the biennial report and/or annual export reports include:

_____	_____	_____	a. A description of the efforts undertaken during the year to reduce the volume and toxicity of waste generated? [Sections 262.41(a)(6); 262.56(a)(5)(i); 264/265.75(h)]
_____	_____	_____	b. A description of the changes in volume and toxicity of waste actually achieved during the year in comparison with previous years? [Sections 262.41(a)(7); 262.56(a)(5)(ii); 264/265.75(i)]
_____	_____	_____	c. Certification by the generator or authorized representative? [Sections 262.56(a)(6) require certification by primary exporter, Sections 264/5.75(j) require certification by the owner/operator or authorized representative.] [Sections 262.41(a)(8); 262.56(a)(6); 264/265.75(j)]
_____	_____	_____	6. For permitted facilities, does the operating record contain a certification by permittee (at least annually) that the

FIGURE 16–11. (*Continued*)

YES	NO	N/A	*Statutory/Regulatory Requirements*
			permittee has a program in place to reduce the volume and toxicity of the hazardous wastes? [Section 264.73.(b)(9)]

YES	NO	N/A	*Permit/Enforcement Requirements*
_____	_____	_____	7. Does the facility's permit contain any waste minimization requirements?

If yes, briefly describe the requirements and indicate if they have been met:

_____

_____

_____

YES	NO	N/A	*Permit/Regulatory Requirements*
_____	_____	_____	8. Are there waste minimization requirements contained in enforcement orders or settlement agreements with the facility?

If yes, briefly describe the requirements and indicate whether they have been met.

_____

_____

Source: Adapted from U.S. Environmental Protection Agency, Solid Waste and Emergency Response, *RCRA Inspection Manual 1993 Edition,* OSWER Directive 9938.02B (Washington, D.C.: U.S. Environmental Protection Agency, 1993).

FIGURE 16–12.   RCRA Transporter Checklist:

Responses			Questions to Ask
YES	**NO**	**N/A**	***General***

___   ___   ___   1. Does transporter have an EPA ID no.?

2. a. If yes, specify EPA ID? _____

YES	**NO**	**N/A**	***Manifest***

___   ___   ___   3. Does transporter use manifests?

___   ___   ___   a. If yes, are manifests signed and dated?

___   ___   ___   b. Does transporter return signed copies of manifests to generators? [Section 263.20(b)]

___   ___   ___   c. Does transporter carry manifests with waste shipments? (Section 263.20)

___   ___   ___   d. Does transporter obtain delivery date and signature of next transporter or owner/operator of designated facility at delivery? [Section 263.20(d)(1)]

___   ___   ___   e. Does transporter retain copies? [Section 263.20(d)(2)]

___   ___   ___   f. Does transporter give remaining copies to accepting transporter or facility? [Section 263.20(d)(3)]

___   ___   ___   g. Is transporter a water (bulk shipment) transporter? [Section 263.20(e)]

___   ___   ___   1) If yes, is waste delivered to receiving facility by water? [Section 263.20(e)(1)]

___   ___   ___   2) Does transporter carry with the waste a shipping paper containing all information required on manifest (excluding EPA ID numbers, generator certification, and signatures)? [Section 263.20(e)(2)]

___   ___   ___   3) Does transporter obtain delivery date and handwritten signature of owner/operator of designated facility on manifest or shipping paper? [Section 263.20(e)(3)]

___   ___   ___   4) Does transporter retain copies of shipping papers or manifests, in accordance with Section 263.22? [Section 263.20(e)(5)]

___   ___   ___   h. Is transporter a rail transporter? [Section 263.30(f)]

___   ___   ___   1) If yes, when accepting waste from a nonrail transporter, does rail transporter sign and date manifest acknowledging acceptance of waste? [Section 263.30(f)(1)(i)]

___   ___   ___   2) Does rail transporter return a signed copy of manifest to nonrail transporter? [Section 263.20(f)(1)(ii)]

FIGURE 16–12. *(Continued)*

YES	NO	N/A	*Manifest*
			3) Does rail transporter forward manifest copies to: [Section 263.20(f)(1)(iii)]
———	———	———	a) The next nonrail transporter? OR
———	———	———	b) Designated receiving facility (if reached by rail)? OR
———	———	———	c) The last rail transporter designated to handle waste in the United States?
———	———	———	4) Does rail transporter retain a copy of manifest? [Section 263.20(f)(1)(iv), 236.22 (c)]
———	———	———	5) Does rail transporter ensure that a shipping paper accompanies the hazardous waste and contains all information required on manifest (excluding EPA ID, generator certification and signatures)? [Section 263.20(f)(2)]
———	———	———	6.) Does rail transporter obtain delivery date and handwritten signature of owner/operator of designated facility or the next nonrail transporter on manifest? [Section 263.20(f)(3), (4)]
———	———	———	7) Does rail transporter retain a copy of manifest or signed shipping paper? [Section 263.20(f)(3, 4)]
———	———	———	i. Does transporter transport waste outside of the United States? [Section 263.20(g)]
			1) If yes, does the transporter:
———	———	———	a) Indicate on manifest the date that shipment left the U.S.?
———	———	———	b) Sign manifest and retain one copy?
———	———	———	c) Return a signed copy of manifest to generator?

YES	NO	N/A	*Compliance with the Manifest (40 CFR Section 263.21)*
			4. Does transporter deliver entire shipment of hazardous waste to: [Section 263.21(a)]
———	———	———	a. Designated facility listed on manifest?
———	———	———	b. Alternate designated facility, if emergency prevents delivery to designated facility?
———	———	———	c. Next designated transporter?
———	———	———	d. Place outside U.S. designated by generator?
———	———	———	e. If no, does transporter contact generator for further directions, and then revise manifest accordingly? [Section 263.21(b)]

FIGURE 16–12.   *(Continued)*

YES	NO	N/A	**Record Keeping (40 CFR Section 263.22)**
____	____	____	5. Does transporter keep a copy of manifest signed by generator, itself, and next designated transporter for 3 years from the date the hazardous waste was accepted by the initial transporter? [Section 263.22(a)]
____	____	____	6. Does water (bulk shipment) transporter retain copy of shipping paper for each shipment delivered by water for 3 years from the date hazardous waste was accepted by the initial transporter? [Section 263.22(b)]
____	____	____	7. Does initial rail transporter keep a copy of a manifest and/or shipping paper for 3 years from the date the hazardous waste was accepted by the initial transporter? (Section 263.22)
____	____	____	8. Does transporter shipping waste outside of the U.S. keep a copy indicating that waste was shipped for 3 years from the date the hazardous waste was accepted by the initial transporter? [Section 263.22(d)]

Source: Adapted from U.S. Environmental Protection Agency, Solid Waste and Emergency Response, *RCRA Inspection Manual 1993 Edition,* OSWER Directive 9938.02B (Washington, D.C.: U.S. Environmental Protection Agency, 1993).

ENDNOTES

[1] Section 261.2 Definition of a Solid Waste

(a) (1) A *solid waste* is any discarded material that is not excluded.

    (2) *Discarded* is abandoned (disposed of, burned, stored, or treated prior to disposal), recycled, or inherently wastelike.

(c) (1) *Recycled materials* are solid wastes if they are used in a manner constituting disposal, such as applied on land or used in products applied on the land if that is not their ordinary manner of use.

    (2) *Recycled materials* are solid waste if they are burned to recover energy or used to produce a fuel.

    (3) The following materials are solid wastes when reclaimed: spent materials, listed sludges, and byproducts, and scrap metal.

    (4) *Recycled materials* are solid wastes if they are accumulated speculatively, except for listed commercial chemical products.

(d)   *Inherently wastelike materials* are solid wastes when recycled in any manner.

(e) (1) *Recycled materials* are not solid waste when they are used as ingredients in an industrial process or as effective substitutes for products, provided that no reclamation is involved.

(f)   Respondents in actions to enforce RCRA Subtitle C regulations must document claims that materials are not solid waste or are conditionally exempt.

[2] Section 261.4 Exclusions:

The following materials are not solid wastes:

(a) (1) Domestic sewage and mixtures of domestic sewage and other wastes;

    (2) Industrial wastewater point source discharges under the Clean Water Act;

    (3) Irrigation return flows;

    (4) Special nuclear or byproduct material;

    (5) Materials subject to *in situ* mining techniques;

    (6) Pulping liquors that are reclaimed in a pulping furnace;

    (7) Spent sulfuric acid used to produce virgin sulfuric acid;

    (8) Secondary materials that are reclaimed and returned to the original process;

    (9) Spent wood-preserving solutions that are reclaimed and reused, and wood-preserving wastewaters reclaimed and reused to treat wood; and

    (10) Toxicity characteristic and listed as hazardous coke byproduct residues that are used in producing coke and coal tar.

The following materials are solid wastes but not hazardous wastes:

(b) (1) Household waste;

    (2) Agricultural waste returned to the soils as fertilizers;

    (3) Mining overburden returned to the mining site;

    (4) Ash from combustion of coal or other fossil fuels;

    (5) Drilling fluids and other wastes associated with the production of crude oil, natural gas, or geothermal energy;

    (6) Wastes failing the toxicity test for chromium if the chromium is in the trivalent state and generated in the leather tanning and finishing industry;

(7) Solid waste from the extraction, beneficiation, and processing of ores and minerals;

(8) Cement kiln dust;

(9) Arsenic-treated wood or wood products failing the toxicity test for arsenic, used as intended and discarded;

(10) Petroleum-contaminated media and debris failing the toxicity test for D018–D043 (newly identified organic constituents only), which are subject to corrective action under the Underground Storage Tank program described in 40 CFR Part 280;

(11) Injected groundwater, failing the toxicity test for D018–D043 only, from hydrocarbon recovery operations undertaken at a certain petroleum industry site until January 25, 1991;

(12) Reclaimed chloroflurocarbon refrigerants; and

(15) Non-terne-plated, used oil filters that are not drained.

(c) Hazardous waste is not regulated if generated in a product or raw material storage tank, or in a manufacturing process unit until it exits the unit or until the unit ceases to operate for more than 90 days.

(d) Samples of solid waste or environmental media tested for characteristics or composition are not subject to hazardous waste regulation.

(e,f) Treatability study samples of hazardous waste are not subject to hazardous waste regulation.

CHAPTER 17

AIR POLLUTION

INTRODUCTION

Air pollution activities in the U.S. are regulated under the Clean Air Act (CAA) of 1970, which was last amended in 1990. Air pollution regulations are complex, and charting a course through them can be a daunting task. At the core of the national and state air pollution control programs is a requirement for sources to obtain permits before *constructing* or modifying air emission sources and before *operating* new emission sources or modified emission sources. Permits are important to the control of air pollution for many reasons. Permits ensure that new sources and existing sources (and their modifications) will meet applicable air pollution standards or additional requirements, such as monitoring and housekeeping. This, in turn, helps prevent the degradation of air quality. Permits provide an enforcement tool if sources do not comply with permit requirements. The information that is recorded on permits is also used by federal and state governments to develop emissions inventories and to track progress toward meeting the goal of attaining clean air.

Because of the CAA Amendments of 1990, sources may be required to meet one of several new standards, in addition to conditions from the older regulations. Sources can approach these new challenges using earlier strategies—create an emissions inventory, determine what regulations apply, and prepare an action plan for compliance. Appendix F contains addresses for U.S. EPA World Wide Web sites (*http://www.epa.gov/oar*) and other resources that can provide helpful information for meeting these challenges.

EMISSIONS INVENTORY

Creating an emissions inventory means that all emissions units and pollutants (solid particles, liquid particles, or gases) must be identified for the entire facility. The inventory should include emissions units for both point sources (released through a single, identifiable source such as a stack) and fugitive sources (emissions not caught by a capture system). The inventory should include both *actual* and *potential* emissions. Actual emissions typically are those released from the emissions unit after controls and during actual operating conditions. Potential emissions are typically those that could be released under worst-case operating conditions (may or may not include control equipment).

SOURCES

Air pollution sources are broadly classified into stationary process sources, stationary combustion sources, materials storage and transport sources, and transportation sources. Process units include a variety of sources such as cookers, grinders, mixers, dryers, distillers, pulpers, coaters, and extruders. Stationary combustion sources include dryers, process heaters, calciners, flares, furnaces, smelters, incinerators, or boilers. Material transport is accomplished by pipes, moving belts, bucket elevators, screw conveyers, and other mechanical devices. Material storage devices include tanks, silos, and other containers.

Fugitive or uncontrolled emissions can be significant sources of emissions, and these are sometimes difficult to control. Tanks and silos typically do not release emissions during routine storage, but filling operations can allow large amounts of product to escape into the air. Fugitive dust or fumes can escape processes, such as spray painting and sandblasting, melting and casting operations, and solvent cleaning operations. Flanges, pumps, and valves are typical release points for fugitive emissions. Fugitive emissions can escape from doors, windows, and roof vents. Additional sources include dust blowing off storage piles, roads, or lots and any point where materials are transferred.

Smaller noncontact combustion sources typically are not required to have emissions controls because the amounts of pollutants released are not large, but these sources should be evaluated to determine whether company policies are followed with respect to operation, maintenance, fuel use, and so forth.

Plant vehicles and stationary engines are additional air pollution point sources. Vehicle emissions may be managed by maintenance and emissions testing programs or through the use of alternate fuels.

CONTROL EQUIPMENT FOR PROCESS UNITS

Air pollution control equipment includes cyclones, baghouses, electrostatic precipitators, absorbers, adsorbers, condensers, and incinerators. Additional control strategies include restricting the sulfur content of fuel and maintaining proper combustion temperatures and oxygen content, wetting roads to minimize particulate releases, and other housekeeping measures.

Cyclones remove medium to coarse particles, and they are typically used as precollectors or for product recovery, especially in the processing of wood, minerals, and grains. The cyclone body consists of a main cylinder that has a centrally located discharge outlet at the top of the body. Waste gases enter a tangential inlet near the top of a cylindrical body. This creates a vortex, which causes the gas stream to spiral downward between the walls of the cylinder and the discharge outlet. As the gas stream reaches the bottom of the cyclone, the vortex changes its direction of flow and forms an inner vortex that carries the gas stream upward to the gas outlet. Particles are removed by inertial impaction on the walls of the cyclone. The removed particles are collected into a hopper at the bottom of the cyclone. Cyclones can be configured as single units or they can be arranged in parallel, with as many as 600 tubes in one housing. Single units can range in size from 0.5-inch diameter to over 80 inches in diameter. Flow rates can range up to 40,000 actual cubic feet per minute (acfm) in a single large cyclone. Efficiencies vary from less than 50 percent to 99 percent, depending on the particle size and size and configuration of the cyclone unit.

Filtration is used to collect solid particles from a variety of industrial sources, including foundries, oil-fired boilers, cement kilns, electric generating plants, and steel-making operations. Filtration is typically used when high collection efficiencies are needed for small-sized particles and high dust loadings exist. The fabric filter system consists of multiple tubular or flat collecting bags suspended inside a housing called a baghouse. A single baghouse may contain several hundred to several thousand bag filters. Depending on the characteristics of the gas stream, the bags may be made of cotton, wool, glass, nylon, or dacron. The gas stream typically enters the unit from the side and flows upward through the bags. As the gas passes through the bags, dust collects on the inside of the bags, and filtered air exits at the top of the baghouse unit. Collected particles are typically cleaned by mechanical shaking or high-pressure reverse jets of air. Flow rates of gas through the collection system are in the range of 10,000–50,000 acfm. Overall efficiencies that are greater than 99% can be achieved.

High-efficiency filters are used to control organic particulates, such as oil mists or resin fumes. In smaller units, the filters may be changed manually in response to high pressure drops, but in larger units the clean filter advances automatically when the pressure drop gets too high.

Electrostatic precipitators are used when high collection efficiencies are needed for a gas stream with liquid and solid particles over a wide size range, including submicron particles. The electrostatic precipitator removes particles by imparting a negative charge to the particles, which are then attracted to positively charged collection plates. The charged particles are retained on the plates until the cleaning cycle allows the particulates to fall into a hopper for final disposal. The cleaning phase can be accomplished by rapping the plates or by washing the plates with water sprays. Collection efficiency is determined primarily by the particle retention time in the electrical field and particle resistivity; efficiencies greater than 99% can be achieved even with high flow rates.

Scrubbing is a technique that removes smaller-sized particles and gases from a waste stream through turbulent contact of the pollutants with a liquid. This process is called absorption. There are many scrubber designs; three of the more common ones are the open spray tower, packed tower, and the venturi. All scrubbers have two basic components: a section where liquid-gas contact occurs and a section where the wetted particles are removed.

The open spray tower is a large chamber in which a scrubbing liquid is sprayed downward through low-pressure nozzles. The waste gas enters tangentially at the bottom of the unit, and as the gas stream moves upward, the particles are removed by the falling droplets and are collected at the bottom of the scrubber. Cleaned gas exists at the top of the unit, typically passing through a demister to remove excess water from the gas stream. Open tower scrubbers have relatively low removal efficiencies (80–90%), and they are typically used as precleaners to remove particles greater than 5 microns in size. The efficiency of the open spray tower can be improved by packing the scrubber with materials (for example, ceramics) in shapes that improve the contact of the waste gas stream with the scrubbing liquid.

The venturi scrubber is used when high collection efficiencies (over 99%) are needed for smaller particles. The venturi scrubber consists of a venturi section (short tube with a taper in the middle) that connects to a cyclone separator. As gases enter the venturi, the velocity of the gas stream increases, and this, in turn, atomizes the scrubbing liquid into many fine droplets. The difference in velocity between the particles and the scrubbing liquid promotes the impaction of the particles into the scrubbing liquid. As the gas stream enters the cyclone, removal continues. A slurry of particles and liquid that requires disposal collects at the bottom of the cyclone.

Incineration is an effective control strategy for oxidizing a variety of solid, liquid, and gaseous materials. In some instances, incineration is the only effective control for highly odorous or volatile emissions. Under ideal conditions (adequate oxygen, dwell time, and sufficient temperatures), carbon dioxide and water vapor are the end products of incineration. More typically, most incineration units produce some volatile products, solids, and ash. When the waste stream includes metals, nitrogen, phosphorous, sulfur, halogens, and other toxic components (other than hydrogen and carbon), additional toxic by-products result (for example, the incineration of halogenated hydrocarbons produces acid gases such as hydrochloric acid). These toxic by-products require additional control equipment. Flares are not permitted for routine control, but they may be allowed in emergency situations.

Adsorbers and condensers are used to control volatile organic hydrocarbons (VOCs) from surface coating operations or bulk liquid transfer/storage operations. Adsorption is used to control operations that produce lower molecular weight VOCs and odor-causing chemicals (for example, dry cleaners, metal degreasing operations, surface coating systems, and processing for rayon and plastics). The gas stream is passed over activated charcoal or other adsorbents packed into canisters or beds. The adsorbent captures and holds the contaminants until the bed is saturated and cleaned. During the purge cycle, the bed is taken out of operation, and the contaminants are removed from the bed by heat, vacuum, or both. The released contaminants are then incinerated, absorbed, or condensed for final disposal. Portable units that are used primarily for odor control are typically discarded after saturation or returned to the manufacturer for recharging. Adsorption efficiencies of 95 to 99% can be achieved, depending on the characteristics of the waste stream and the adsorption system.

Condensation is used to control VOCs when recovery is desirable (for example, gasoline vapors). The gas stream passes across chilled plates onto which the vapor condenses. A periodic warming cycle is needed to drain the condensed liquid off the plates. The warming cycle should occur after process shutdown unless short periods without control are acceptable.

QUANTIFYING EMISSIONS

After sources are identified, the actual and potential emissions should be quantified from the identified release points. Any of four basic methods can be used to estimate emissions. These include: mass balances, emissions factors, engineering calculations, and stack or other analytical data. Stack or analytical data must be obtained from operational sources whenever other information is unavailable or insufficient for making a good estimate. Taking time to obtain emissions estimates is important and can save considerable time and resources later in the process.

Once emissions are quantified, it is important to maintain accurate records of all calculations and other information used to obtain the estimates. These calculations should include sources of information, emissions factors, and all assumptions used in the calculations and the identity of individuals making the estimates.

DETERMINING APPLICABLE REQUIREMENTS

Determining applicable requirements can be a complex and confusing process. Sources must evaluate their emissions with respect to impact on the National Ambient Air Quality Standards (NAAQS), attainment and nonattainment areas, hazardous air pollutant emissions (HAPs), accidental release prevention, stratospheric ozone depletion, the Title V operating permit program, and mobile sources emissions. The particular requirements applicable to a stationary source will depend on whether the source is *major* or *minor* and *new* or *existing*. These requirements are discussed in the following sections.

Determining applicability may require the source to model the impact of its emissions on air quality. Often relatively simple screening models can be used to determine applicability, but in other cases more complex models (and, perhaps, consulting services) are needed. The EPA and state agencies can provide considerable guidance for the selection and use of air quality models.

PREPARING AN ACTION PLAN

An action plan to bring the facility into compliance can be developed based on the identified applicable requirements and findings from a facility audit. Because of the complexity of the air pollution regulations, a single facility may have a number of issues to address. It is important to rank and prioritize the individual compliance requirements. For example, the facility may need to acquire additional permits or pollution control equipment, increase training, submit reports or develop/modify other documentation systems, change processes or other aspects of production, or implement other changes. Once the prioritized list is produced, the resources needed for compliance can be identified and quantified to facilitate decision making.

APPLICABLE REQUIREMENTS

TITLE V PERMIT PROGRAM

Prior to the 1990 amendments of the CAA, the regulatory framework for allowing sources to construct and operate facilities required a variety of federal and state permits, depending on the classification and anticipated emissions from the sources. To meet federal requirements, a preconstruction permit (discussed above) was required for new sources and modifications to existing sources. Federal source-specific operating permits were not required for existing sources. The operation of existing sources was further regulated through provisions established in the State Implementation Plans (SIPs).

The Title V permit program was developed because the existing regulatory framework was not working effectively. The fragmentation and complexity of the pre-1990 permitting scheme led to confusion about compliance responsibilities, ineffective oversight by understaffed and inadequately funded state agencies, and a lack of public participation in the process. The Title V provisions of the 1990 amendments establish a new Part 70 operating permit program, which is patterned after the successful NPDES water permit program. The Title V program alleviates many problems because it results in *one comprehensive operating permit* for a single source. The permit, which contains all of the federal and state requirements that apply to a single source, has a maximum term of 5 years, and for most purposes an entire plant constitutes a single source.

Under Title V, the permit program is administered by the states, but the EPA has the authority to review and approve the state programs and each permit issued by the state. After the initial phase-in of the Title V permits, any first-time applicant must submit a complete permit application to the state within 12 months after becoming subject to the Part 70 operating permit program.

WHO NEEDS A TITLE V PERMIT?

A facility is subject to the Part 70 requirements if it is listed or meets the definition of a *major* stationary source (see Figure 17–1). A major source is one that emits or has the *potential to emit* (PTE) pollutants above allowed levels. The PTE means the maximum capacity of a stationary source to emit any air pollutant under its physical and operational design. Any physical or operational limitation on the capacity of a source to emit an air pollutant, including air pollution control equipment, restrictions on hours of operation, or type or amount of material combusted, stored, or processed shall be treated as part of its design if the limitation is enforceable by the EPA. The *unrestricted* PTE means the maximum capacity of the source to emit. In most cases, this means the emissions that would result from the potential worst-case operating scenario (running at maximum capacity without controls, 24 hours per day, 365 days per year). The unrestricted PTE is the basis for determining the applicability of the Title V permit program. Fugitive emissions (emissions that do not pass through a stack, chimney, vent, or equivalent opening) are not considered in determining actual emissions or PTE, unless the source is one that is listed in Figure 17–2. The unrestricted PTE can be determined using mass balance calculations, stack tests, or manufacturer or vendor information on design ratings or process capacity with published emissions factors (see Appendix F for resources).

If a source is not listed, a three-step approach should be used to determine whether or not the source is a major source under Part 70. First, identify each of the emissions units at the source. An emission unit as defined in Part 70 includes any part or activity of a stationary source that emits or has the potential to emit

FIGURE 17–1. Title V Source Categories:

Category 1 Sources. All facilities that are major sources are required to comply with the Title V permit system. A major source is one that has the potential to emit:

- 100 ton/yr or more of criteria pollutants (includes ozone, O_3; nitrogen dioxide, NO_2; particulate matter, PM10; sulfur dioxide, SO_2; lead, Pb; and carbon monoxide, CO) and their precursors (nitrogen oxides, NO_x, and certain volatile organic compounds, VOCs)

- 10 ton/yr or more of any one HAP, or 25 ton/yr or more of any combination of HAPs [see Appendix E.1 for a list of the HAPS and Appendix E.2 for a list of maximum achieveable control technology[1] (MACT) source categories and timetables]

- 10–70 ton/yr for some criteria pollutants in nonattainment areas

- other pollutants regulated under Title VI (see Appendix E.3 for a list of stratospheric ozone-depleting pollutants)[2]

Other sources that need a Title V permit include sources subject to:

- National Emissions Standards for Hazardous Air Pollutants (NESHAPS) (see Appendix E.4 for a list of these pollutants)

- New Source Performance Standards (NSPS) (see Appendix E.5 for a list of sources subject to these standards)

- Prevention of Significant Deterioration (PSD) permits

- nonattainment New Source Review (NSR) permits

Under the federal regulations, the states may choose to exempt nonmajor NESHAPS and NSPS sources. These program requirements are outlined below.

Category 2 Sources. There are two types of sources that must obtain a Title V permit regardless of their PTE. One type is an "affected source," which is a source that is *affected* by the Acid Rain/Deposition Control requirements of Title IV of the CAA. Affected sources include utilities and other facilities that combust fossil fuel and generate electricity for wholesale or retail sale. The second type is sources that are solid waste incineration units that are required to obtain a Part 70 permit under Section 129 of the CAA.

Category 3 Sources. The EPA has currently deferred regulation of the following nonmajor sources, which are included in the Part 70 permit program: area sources subject to NSPS under Section 111 of the CAA and area sources subject to a HAP standard under Section 112 of the CAA. Small businesses such as filling stations, dry cleaners, and printers that have emissions under the 10/25 ton/yr levels for major HAP sources would fit into this category of area sources because any emission of HAPs makes them subject to this currently deferred portion of Part 70.

any regulated air pollutant (for example, boilers, spray booths, coal storage piles, and so forth). The emission point is the place where the emission is released to the environment (for example, stacks, roof vents, exhaust fans, and so forth). Second, estimate the unrestricted PTE for regulated air pollutants for each emission source. Third, compute the total unrestricted PTE for all emission units at the source and compare the total against the applicability levels in Part 70.

Once the unrestricted PTE has been calculated, there are three potential outcomes: the source is clearly not subject to Part 70 (unrestricted PTE is below applicability levels), the source is clearly subject to Part 70 because the unrestricted PTE is above applicability levels, or the unrestricted PTE is greater than the applicability levels, but the actual emissions either fall under the applicability levels

FIGURE 17–2.   Major Stationary Sources for which Fugitive Emissions Count
in Determining the Potential to Emit Under Title V:

- Coal-cleaning plants (with thermal dryers)
- Kraft pulp mills
- Portland cement plants
- Primary zinc smelters
- Iron and steel mills
- Primary aluminum ore reduction plants
- Primary copper smelters
- Municipal incinerators capable of charging more than 250 tons of refuse per day
- Hydrofluoric, sulfuric, or nitric acid plants
- Petroleum refineries
- Lime plants
- Phosphate rock-processing plants
- Coke oven batteries
- Sulfur recovery plants
- Carbon black plants (furnace process)
- Primary lead smelters
- Fuel conversion plants
- Sintering plants
- Secondary metal production plants
- Chemical process plants
- Fossil fuel boilers (or combination thereof) totaling more than 250 million BTU/hour heat input
- Petroleum storage and transfer units with a total storage capacity exceeding 300,000 barrels
- Taconite ore processing plants
- Glass fiber processing plants
- Charcoal production plants
- Fossil fuel-fired steam electric plants of more than 250 million BTU/hour heat input
- All other stationary source categories regulated by a standard promulgated under Sections 111 or 112 of the CAA, but only with respect to those air pollutants that have been regulated for that category.

or could reasonably be maintained below those levels. If a source falls into the last category, it may be able to establish that its PTE with applicable limits (for example, those established through the NSPS[3] NESHAPS[4], or SIP restrictions) may reduce its emissions below those that trigger Part 70 applicability. In addition, it may be possible for a source that remains above the applicability limits to consider limiting emissions so that the source becomes a "synthetic minor" source. These sources can opt out of a Part 70 operating permit by applying for and agreeing to operate within the limits given in a *federally enforceable state operating permit* (FESOP) or perhaps a *state source-specific operating agreement* (SSOA). The FESOP and SSOA permits may be attractive to a facility because of reduced fees, reduced or expedited paperwork, or less burdensome record keeping, monitoring, and reporting requirements.

AGENCY REVIEW

Once the agency receives an application, it reviews the application for completeness and accuracy. Incomplete applications cannot be evaluated, and they are returned to the source for additional information. After the agency evaluates the emissions estimates and other technical information of a completed application, it makes a series of determinations. Will the source comply with all applicable regulations? Will the source cause any adverse air quality impacts? Next, the agency drafts a permit and technical support document. The agency gives notice of the draft permit and provides a public comment period, which may include a public hearing. After the agency responds to comments, a final permit is issued.

NEW SOURCE REVIEW PROGRAM

One of the foundations of the CAA is a requirement to review new sources before they are constructed and existing sources before they are modified or reconstructed to ensure that they will comply with all applicable requirements. New sources and changes to existing sources are required to undergo review and to obtain a construction permit that authorizes the source to construct or modify facilities that have the potential for increased air pollution and to obtain an operating permit that authorizes a source to conduct the day-to-day operations that release allowable amounts of emissions to the air. State and local agencies issue construction permits to implement the minor New Source Review (NSR) program required by the 1970 CAA and the major NSR program required by the CAA Amendments of 1977.

MAJOR NSR

Major NSR applies to the construction of *major stationary sources* and *major modifications* to existing major stationary sources. The nature of the required review depends on whether the source plans to locate in an area that is attainment or nonattainment for the NAAQS for the specific pollutant(s) released by the facility. The Prevention of Significant Deterioration (PSD) provisions of the Act apply to sources that locate in attainment areas, whereas sources that locate in nonattainment areas are subject to Nonattainment NSR. Both programs require the source to be reviewed and to obtain a permit before construction can begin.

The steps in a major NSR consist of:

- determining applicability
- technology analysis
- air quality analysis
- other impact analysis.

The NSR program definitions refer to "actual" and "potential" emissions. *Actual emissions* means the average emission rate over the 2 years preceding the proposed addition/change. *Potential emissions* means the emissions based on operating at the maximum rated capacity for 8,760 hours per year, unless a federally enforceable emission limitation or restriction on production of hours of operation applies.

For the PSD program, a major stationary source is a new plant site or an addition to an existing nonmajor plant site that has potential emissions greater than 100 ton/yr for sources belonging to a list of 28 groups or 250 ton/yr for nonlisted sources. A major modification under PSD is an addition or change to an existing major stationary source where the potential emissions of the plant after the addition/change exceed the actual emissions by a "significant" amount. The owner of a source subject to a PSD permit agrees to comply with the PSD increments and to use best available control technology (BACT) for each regulated pollutant that it will emit in "significant" amounts. BACT means an emission limitation based on the maximum degree of reduction that can be achieved, taking into account economic, energy, and environmental factors. The control technology derived from the BACT analysis must be as least as stringent as any NSPS applicable to the source category and applicable NESHAPS.

The 1990 Amendments, however, specifically state that the hazardous air pollutants listed in the Amendments are not subject to the PSD program. Individual states, however, may elect to include these contaminants in the PSD analysis. The air quality analysis required by the PSD program must demonstrate that the new major source or major modification does not cause the NAAQS to be exceeded and must ensure that all of the available PSD increment is not consumed. In addition, the permitting agency must review the potential for other impacts, including growth-related impacts, on soil, vegetation, and visibility.

For the nonattainment NSR program, a major stationary source is a new plant site or an addition to an existing nonmajor plant site that has potential emissions greater than 100 ton/yr of the nonattainment pollutant. The definition of a major modification under the nonattainment NSR program is equivalent to the definition under the PSD program. The nonattainment NSR program requires sources to commit to achieving the Lowest Achievable Emission Rate (LAER). LAER means the most stringent of the following emission limitations: the most stringent emission limitation that is contained SIP of any state or one that is achieved in practice by the same or similar class or category of source. If the owner or operator of the proposed source or major modification can demonstrate that the most stringent technology is not feasible, the next most stringent limitation can be established as LAER. The air quality analysis for the nonattainment NSR program requires the source to demonstrate that it will not cause the NAAQS to be exceeded. The source is also required to obtain an offset in its nonattainment emissions from nearby facilities and to ensure that the offsets provide a net air quality benefit. This means that nearby facilities must agree to reduce their emissions at a greater than one-to-one ratio. After LAER is established, the source must certify that its other sources are in compliance or on a schedule to comply with all applicable air quality requirements. A final condition to obtaining the permit requires the source to certify that the benefits of the proposed facility outweigh its environmental and social costs.

MINOR NSR

Minor new source review permits/registrations issued by the states are applicable to facilities or modifications that do not qualify as major sources or major modifications. Activities that trigger the need for a minor NSR include new plant sites, expansion of an existing facility, addition of new equipment or replacement of existing equipment, reconstruction or modification of equipment, changes in raw materials, or changes in the method of operation.

To determine the type of permit needed, the source must determine whether the project is construction, reconstruction, or modification. Sources that meet the test of construction or reconstruction activities are required to evaluate only the emissions associated with the construction/reconstruction equipment. If the source qualifies as an equipment modification, raw material change, or change in method of operation, the source must evaluate the increase in emissions caused by the change. The minor NSR permit application typically requires general information about the facility, location, raw materials, explanation of the process (flow diagram), types and quantity of equipment, capacity of equipment, production rate, emission estimates and justification, air pollution control equipment (design parameters and efficiency), stack parameters (height, diameter, exhaust flow rate), applicable rules and the methods used for compliance, and background information (equipment specifications, emission calculations, assumptions).

AGENCY REVIEW

Once the agency receives an application, it reviews the application for completeness and accuracy. Incomplete applications cannot be evaluated, and they are returned to the source for additional information. After the agency evaluates the emissions estimates and other technical information of a completed application, it makes a series of determinations. Will the source comply with all applicable regulations? Will the source cause any adverse air quality impacts? Next, the agency drafts a permit and technical support document. The agency gives notice of the draft permit and provides a public comment period, which may include a public hearing. After the agency responds to comments, a final permit is issued.

ACCIDENTAL RELEASE PREVENTION

Under the accident prevention provisions of Section 112(r) of the CAA Amendments of 1990, some facilities may be required to develop accidental release prevention plans. Section 112(r) requires covered facilities to develop risk management programs (RMP). The rule applies to those facilities that have quantities of regulated hazardous substances present in amounts that are greater than threshold quantities (see Appendix E.6 for a list of these substances and the threshold quantitites).[5] The rule applies to most manufacturers, chemical wholesalers, utilities, public drinking water and wastewater treatment systems, propane retailers, and cold-storage facilities using ammonia as a refrigerant. Some requirements of the rule are that companies must design and maintain a safe facility, describe ways that accidents will be prevented, identify worst-case hazards that may result from unplanned releases, and minimize the consequences of releases when they occur.

The RMP must be registered with the EPA prior to the effective date of the regulation, and registrations must be updated within 60 days if the information becomes inaccurate. The employer must conduct a safety audit once every 3 years, must document the audit in a report that includes findings and recommendations, and must document management's response to the findings. The employer is required to conduct safety inspections to prevent leaks and other episodic releases. Employers must also develop and document a training program, train employees in applicable and appropriate operating procedures, and provide refresher training at least once every 3 years. The employer must evaluate the effectiveness of training and develop a schedule for reviewing and revising the training.

The regulatory requirements under Section 112(r) overlap with Occupational Safety and Health Administration (OSHA) Process Safety Management Standard (29 CFR Section 1910.119),[6] the Hazardous Waste Site Operations (HAZWOPER) standard (29 CFR Section 1910.120) and emergency action plan regulation [(29 CFR Section 1910.38(a)], and the facility emergency response plan required under the Emergency Planning and Community Right-to-Know Act (EPCRA) for chemical releases. Each of these regulatory programs should be evaluated for applicability.

THE AIR POLLUTION AUDIT PROCESS

Prior to conducting the audit, the audit team will gain an understanding of the facility's processes, the types of air pollution sources, control equipment, and monitoring equipment used to control air pollution. Two important sources of information about specific industrial processes are the NSPS for the industry and the Air Pollution Engineering Manual (AP-42). The NSPS typically describes all the process units specific to that industry and defines the types of controls that meet the BACT requirements for that source. AP-42 provides a compilation of source-specific factors related to controlled and uncontrolled emissions. Appendix F provides information on accessing AP-42 information and other EPA resources for air pollution. In addition, the audit team will review the regulations (federal, state, or local) that govern the facility, and it will probably visit the applicable regulatory authority to review the compliance history of the facility. Typical areas of the air pollution audit are shown in Figure 17–3.

It is important for the facility to have permits and other documentation available in an organized, easy-to-use format. The auditor will examine the air permit file to complete the inventory forms, and to identify citations, notices of violation (NOVs), agency inspection reports, source test reports, and source inventory reports that have been submitted for regulatory purposes. The auditor will note any permit requirements, such as periodic source testing, or required reports, such as submission of fuel logs or operating logs.

Prior to the in-depth inspection, the audit team will use available data to construct its working inventory of sources to facilitate the identification of sources and to evaluate compliance with permit conditions. The Title V or FESOP permit application forms provide a listing of significant and insignificant sources that can form the basis of the inventory. Alternatively, Figure 17–4 shows an example of a source inventory form that could be used for the audit. The sources from the inventory forms or permit application can be collated into a summary that provides a quick review of sources, controls, and permits. The auditor can place the source numbers on a copy of the plot plant to facilitate the plant tour

FIGURE 17–3.   Typical Elements of an Air Pollution Audit:

Areas to Consider

- stationary combustion sources
- pollution control devices
- VOC sources
- hazardous air pollutants
- accidental release prevention and emergency response
- fugitive emissions
- mobile sources
- noise
- odors

Records to Review for Process Air Emissions

- source emission inventory
- stack/vent operating certificates
- permits
- operation and maintenance logs
- reporting of unplanned releases
- training

Records to Review for Petroleum Storage

- source emission inventory
- vapor control systems
- permits
- operation and maintenance logs
- reporting of unplanned releases
- training

Records to Review for Boiler Operations

- permits
- daily boiler logs
- boiler operation inspection records
- fuel use records
- opacity and stack emission test records
- training

Records to Review for Asbestos Management

- asbestos inventory
- reporting of abatement or demolition activities
- monitoring
- waste disposal
- training

FIGURE 17–4.   Example of an Air Pollution Source Inventory Form:

ABC Company
Air Pollution Source Inventory Form

Plant: _____   Auditor _____

Plant Contact: _____   Date: _____

Source No.: _____

Operational Date: _____

Source Description: _____

_____

Has source been modified/reconstructed?   _____ yes   _____ no

If yes, explain: _____

_____

Location: _____

Fuel: _____   Rate: _____

Rated Capacity: _____   Operational Capacity: _____

Operation Schedule: _____ hr/day _____ days/wk _____ wk/yr

Exhaust Gas: _____   CFM @ _____ °F

Permitted Emission Rates:

  SO_2 _____ NO_2 _____

  CO _____   PM _____

  VOCs _____   Pb _____

  Other _____

Monitoring Data:

  a. _____ in-stack

  real-time concentrations: _____ ave during _____

                            _____ peak during _____

  b. _____ stack test   _____ date

  results _____

  _____

  c. _____ upwind/downwind _____ date

  results _____

  _____

Comments: _____

_____

_____

Source Control Equipment:

#1 _____

  Location: _____

FIGURE 17–4. *(Continued)*

Date of Operation: _____
Permit Number: _____ Expiration Date _____
Observations: _____
_____
#2 _____
Location: _____
Date of Operation: _____
Permit Number: _____ Expiration Date: _____
Observations: _____
_____
#3 _____
Location: _____
Date of Operation: _____
Permit Number: _____ Expiration Date: _____
Observations: _____
_____
Comments: _____
_____
_____
_____
Recommendations: _____
_____
_____
_____
Rev. 1   FWP
9/15/96   Page ___ of ___

and final report preparation. The plot plan also gives the auditor an easy way to identify sources that were omitted from the initial inventory. Omitted sources will be added to inventory forms during the audit.

The audit begins with a briefing of site management and of the environmental staff. The audit team obtains information about safety requirements and participates in an orientation tour. During the orientation tour, the audit team becomes better acquainted with the physical layout of the facility and its operations. The orientation tour is conducted with the facility representative, typically, the environmental coordinator. During the tour, the audit team becomes oriented to the facility and becomes acquainted with the processes that generate point source and fugitive emissions. The tour should include the entire facility and its grounds.

After the orientation tour, the audit team reviews records and management systems, interviews site personnel, and conducts an in-depth inspection of the facility. Previously developed checklists will be used by the audit team to conduct the in-depth evaluation of the facility. Figure 17–5 at the end of this chapter shows an example of a checklist that could be used to audit a facility. During the in-depth inspection, the auditor's objectives are to find all sources of stack and fugitive emissions, inspect control devices, and note apparent problems. The auditor will take photographs to provide for unanswered questions or documentation for problems, and the auditor may ask for facility staff to explain some processes. The inventory of sources will be completed during this phase of the audit.

Locating all emissions sources requires the auditor to examine the interior and exterior of all buildings, storage facilities, and other structures. The auditor will look for building vents, ventilators, clerestories, stacks, exhaust ports, and fans, and will examine these for evidence of visible emissions. The auditor will also look for fugitive emissions from piping systems, transfer operations, materials storage, and roads, and will determine whether control strategies are operational and effective.

The auditor will check pollution control equipment against the inventory and determine the operating condition of the equipment. Continuous monitors and stack probes will be noted. The auditor will not conduct stack tests but will observe plumes for opacity (visual opacity checks should be done only by a certified smoke tester). The auditor will not inspect each piece of control equipment in depth. At a minimum, monitoring equipment and gauges will be evaluated for proper operation, and daily/weekly operation logs will be checked to determine whether they are current.

Figure 17–6 shows minimum monitoring and record-keeping practices for selected types of air pollution control equipment that are likely to be evaluated by an auditor. In addition to these monitoring and record-keeping checks, the auditor will look for visible evidence of poor control of emissions and condition of equipment. Some examples of checks that the auditor might make on control equipment include checking cyclones for noticeable emissions or settled deposits around the outlet, rate of discharge, and pressure drop across the unit. Baghouses can be checked for visible emissions and signs of poor dust control or reduced solids removal, and pressure drop records can be evaluated. Filters used to control particles can be evaluated for proper changing of the filters by examining pressure drop records and change cycles. The auditor can check filtration units for signs of inadequate capacity or malfunctions by looking for evidence of incomplete fume clearance in the work area or oil stains around the outlet stack. Because the filters are discarded after use, the disposal method for the filters should also be evaluated.

There are many sources of potential problems with the electrostatic precipitator, including plugging of the prefilter, inadequate cleaning of the plates, improper voltage differential, and undersized units. The auditor can look for evidence of problems with electrostatic precipitators, such as insufficient flow and opacity. It may be possible for the auditor to hear arcing outside of the unit, and, if so, the rate can be counted and compared to manufacturer specifications as an indicator of voltage differential. Similarly, the frequency of rapping can be noted. Proper functioning of the electrostatic precipitator requires a complete inspection by a qualified technician, and maintenance should be checked to evaluate routine problems and the last date of inspection.

Scrubbers typically have common problems that include plugging, corrosion, and pump malfunctions. Scrubbers can be examined for visible emissions,

FIGURE 17–6.   Generally Accepted Record-Keeping and Monitoring Practices for Selected Types of Air Pollution Control Equipment:

Type of Pollution Control Equipment	Monitoring	Record Keeping
Baghouse (fabric filter)	• pressure drop	• record pressure drop at least once/day when in operation • record corrective action taken when pressure drop or opacity readings indicate there is a problem; includes recording when each bag is replaced
Cyclone/Rotoclone	• pressure drop; and • level indicator readings (if applicable) • water pressure for rotoclones only	• record pressure drop, level indicator reading, and water pressure at least once/day when in operation • record corrective action taken when pressure drop, level indicator readings, or opacity readings indicate there is a problem
ESP (Electrostatic Precipitator)	• primary and secondary voltage • primary and secondary current • sparking rate • number of fields on line	• record each parameter at least once every calendar day when in operation • record corrective action taken when monitored parameters indicate that there is a problem
Wet Scrubber	• pressure drop across the absorber and demister in inches of water • additive concentration, if applicable • pH of liquid, if applicable • liquid flow rate in gpm	• record each parameter at least once/day when in operation • record corrective action taken when monitored parameters indicate that there is a problem
Thermal Oxidizer	• combustion temperature; or • inlet temperature and outlet temperature	• continuous hard copy readout; or • manual readings every 15 minutes when in operation • record corrective action taken when combustion temperature/inlet and outlet temperature indicate(s) that there is a problem
Catalytic Oxidizer	• inlet temperature and outlet temperature across the bed	• continuous hard copy readout of temperatures, or • manual readings every 15 minutes when in operation • record the quarterly results of the catalyst reactivity testing and any corrective action taken as a result of the testing • record corrective action taken when temperatures indicate that there is a problem
Adsorber	• operational temperature, where appropriate • amount of solvent adsorbed or • VOC breakthrough monitor	• record temperature once/day when in operation • record each time the adsorber is cleaned or replaced • record the results for the adsorbability and retentivity tests every quarter • measure and record solvent adsorbed monthly • record all alarms from the VOC breakthrough monitor and the resulting corrective action
Wall Filter	• condition of the filters (alignment, saturation, tears, holes, and so forth)	• inspect the filters at least once/day when in operation • record the condition of the filters and the corrective action taken

Source: Adapted from Indiana Department of Environmental Management, Office of Air Management, *Operating Permit Application Forms & Instructions Part 70/Enhanced NSR/FESOP* (Indianapolis, IN: Indiana Department of Environmental Management, 1995).

pressure drop, operation of pumps, signs of corrosion on plumbing lines, and operating recirculating pumps. The disposal of sludge should be evaluated.

It is difficult for an auditor to perform visual checks to determine whether adsorbers and condensation units are operating properly. The vapors are not visible, and odors are difficult to evaluate. The auditor will examine maintenance and operation records and will recommend control efficiency tests for indications of problems with the units. The auditor will check the combustion temperature of incinerators and compare them to the permit or design specifications.

The auditor will examine tanks and silos for evidence of fugitive emissions and will note the location and condition of control equipment, vents, vapor collection systems, or balance filling systems. Moving belts, screw conveyors, bucket elevators, and other mechanical devices that move or transfer solids will also be evaluated for fugitive emissions. Fugitive emissions from liquid transfer systems are also important, especially in refineries and chemical plants. Facilities with the potential for these fugitive emissions should have well-established programs for routinely checking and correcting piping leaks. The auditor will examine the survey records and calibration records for the instruments that are used to measure the contaminants. The auditor will look for visible evidence of emissions from process operations by checking for dusty emissions on walls or roofs of buildings and from windows and doors. The release of emissions from roadways and storage areas will be evaluated by looking for evidence of blowoff from these sources at the time of the evaluation and evidence of past releases to adjacent areas. The auditor will also examine records for evidence of required dust suppression activities.

The auditor will examine records related to the maintenance and operation of control equipment. Proactive maintenance programs demonstrate a commitment to minimize environmental impacts by keeping equipment in top operational condition. A proactive maintenance program will include a schedule of daily and periodic maintenance activities and relevant procedures, a spare parts inventory, and a maintenance log for each piece of control equipment. Operational procedures should be documented, and records of daily checks and data collection should be current and available for inspection. The auditor will also review the plant's procedures for reporting upset conditions.

Similarly, the auditor will determine whether monitoring equipment is operational and meets performance requirements. Checks will include a review of procedures and records for operation, maintenance, and calibration activities; frequency of data collection and calibration; percentage of downtime; data reporting requirements; and other quality assurance/quality control (QA/QC) aspects. If outside contractors are employed for monitoring and/or analysis, the auditor will determine whether the contractor complies with applicable requirements.

At the end of the audit, the audit team debriefs the management of the facility. Final activities include drafting the audit report, responding to review comments, and submitting the final report. Facility or site management will follow up on the audit findings, initiate corrective actions, determine the effectiveness of the actions, and report back to the audit team. Figure 17–7 gives examples of typical problems that might be noted during the air pollution audit.

FIGURE 17–7.   Typical Problems Encountered During Air Pollution Audits:

Problems Related to Sources and Control Equipment

- missing or incomplete inventory
- air pollution sources not identified
- permits not obtained or outdated
- exemption letters not available
- missing or incomplete operation and maintenance procedures and records
- performance test records not available
- sulfur content of fuel not in compliance with regulatory limits
- sulfur content of fuel certificate missing
- incinerator operations not maintained at correct temperature requirements and charging rates
- incomplete or missing vapor controls
- excessive unplanned releases and downtime
- visible source and fugitive emissions

Problems Related to Monitoring Equipment

- missing or incomplete operation, maintenance, and calibration procedures and records
- excessive downtime
- missing or incomplete QA/QC program on the part of facility or contractors

Problems Related to Reporting and Management

- missing or inadequate Accidental Release Prevention Plan
- air pollution alert and emergency plan not available or incomplete
- modifications to air sources not reported to regulators
- asbestos abatement or demolition not reported to regulators prior to the initiation of work
- upset conditions not reported or improperly reported to regulators
- missing or incomplete personnel training program and records
- missing or inadequate corrective action program

SUMMARY

The CAA regulatory framework for air pollutants (and good management practices) is directed toward ensuring that releases of contaminants to the outdoor air are kept as low as practicable. Air pollution sources must meet a variety of regulations for stationary stack and fugitive sources, and for mobile sources. The Title V amendments to the CAA establish a new national permit system that coordinates all requirements applicable to a source in a single permit. Attaining and maintaining compliance with air pollution regulations requires a management system that allows sources and contaminants to be identified, ensures good record keeping and document control, provides adequate operational controls, and ensures competent laboratory measurement and support.

FIGURE 17–5. Air Pollution Checklist:

	Response		Questions to Ask
YES	**NO**	**N/A**	***General Requirements: Emissions Inventory, Control Equipment, and Monitoring Equipment***

_____ _____ _____ 1. Does the facility have an emissions inventory?

_____ _____ _____ 2. Is the inventory complete?

_____ _____ _____ 3. Is the inventory current?

_____ _____ _____ 4. Are there any changes in emissions levels since the inventory was last updated?

_____ date of last update

Comments: _____

_____

_____

_____ _____ _____ 5. Does the facility have a system for regularly inspecting and maintaining control devices?

_____ _____ _____ 6. Is the inspection and maintenance program documented?

_____ _____ _____ 7. Does the system include a plan for corrective action?

Comments: _____

_____

_____

8. Does the facility have a system for regularly inspecting and maintaining instruments and equipment for monitoring:

_____ _____ _____ ambient air?

_____ _____ _____ stack tests?

_____ _____ _____ in-stack monitors?

Comments: _____

_____

_____

9. Does the monitoring system include QA/QC for:

_____ _____ _____ laboratory analyses?

_____ _____ _____ field data collection?

Comments: _____

_____

_____

| **YES** | **NO** | **N/A** | ***Permits*** |

10. List the permits held by the facility and the major requirements of the permits. Use attached page to describe.

11. Does the facility meet its:

_____ _____ _____ emission limits?

FIGURE 17–5. (*Continued*)

YES	NO	N/A	*Permits*
_____	_____	_____	monitoring requirements?
_____	_____	_____	data reporting requirements?
_____	_____	_____	upset conditions/startup/shutdown reporting requirements?

Comments: _____

_____

_____

| _____ | _____ | _____ | 12. Are all emission points, processes, and appropriate equipment covered by permits or exemptions? |

Comments: _____

_____

_____

| _____ | _____ | _____ | 13. Are copies of all permits and exemptions readily available at the facility? |
| _____ | _____ | _____ | 14. Are the facility's permits current? |

If no, list any that have expired: _____

_____

_____

| _____ | _____ | _____ | 15. Is there a system to ensure timely renewal of air permits and associated fees? |
| _____ | _____ | _____ | 16. Have any emission levels changed significantly since the permit/exemption applications and/or approvals? |

Comments: _____

_____

_____

| _____ | _____ | _____ | 17. Have any operating practices changed signficantly since the permit/exemption applications and/or approvals? |

If yes, describe: _____

_____

_____

_____	_____	_____	18. Has the facility had any air pollution-related inspections during the last calendar year?
_____	_____	_____	If yes, were any deficiencies noted? (Attach list of deficiencies.)
_____	_____	_____	Were deficiencies corrected? (Attach list of corrective actions.)

FIGURE 17–5. *(Continued)*

YES	NO	N/A	*Permits*
_____	_____	_____	19. Is the facility operating under a Consent Decree and/or compliance schedule?
_____	_____	_____	If yes, is the facility on schedule?
			Comments: _____
			_____
			_____
_____	_____	_____	20. Does the facility have future plans that would require new permits?
			If yes, describe: _____
_____	_____	_____	21. Which of the following regulatory programs apply to the facility? For each regulatory program that applies, attach a sheet that describes the status of the facility's operations for each program area.
_____	_____	_____	NESHAPS?
_____	_____	_____	NSPS?
_____	_____	_____	NSR in a nonattainment area?
_____	_____	_____	NSR in an attainment area?
_____	_____	_____	NAAQS, increment and visibility requirements?
_____	_____	_____	Title V acid rain program?
_____	_____	_____	Stratospheric Ozone Protection?
_____	_____	_____	HAPs?
_____	_____	_____	Accident Prevention Program?
_____	_____	_____	Title V mobile sources?

YES	NO	N/A	*Accident Prevention Program*
_____	_____	_____	22. Does the facility have a plan to prevent accidental releases?
_____	_____	_____	23. Does the plan list hazardous substances and quantities on site?
_____	_____	_____	24. Does the plan include a program for training employees in appropriate operating procedures?
_____	_____	_____	25. Does the training program include refresher training at least once every 3 years?
_____	_____	_____	26. Is training documented?
_____	_____	_____	27. Is the effectiveness of training evaluated?
_____	_____	_____	28. Does the plan contain provisions for leak detection and preventing accidental releases?
_____	_____	_____	29. Are leak-detection and accidental release prevention measures documented?

FIGURE 17–5. *(Continued)*

YES	NO	N/A	*Permits*
———	———	———	30. Is there a system of corrective action when leaks or accidental releases are detected?

YES	NO	N/A	*Air Pollution Incident Plan*
———	———	———	31. Does the facility have an air pollution incident plan?
———	———	———	32. Is the plan reviewed and updated on a regular basis?
———	———	———	33. Are copies available at the facility for review and inspection?
———	———	———	34. Has the plan ever been activated by the state?

If so, identify the level and frequency

_____ Alert

_____ Warning

_____ Emergency

———	———	———	35. Are scheduled briefings for all staff conducted frequently enough to ensure that all staff understand the plan and the actions required by the plan?

ENDNOTES

[1] Under the CAA Amendments, the EPA must promulgate standards that require the installation of control technology that will result in the maximum degree of reductions of HAPs that it determines is achievable. This requirement is called the *maximum achievable control technology* (MACT) standard. The MACT is to be based on the best technology currently available for a given source category. For new sources, the MACT standard must also be at least as stringent as the level achieved in practice by the best controlled source in the source category. For existing sources, the MACT standard must be at least as stringent as the level achieved in practice by the best performing group of sources in the source category. Existing sources may be exempt from the MACT or may be eligible for extensions of deadlines under some conditions. Appendix E.2 provides a listing of the source categories and scheduled promulgation dates for the MACT standard in each category.

[2] Stratospheric ozone-depleting chemicals are grouped into Class I substances (chlorofluorocarbons, CFCs) and Class II substances (hydrochlorofluorocarbons, HCFCs). The production of all Class I substances is to be phased out by the year 2000 (2002 for methyl chloroform). The use and production of all Class II substances is scheduled to be phased out by 2030. Appendix E.3 provides a listing of the Class I and Class II substances.

[3] The CAA of 1970 required new or significantly modified sources to comply with the NSPS (New Source Performance Standards). The NSPS are emissions standards established by the EPA, and they apply to all new sources in designated source categories, regardless of the location of the source.

[4] The 1970 CAA established a program to regulate hazardous or toxic pollutants through emissions standards. As originally conceived, NESHAPs (National Emission Standards for Hazardous Air Pollutants) could be implemented more quickly than could a new NAAQS. In practice, the designation of NESHAPS was very slow (only seven substances were regulated in 20 years by the EPA). The 1990 CAA Amendments substantially increased the number of pollutants regulated as hazardous. Appendix E.4 provides a list of the NESHAPs pollutants and source categories.

[5] The threshold quantities are listed and updated in 40 CFR Part 68. An EPA publication, *Title III List of Lists,* can help facilities handling chemicals to determine whether they are subject to regulatory requirements under Sections 302, 304, or 313 of SARA Title III [EPCRA or Section 112(r) of the CAA Amendments]. In addition, this publication provides lists of Resource Conservation and Recovery Act (RCRA) waste streams and unlisted hazardous wastes and radionuclides reportable under the Comprehensive Environmental Response, Compensation and Liability Act (CERCLA). The list document can be obtained from: EPCRA Document Distribution Center, Attention: NCIPI, 11029 Kenwood Road, Cincinnati, OH 45412 (ask for document number 740-R-95-001 or its updates). A dBase version is also available on disk or through the Internet (through the EPA Public Access Gopher server). The Internet address is **gopher.epa.gov**, or the list can be accessed through dial-in access **(919) 558-0335**. The file can be found by following these menu options: EPA Offices and Regions; Office of Solid Waste and Emergency Response; OSWER publications; *Title III List of Lists.*

www

[6] The requirements under Section 112(r) are similar to OSHA's Process Safety Management Standard (29 CFR Section 1910.119), but there are differences between the two programs in the areas of facilities covered, lists of chemicals, hazard assessment, emergency response requirements, and registration/submission. In addition, the EPA rule is intended to protect public health and environment off site whereas the focus of the OSHA rule is to protect workers from chemical accidents at facilities that use highly toxic, reactive, flammable, or explosive substances.

APPENDIX A

PLANNING AND IMPLEMENTATION QUESTIONNAIRES

A.1 ENVIRONMENTAL MANAGEMENT SYSTEM PLANNING QUESTIONNAIRE

A.2 ENVIRONMENTAL MANAGEMENT SYSTEM IMPLEMENTATION QUESTIONNAIRE

The following questionnaires will help you identify areas of strength and weakness in the existing Environmental Management System (EMS). The planning questionnaire focuses on the planning activities needed to develop the EMS whereas the implementation questionnaire more directly targets whether or not the developed procedures and practices are part of the organization's operational activities.

A.1 ENVIRONMENTAL MANAGEMENT SYSTEM PLANNING QUESTIONNAIRE

The following questions will help you identify areas of strength and weakness in the existing EMS. A "no" means that you may need to develop or more completely characterize a particular element of the EMS. Refer to Chapters 3, 4, and 5 for more information about specific topic areas for planning the EMS and preparing documentation.

Response			Questions to Ask
YES	**NO**	**N/A**	***Environmental Management Policy and Commitment***
_____	_____	_____	1. Does the organization have an environmental policy?
_____	_____	_____	2. Has the policy been approved by top management?
_____	_____	_____	3. Does the policy make it clear that environmental management is one of the highest corporate priorities, and is it incorporated into the highest level of decision making (for example, the board of directors)?
_____	_____	_____	4. Does the policy make it clear to all management, staff, and workers that compliance with the organization's environmental management policy and the provisions of the EMS is critically important?
_____	_____	_____	5. Does the environmental policy include the organization's vision, mission, and core values?
_____	_____	_____	6. Is the policy based on a set of guiding principles?
_____	_____	_____	7. Is the corporate environmental policy appropriate to the activities, products, and services of the organization?
_____	_____	_____	8. Does the policy allow for the creation of a high-level management system that has the objective of reducing pollution and its associated costs (environmental remediation, liability, energy use, and so forth)?
_____	_____	_____	9. Does the policy ensure that new products, processes, and activities will be developed in an environmentally efficient manner?
_____	_____	_____	10. Does the policy provide for employee training?
_____	_____	_____	11. Does the policy include a commitment to regulatory compliance as a minimum?
_____	_____	_____	12. Does the policy include a commitment to the prevention of pollution?
_____	_____	_____	13. Does the policy include a commitment to reduce energy consumption whenever possible?

(*Continued*)

YES	NO	N/A	*Environmental Management Policy and Commitment*
____	____	____	14. Does the policy encourage the development and use of appropriate environmental performance indicators, such as quantity of emissions, waste produced per quantity of finished product, and so forth?
____	____	____	15. Does the policy encourage the use of life-cycle thinking about products?
____	____	____	16. Does the policy address health and safety concerns?
____	____	____	17. Does the policy consider factors specific to the locality?
____	____	____	18. Does the policy encourage movement toward sustainable development?
____	____	____	19. Does the policy include a commitment to continual improvement?
____	____	____	20. Does the policy consider the needs of interested parties, such as community groups?
____	____	____	21. Does the policy include a commitment to a continuing communication program with all stakeholders, both internal and external?
____	____	____	22. Does the policy encourage the development and use of an EMS by all contractors and suppliers?

YES	NO	N/A	*Training*
____	____	____	23. Does the organization have an environmental training procedure that covers all employees and all aspects of the operation?
____	____	____	24. Are all employees trained on the environmental policy?
____	____	____	25. Does this training program include the EMS, corporate policies and procedures; environmental objectives and targets; environmental aspects of specific jobs; pollution prevention; waste minimization; pertinent regulations; health, safety, and worker protection; emergency response; spill prevention and cleanup; roles, responsibilities, and accountability; measurement and evaluation; continual improvement; and other areas of the EMS?
____	____	____	26. Are trainers adequately trained for their training activities?

YES	NO	N/A	*Environmental Aspects and Impacts*
____	____	____	27. Does the organization have a procedure for identifying environmental aspects and impacts?

(Continued)

YES	NO	N/A	Environmental Aspects and Impacts
____	____	____	28. Have the environmental aspects and impacts of the organization's products, processes, services, and activities been evaluated?
____	____	____	29. Have administrative and service activities been evaluated for environmental impacts?
____	____	____	30. Are there any negative environmental impacts from the organization's operations?
____	____	____	31. Are there positive environmental impacts from the organization's operations?
____	____	____	32. When new products, processes, or services are planned, is there a system in place to identify and evaluate the environmental aspects of these new products and processes?
____	____	____	33. When new products, processes, or services are planned, is there a system in place to identify and evaluate the use of energy and natural resources of these new products and processes?
____	____	____	34. Have unique environmental aspects been identified for the organization (for example, adjacent to a wetland, desert location, near a national park, and so forth)?
____	____	____	35. Has potential impact and its likelihood been identified in the event of a process or control mechanism failure?
____	____	____	36. Is there a system in place to prioritize impacts for mitigation and to implement actions to reduce the number and severity of impacts?
____	____	____	37. Is there a system in place to assess environmental impacts before decommissioning a facility or leaving a site?
____	____	____	38. Are employees trained to recognize environmental impacts?

YES	NO	N/A	Legal and Other Requirements
____	____	____	39. Does the organization have a procedure for identifying legal and other requirements?
____	____	____	40. Are requirements from local, state, and federal agencies all evaluated?
____	____	____	41. Are there foreign regulatory requirements that apply to the organization?
____	____	____	42. Does the organization determine what legal and regulatory requirements apply and ensure that they are updated regularly?

(Continued)

YES	NO	N/A	*Legal and Other Requirements*
———	———	———	43. Are there requirements from industry organizations or other nongovernmental entities and are they updated regularly?
———	———	———	44. Do managers and staff have access to all legal and other requirements through an on-line system or other references?
———	———	———	45. Is there a person with responsibility for ensuring that requirements are determined, updated, and communicated?
———	———	———	46. Has the organization identified and complied with regulatory training requirements?
———	———	———	47. Does the organization maintain communication with government agencies regarding regulatory development?
———	———	———	48. Has the organization evaluated local emergency planning regulations and coordinated with the appropriate emergency planning entities?
———	———	———	49. Has the organization evaluated the requirements of the emergency health care provider?

YES	NO	N/A	*Environmental Objectives and Targets*
———	———	———	50. Does the organization have a procedure for establishing environmental objectives and targets?
———	———	———	51. Have environmental objectives and targets been established?
———	———	———	52. Are the established environmental objectives and targets appropriate for the organization's environmental policy?
———	———	———	53. Are the environmental objectives and targets appropriate when the environmental aspects and impacts of the organization are considered?
———	———	———	54. Have employees at all levels of the organization had input into the development of the environmental objectives and targets?
———	———	———	55. Have external affected parties been considered in the development of the environmental objectives and targets?
———	———	———	56. Have specific environmental performance indicators been developed that can help evaluate progress toward environmental objectives and targets?
———	———	———	57. Have responsibilities been assigned for achievement of the environmental objectives and targets?

(Continued)

YES	NO	N/A	**Environmental Objectives and Targets**
____	____	____	58. Is there a system in place to hold management accountable for the achievement of environmental objectives and targets?
____	____	____	59. Are environmental objectives and targets reviewed and updated regularly?
____	____	____	60. Is there a continual improvement program in place that ensures that environmental objectives and targets are achieved and improved as needed?

YES	NO	N/A	**Environmental Management Program/Management System**
____	____	____	61. Has a process been established for development of an environmental management program/management system to achieve the organization's objectives and targets?
____	____	____	62. Does the process involve all people with responsibility for environmental protection, including line workers?
____	____	____	63. Are resources, including staff time and funding, available for development and implementation of the environmental management program/management system?
____	____	____	64. Is the environmental management program/management system responsive to the environmental policy?
____	____	____	65. Is the environmental management program/management system integrated into the quality program and other corporate management programs?
____	____	____	66. Is the environmental management program periodically reviewed?
____	____	____	67. Does the environmental management program/management system include a process of internal audit and management review?
____	____	____	68. Is continual improvement a part of the review process?
____	____	____	69. Does the organization have internal performance criteria for those areas not governed by external criteria or when external criteria are inadequate?
____	____	____	70. Have environmental teams that include management, workers, and other affected parties (all organizational levels) been established to review and design processes?
____	____	____	71. Have similar teams been established for environmental health and safety policy development?

(Continued)

YES	NO	N/A	*Environmental Management Program/Management System*
_____	_____	_____	72. Have teams been created to help solve environmental problems?
_____	_____	_____	73. Does the environmental management program identify responsible parties for each activity?
_____	_____	_____	74. Have adequate resources been made available to those with responsibility for the program?
_____	_____	_____	75. Is there a system for prioritization of environmental activities?
_____	_____	_____	76. Have all legislative and regulatory requirements been identified?
_____	_____	_____	77. Is there a logical, thorough, and responsible process for establishing environmental performance levels?
_____	_____	_____	78. Has a mechanism been established for achieving appropriate performance levels that are directly related to environmental objectives and targets?
_____	_____	_____	79. Is there an internal process that reviews progress toward environmental targets and recommends any necessary changes to procedures?

YES	NO	N/A	*Employee Responsibility and Accountability*
_____	_____	_____	80. Have the roles and responsibilities of each employee been established and communicated?
_____	_____	_____	81. Has performance of the EMS been incorporated into evaluation criteria for managers and employees with responsibilities in the EMS?
_____	_____	_____	82. Is there a system for evaluating employee ideas and incorporating them into the EMS?
_____	_____	_____	83. Do employees at all levels of the organization feel they are part of the EMS, that their ideas are important, and that they have a responsibility to ensure that the organization's objectives and targets are met?
_____	_____	_____	84. Is there a system of accountability for environmental protection that has clearly defined roles and responsibilities?
_____	_____	_____	85. Do employees understand that the environmental policy makes it their responsibility to prevent and stop environmental incidents, even if it interferes with production?

YES	NO	N/A	*Emergency Preparedness and Response*
_____	_____	_____	86. Have emergency preparedness and response procedures been established?

(Continued)

YES	NO	N/A	*Emergency Preparedness and Response*
_____	_____	_____	87. Is there an internal team with responsibility for coordinating emergency response activities?
_____	_____	_____	88. Has the emergency response team been adequately trained in all topics necessary for effective emergency response at the facilities?
_____	_____	_____	89. Have all necessary emergency and spill response supplies been obtained?
_____	_____	_____	90. Is appropriate emergency response equipment on hand?
_____	_____	_____	91. Has staff been trained in the use of emergency response equipment and supplies?
_____	_____	_____	92. Have emergency communications systems been established?
_____	_____	_____	93. Have procedures for emergency evacuations been developed?

YES	NO	N/A	*Prevention of Pollution, Waste Minimization, and Natural Resource Consumption*
_____	_____	_____	94. Does the organization have a procedure to prevent pollution, minimize wastes, and reduce natural resource consumption?
_____	_____	_____	95. Does the pollution prevention policy include a commitment to reduce the use of hazardous chemicals and to reuse, recover, and recycle materials instead of disposing of them whenever possible?
_____	_____	_____	96. Does the organization have a program in place to minimize waste production?
_____	_____	_____	97. Does the waste minimization program include reuse, recycling, and redesign of processes?
_____	_____	_____	98. Does the waste minimization program focus on reducing costs rather than simple regulatory compliance?
_____	_____	_____	99. Does the organization search out new technologies that will improve the EMS?
_____	_____	_____	100. Is the purchasing system of the organization designed to reduce pollution, minimize waste, and conserve natural resources?
_____	_____	_____	101. Is there a system in place that allows each department (including administrative departments) to measure waste and pollution that is produced?

(Continued)

YES	NO	N/A	**Prevention of Pollution, Waste Minimization, and Natural Resource Consumption**
_____	_____	_____	102. Is each department head responsible for reducing those wastes to the lowest possible level?
_____	_____	_____	103. Has the organization explored the sale of by-products or wastes to another organization that might use them as raw materials?
_____	_____	_____	104. Are measures in place to evaluate energy use and to implement innovative methods for reduction of energy use?
_____	_____	_____	105. Does the organization participate in voluntary programs such as U.S. EPA's 33-50 or Green Lights programs?

YES	NO	N/A	**Procurement, Contractors, and Vendors**
_____	_____	_____	106. Does the organization have a procedure to review the EMSs of contractors and suppliers?
_____	_____	_____	107. Are contractors and suppliers encouraged to establish or improve their EMS?
_____	_____	_____	108. Has the organization developed an environmental partnership with customers and suppliers to meet the needs of all and produce collaborative efforts that minimize environmental impacts while reducing costs?
_____	_____	_____	109. Does the organization consider environmental issues in developing capital projects?

YES	NO	N/A	**New Products/Processes**
_____	_____	_____	110. Does the organization have a procedure for identifying environmental aspects for new products or processes?
_____	_____	_____	111. Is there a commitment to minimize the environmental impact of new products or processes?
_____	_____	_____	112. Does the organization use design for environment or life-cycle principles when designing new products/processes?

YES	NO	N/A	**Internal and External Communications**
_____	_____	_____	113. Does the organization have a procedure for internal and external communications?
_____	_____	_____	114. Is there a system of environmental communications in place that meets the needs of internal and external interested parties?

(Continued)

YES	NO	N/A	*Environmental Measurement and Evaluation*
_____	_____	_____	115. Has the organization developed and implemented environmental measurement procedures and work instructions to ensure the gathering of high-quality data?
_____	_____	_____	116. Do the environmental procedures and work instructions include both field and laboratory activities?
_____	_____	_____	117. Do the procedures and work instructions include operation, maintenance, and calibration of equipment?
_____	_____	_____	118. Do the procedures and work instructions include data reduction and disposal of waste?
_____	_____	_____	119. Is preventive maintenance scheduling used?
_____	_____	_____	120. Is there an inventory control program in place?
_____	_____	_____	121. Is there a quality assurance program for field and laboratory data?
_____	_____	_____	122. Does the quality assurance program include quality control parameters to include precision and accuracy?
_____	_____	_____	123. Does the environmental measurement program include a field laboratory chemical hygiene and safety plan?

YES	NO	N/A	*Property Management and Transfers*
_____	_____	_____	124. Does the EMS include procedures for evaluating the environmental hazards of property to be purchased?
_____	_____	_____	125. Does the EMS include procedures for evaluation and disclosure of environmental hazards at the time of property sale?
_____	_____	_____	126. Does the EMS incorporate environmental considerations into property management practices?

YES	NO	N/A	*Media-Specific Procedures*
_____	_____	_____	127. Have transport and management procedures been developed and implemented for hazardous materials?
_____	_____	_____	128. Are hazardous waste management practices in place and regularly updated?
_____	_____	_____	129. Have procedures to protect the health and safety of workers been developed and implemented?
_____	_____	_____	130. Are air management practices in place and regularly updated (efficient combustion, sulfur management,

(Continued)

YES	NO	N/A	*Media-Specific Procedures*
			fuel use minimization, local exhaust ventilation, scrubbers, bag houses, and so forth)?
―――	―――	―――	131. Are water management practices in place and regularly updated (reuse, treatment, storm water management, and so forth)?
―――	―――	―――	132. Are there management practices in place to minimize the risks associated with aboveground and underground storage tanks?
―――	―――	―――	133. Is there a risk reduction process that works to protect workers, the community, and the environment?

A.2 ENVIRONMENTAL MANAGEMENT SYSTEM IMPLEMENTATION QUESTIONNAIRE

This questionnaire will help you identify areas of strength and weakness in the existing Environmental Management System (EMS). A "no" means that you may need to develop or more completely characterize a particular element of the EMS. Refer to Chapters 6 and 7 for more information about implementation, Chapter 8 for checking and corrective action, Chapter 9 for management review and continual improvement, and Chapter 10 for auditing.

Response			Questions to Ask
YES	**NO**	**N/A**	***Ensuring Capability***
_____	_____	_____	1. Does the organization utilize a systematic process for allocating human, physical, technical, and financial resources that facilitates meeting objectives and targets?
_____	_____	_____	2. Are adequate resources allocated to new project planning to allow the consideration of environmental aspects in design of new equipment and processes?
_____	_____	_____	3. Is there a total-cost accounting program in place to track and evaluate the costs (both negative and positive) and benefits of environmental activities?
_____	_____	_____	4. Have cooperative efforts with other industries, trade organizations, or universities been investigated for applicability to the organization?
YES	**NO**	**N/A**	***EMS Alignment and Integration***
_____	_____	_____	5. Does the organization integrate environmental factors into other management systems?
_____	_____	_____	6. Are environmental policies integrated into other organizational policies?
_____	_____	_____	7. Is the allocation of environmental resources included with the budgeting process for other activities?
_____	_____	_____	8. Are waste disposal resources tracked and allocated for each department?
_____	_____	_____	9. Are environmental procedures incorporated into operation manuals for all processes?
_____	_____	_____	10. Have environmental information systems been established?
_____	_____	_____	11. Are environmental factors incorporated into orientation and job training?
_____	_____	_____	12. Does the organizational structure identify employees with environmental responsibility?

(Continued)

YES	NO	N/A	**EMS Alignment and Integration**
____	____	____	13. Is there a system of accountability for accomplishing environmental objectives and targets that is built into the accountability system for other activities?
____	____	____	14. Are employees rewarded for contributions to the environmental management program?
____	____	____	15. Has the performance appraisal system been designed to include performance of environmental responsibilities as evaluation criteria?
____	____	____	16. Are systems in place for measuring environmental performance?
____	____	____	17. Does the organization communication system include provisions for environmental communications (informational, educational, reporting)?
____	____	____	18. Is a conflict resolution process used when environmental factors conflict with other business practices?

YES	NO	N/A	**Accountability and Responsibility**
____	____	____	19. Has responsibility for the effectiveness of the EMS been assigned to a senior manager with sufficient authority, competence, and resources to accomplish objectives and targets?
____	____	____	20. Do all managers and employees understand their accountability and responsibilities for implementation of the EMS?
____	____	____	21. Is there a person responsible for ensuring that sufficient and appropriate training (including emergency response) has been conducted for a successful implementation?
____	____	____	22. Have responsible individuals been identified for environmental compliance activities?
____	____	____	23. Is there a team in place with responsibility for recognition of existing or potential environmental problems and development of solutions to those problems?
____	____	____	24. Is there an individual responsible for ensuring that solutions to problems are implemented?
____	____	____	25. Is there an individual responsible for controlling activities until a solution can be developed when environmental problems are found?
____	____	____	26. Do employees understand that they are responsible for taking the initiative to report problems and recommend solutions?

(Continued)

YES	NO	N/A	*Environmental Awareness and Motivation*
____	____	____	27. Is top management involved in building awareness and motivating employees to incorporate the organization's environmental policy and values into their daily work?
____	____	____	28. Are systems in place to foster employee commitment to the goals of the EMS?
____	____	____	29. Do all members of the organization understand the environmental objectives and targets for which they are responsible?
____	____	____	30. Are employees recognized for achieving environmental objectives and targets?
____	____	____	31. Are employees encouraged to make suggestions that lead to improved environmental performance?

YES	NO	N/A	*Knowledge, Skills, and Training*
____	____	____	32. Are the knowledge and skills necessary for achievement of environmental objectives and targets identified?
____	____	____	33. Are these skills considered when hiring new personnel?
____	____	____	34. Have these skills been built into the organization's training program?
____	____	____	35. Is there a system for updating the skills and knowledge requirements for each position?
____	____	____	36. Is there a system for ensuring that contractors have the required skills and knowledge?
			37. Does the organization training program include:
____	____	____	a. identification of employee training needs?
____	____	____	b. development of a training plan to address those needs?
____	____	____	c. verification that the training program meets regulatory requirements?
____	____	____	d. training of target employee groups?
____	____	____	e. documentation of training received?
____	____	____	f. evaluation of training received?
____	____	____	38. Is there an effective system to track training?

YES	NO	N/A	*Communicating and Reporting*
____	____	____	39. Is there a procedure for internal and external communication?

(Continued)

YES	NO	N/A	**Communicating and Reporting**
——	——	——	40. Is management's commitment to the environmental policy communicated?
——	——	——	41. Is there a feedback mechanism to answer questions and concerns about the organization's environmental performance?
——	——	——	42. Does the communications program include the environmental policy, aspects, impacts, objectives, and targets?
——	——	——	43. Are interested internal and external people informed about the organization's EMS?
——	——	——	44. Are the results of audits and other reviews communicated to those with responsibility for environmental performance?
——	——	——	45. Does the communication program facilitate continual improvement?
——	——	——	46. Is environmental information included in the organization's annual report?
——	——	——	47. Are all required regulatory reports submitted in a timely fashion?

YES	NO	N/A	*EMS Documentation*
——	——	——	48. Are operational processes and procedures defined, documented, and updated as necessary?
——	——	——	49. Is environmental documentation integrated into the organization's overall management system?
			50. Does the documentation of the EMS:
——	——	——	a. have in easily accessible form the environmental policy, objectives, and targets?
——	——	——	b. describe the means of achieving environmental objectives and targets?
——	——	——	c. document the key roles, responsibilities, and procedures?
——	——	——	d. provide direction to related documentation?
——	——	——	e. describe other elements of the organization's management system?
——	——	——	f. demonstrate that the EMS is implemented?
——	——	——	51. Is there a document control system in place that facilitates version control and removal of outdated documents, including date of revision, retention period, division, activity, function, and contact person?

(Continued)

YES	NO	N/A	EMS Documentation
_____	_____	_____	52. Is the documentation process clear and communicated to all?
_____	_____	_____	53. Are there clear procedures for employees to access needed documentation?
_____	_____	_____	54. Are documents periodically reviewed for accuracy and appropriateness, and revised as necessary?
_____	_____	_____	55. Are documents located where they are needed?
_____	_____	_____	56. Are documents converted to electronic format where appropriate?

YES	NO	N/A	Operational Control
_____	_____	_____	57. Are environmental considerations included in design and engineering processes?
_____	_____	_____	58. Are the appropriate people involved to assist with decision making (environmental specialists, industrial hygienists, medical staff, ventilation engineers, and so forth)?
_____	_____	_____	59. Do purchasing procedures include a mechanism for evaluating the environmental impacts of suppliers, vendors, and contractors?
_____	_____	_____	60. Are there procedures in place to minimize the environmental impact of raw materials handling and storage?
_____	_____	_____	61. Are production processes evaluated to determine whether there are environmental impacts that could be reduced?
_____	_____	_____	62. Does the maintenance program have as one of its focuses an emphasis on preventive maintenance to eliminate environmental impacts?
_____	_____	_____	63. Do laboratories and other units that use hazardous materials have laboratory chemical hygiene and safety plans?
_____	_____	_____	64. Are the plans implemented?
_____	_____	_____	65. Do laboratories and other units properly dispose of waste?
_____	_____	_____	66. Are raw materials and finished products stored to minimize waste and prevent environmental impacts?
_____	_____	_____	67. Are regulatory requirements met for hazardous materials transportation?
_____	_____	_____	68. Are fuel use minimization procedures in use?
_____	_____	_____	69. Is energy use minimized for all operations?

(Continued)

YES	NO	N/A	*Operational Control*
____	____	____	70. Does the property acquisition program require an environmental evaluation before property is purchased?
____	____	____	71. Are construction programs designed to minimize environmental impacts?
____	____	____	72. Is the use of packaging materials minimized?
____	____	____	73. Are operational procedures reviewed periodically for updating to meet new requirements?

YES	NO	N/A	*Emergency Preparedness and Response*
____	____	____	74. Has a thorough emergency preparedness plan been developed and implemented?
____	____	____	75. Does the plan consider all of the organization's potential environmental impacts under normal operating conditions, abnormal operating conditions, and potential emergency conditions?
____	____	____	76. Is the organization's structure and responsibility for emergency response clearly explained in the plan?
____	____	____	77. Does the plan include a list of key personnel, including telephone numbers and addresses?
____	____	____	78. Have all of the necessary arrangements been made with emergency services such as ambulance services, hospitals, local emergency response groups, police, and fire departments?
____	____	____	79. Is there a procedure for notifying the utility companies?
____	____	____	80. Is there a procedure for limiting off-site impacts from spills?
____	____	____	81. Is there an evacuation plan for the facility?
____	____	____	82. Is there an evacuation plan for nearby facilities?
____	____	____	83. Is information on all hazardous materials readily available during an emergency situation?
____	____	____	84. Does the training program cover emergency preparedness, and is it adequate?
____	____	____	85. Is annual refresher training conducted?
____	____	____	86. Are emergency response drills conducted?
____	____	____	87. Is there an adequate amount of supplies and equipment to contain and clean up any type of spill that might happen?
____	____	____	88. Is proper personal protective equipment inspected, maintained, and available in adequate quantity to deal with any emergency that might be anticipated?

(Continued)

YES	NO	N/A	**Emergency Preparedness and Response**
_____	_____	_____	89. Is there a written personal protective equipment program?
_____	_____	_____	90. Is there a written respiratory protective equipment program?
_____	_____	_____	91. Are there existing arrangements with a cleanup contractor for removal of contaminated soil, water, or materials?

YES	NO	N/A	**Auditing**
_____	_____	_____	92. Does the organization audit the EMS on a regular basis?
_____	_____	_____	93. Is the audit conducted according to an audit plan and protocol?
_____	_____	_____	94. Is the frequency of audits determined by the nature of the operation and the related environmental aspects and impacts?
_____	_____	_____	95. Are the results of the previous audits considered in determining the frequency of the audits?
_____	_____	_____	96. Are the audit results communicated according to the audit plan?
_____	_____	_____	97. Does management receive a copy of the audit results?
_____	_____	_____	98. Do the audits trigger corrective and and preventive actions?

YES	NO	N/A	**Continual Improvement**
_____	_____	_____	99. Has the organization integrated a continual improvement process into all aspects of the organization?
_____	_____	_____	100. Does the continual improvement process determine the cause(s) of nonconformances?
_____	_____	_____	101. Are corrective and preventive action plans developed and implemented to resolve nonconformance?
_____	_____	_____	102. Is the effectiveness of corrective and preventive action verified?
_____	_____	_____	103. Are corrective and preventive actions and results documented?
_____	_____	_____	104. Is a process of checking objectives and targets integrated into the continual improvement activities?

(*Continued*)

YES	NO	N/A	*Management Review*
_____	_____	_____	105. Does the organization's management periodically review the EMS to ensure its suitability and effectiveness?
			106. Does the EMS review include:
_____	_____	_____	a. a review of environmental objectives, targets, and environmental performance?
_____	_____	_____	b. findings of the EMS audits?
_____	_____	_____	c. an evaluation of its effectiveness?
_____	_____	_____	d. an evaluation of the suitability of the environmental policy?
_____	_____	_____	e. an evaluation of needed changes based on:
_____	_____	_____	1) changing legislation and regulation?
_____	_____	_____	2) changing needs of interested parties?
_____	_____	_____	3) changes in products or activities of the organization?
_____	_____	_____	4) advances in science and technology?
_____	_____	_____	5) knowledge gained from environmental incidents?
_____	_____	_____	6) market preferences?
_____	_____	_____	7) reporting and communications?

EXAMPLE OF AN ENVIRONMENTAL MANAGEMENT POLICY MANUAL

The environmental policy manual that follows is intended to provide an example of how a manual might be structured. Other ways of presenting and organizing the documentation for the EMS are also acceptable. This example manual is not intended to be used in its present form. Each organization will need to develop its own environmental policy and supporting documentation based on how it operates.

Environmental Management Policy Manual

The ABC Food Company

8692 Farlake Road

Scansone Lake, Indiana 46644

Telephone: (812) 555-6362, Fax: (812) 555-6367

November, 1996

ABC Food Company

Rev. 1, 11/96   Page __ of __

Environmental Management Policy Manual

Authorization Statement

This Environmental Management Policy Manual is a statement of the organization, responsibility, procedures, and systems that have been implemented to maintain conformance to the requirements of the ISO 14001 standard and the Company's environmental management program. The ABC Food Company takes its environmental responsibilities seriously and expects all employees to do the same. The Environmental Policy is the document that lays the foundation of our environmental philosophy. Each employee is expected to understand and comply with its provisions.

The requirements of the Environmental Management Policy Manual apply to all employees of the corporation who have been given responsibility for the implementation of the Environmental Management System (EMS) and all the procedures given or referenced in this manual. The relationship between the Procedures Manual and ISO 14001 is given in Section 7.0 of this manual.

The contents of this manual have been reviewed and approved by the senior management team. The contents of the manual are authorized and approved by the Chief Executive of the organization.

This Environmental Management Policy Manual is the property of the ABC Food Company. The manual must not be copied in whole or in part without written authorization from the Quality Assurance Officer of the Company. The manual must be returned to the Company when requested.

Colleen Halpine
Chief Executive Officer

November, 1996

Manual No. 03
Issued To:___ B. Norden ___

Environmental Management Policy Manual

Table of Contents

ABC Food Company

Rev. 1, 11/96   Page 3 of 8

1.0   Scope
      This manual applies to all parts of the organization, its facilities, contractors, suppliers, and vendors. The provisions of the EMS will be applied to process design, operations, transportation, storage, waste management, and administration.

2.0   Environmental Management Responsibility
      The Vice President for Environmental Affairs has overall responsibility for the Company's environmental management program. The Vice President for Environmental Affairs is responsible for the development, implementation, and continual improvement of the EMS and for ensuring that all employees adhere to all its requirements.

      Each employee is responsible for carrying out his or her responsibilities in conformance with the policies, procedures, and work instructions for activities. The Company's policies are found in this manual, procedures in the Procedures Manual, and work instructions in the Work Instructions Manual.

      The Procedures Manual contains the Company's organizational structure, which shows the relationship among members of the Company, as well as responsibilities and authorities for the Environmental Management Program.

3.0   Vision, Mission, and Values
      3.1   Vision
            The ABC Food Company will work to continually improve the price and quality of our products while helping to create a cleaner, healthier environment for the people in our community.
      3.2   Mission
            By holding true to our vision, we will create a safe and healthy work environment in which our people can utilize their abilities to the fullest and can provide high-quality food products to our customers in a cost-effective way.
      3.3   Values
            The values under which The ABC Food Company operates are:
            1) honesty
            2) persistence
            3) patience
            4) community development
            5) cleanliness
            6) determination
            7) customer focus
            8) continual improvement
4.0   Environmental Policy
       As the leading producer of eggs and chicken products in this state, we at the ABC Food Company feel an obligation to be responsible stewards of the environment. We hereby certify that we manage our affairs

ABC Food Company

so that our activities produce minimal negative environmental or health impacts and as many positive impacts as is possible.

We work at all times to prevent pollution and continually strive to improve our operations. We process wastes into useful products or apply those wastes to farm land in ways that improve the fertility of the land without causing runoff that impacts water bodies.

We make sincere efforts in all of our activities to comply with relevant environmental regulations and the environmental policies of the Processed Food Association of America. If unforeseen circumstances cause a temporary violation of applicable standards, we make all efforts to return to compliance in the shortest time possible.

Our management staff annually reviews our environmental performance to evaluate our success in achieving environmental objectives and targets. These managers work with the appropriate staff or other professionals to develop programmatic improvements that continually reduce environmental emissions.

All staff and managers are responsible for creating and keeping appropriate records so that an auditor or manager can easily assess the environmental performance of the company.

This policy is communicated to all employees of the ABC Food Company, and copies are available to interested parties upon request. This policy is discussed in all employee orientation classes.

This policy and all records that do not compromise trade secrets are made available to the public upon request during normal business office hours.

Edward Culver

Chairman, Board of Directors
April 1996

Colleen Halpine

Chief Executive Officer
April 1996

5.0   Supporting Policies
   5.1   Sanitation
      5.1.1   It is our policy to maintain the highest level of sanitation in all of our facilities to prevent the transmission of disease to our workers and customers.
      5.1.2   No product is left in the processing system if it has come in contact with the floor or any other source of contamination.
      5.1.3   No delivery of an animal is accepted if there is evidence that it may be diseased or may have been raised in unsanitary conditions.
      5.1.4   The internal organs of all animals are removed in a way that minimizes the possibility of salmonella bacteria coming in contact with the meat.
      5.1.5   All production and packaging equipment is cleaned and sanitized each work shift. All floors are cleaned and sanitized several times during each work shift.
   5.2   Wastewater Treatment
      5.2.1   It is our policy to treat all wastewater, including process and cleaning waste, at our wastewater treatment facility and to release treated wastewater that exceeds the requirements for quality specified in our permits to operate.

ABC Food Company

    5.2.2 If the treatment facility must be shut down for repairs, all wastewater is held in our retention basin until the treatment facility is operational, at which time the wastewater is pumped to the treatment facility.

    5.2.3 Each time it is necessary to use the retention basin, it is inspected and cleaned upon being emptied. Rainwater that collects in the retention basin when it is clean is pumped into the city storm water system.

  5.3 Employee Health and Safety

    5.3.1 It is our policy to take all appropriate measures to ensure that The ABC Food Company provides a safe and healthful workplace for its employees.

    5.3.2 We strive to exceed all regulatory requirements for health and safety. Emergency preparedness and evacuation procedures are communicated to all employees.

    5.3.4 At no time are safety devices or emergency escape doors removed or compromised.

The Director of Environmental Health and Safety is responsible for establishing and implementing worker health and safety programs.

6.0 Environmental Management System Elements

  6.1 Elements of Planning

    6.1.1 Environmental Aspects

The Company uses and maintains the *Procedure for Identifying Environmental Aspects and Impacts* that ensures that all environmental aspects and impacts are identified for our products and activities. The Environmental Department uses this procedure to perform a risk-based prioritization of environmental aspects and impacts so that management can decide how to apply environmental resources to achieve the most value.

    6.1.2 Legal and Other Requirements

The Company uses and maintains the *Procedure for Identification of Legal and Other Requirements* to guarantee that we have current and correct knowledge of all requirements that we are obliged to meet. This procedure allows us to fulfill our commitment to regulatory compliance and to compliance with industry best practices.

    6.1.3 Objectives and Targets

The Company uses and maintains the *Procedure for Developing and Achieving Objectives and Targets* to ensure that we identify and achieve objectives and targets that accomplish the goals of the Environmental Policy. At a minimum, our objectives and targets comply with regulatory requirements, but we strive whenever possible to exceed these minimum requirements.

    6.1.4 Environmental Management Program

We are committed to the continual improvement of our environmental management program and to keeping our program among the best in the nation. Our environmental management program continually improves through the use of our procedures, industry best practices, and the education and experience of the staff in the Environmental Department.

ABC Food Company

6.2 Elements of Implementation and Operation

    6.2.1 Structure and Responsibility

        The Company uses and maintains the *Procedure for Documenting Roles, Responsibilities, and Authority* to ensure that the structure and system of accountability for implementation of the EMS remains current and clear. Each individual knows what his or her responsibilities are and the authority they are empowered with to carry out those responsibilities.

    6.2.3 Training, Awareness, and Competence

        The Company uses and maintains the *Procedure for Environmental Training* to ensure that all staff members are aware of environmental issues and have adequate training to deal with the responsibilities they are given. We are firmly committed to providing all necessary training to employees.

    6.2.4 Communication

        The Company uses and maintains the *Procedure for Internal and External Communication,* which provides the mechanism whereby two-way communications are maintained between management, supervisors, employees, regulatory agencies, and the public. We believe that maintaining a high level of effective communication is good for the company and the community.

    6.2.5 Environmental Management System Documentation

        The Company uses and maintains the *Procedure for Document and Record Maintenance and Control* to ensure that EMS documentation is adequate, updated, and controlled. In general, these documents are available for inspection by interested parties.

    6.2.6 Document Control

        The Company uses and maintains the *Procedure for Document and Record Maintenance Control* to guarantee that only current documents circulate within the Company. This procedure ensures the security of privileged information and archived documents.

    6.2.7 Operational Control

        The Company uses and maintains the *Procedure for Operational Control and Maintenance* to ensure that the environmental aspects of our operation do not cause undue environmental impacts. Operational control measures are designed into all new equipment and operations. Contractors, vendors, and suppliers are encouraged to implement operational controls to minimize their environmental impacts.

    6.2.8 Emergency Preparedness and Response

        The Company uses and maintains the *Procedure for Emergency Response Planning.* This procedure requires regular updates of the Emergency Response Plan, training in excess of the legal minimums for all response staff, coordination with other response agencies and affected parties, and the use of proper equipment and control measures. The focus of our com-

ABC Food Company

pany is to prevent emergency situations. When that is impossible, we immediately respond to minimize any impacts on workers, the community, or the environment.

6.3 Elements of Checking and Corrective Action

    6.3.1 Monitoring and Measurement

    The Company has and maintains the *Procedure for Monitoring and Measurement* to verify that operational control measures are controlling environmental impacts and are leading to progress toward environmental objectives and targets. Monitoring equipment is operated properly, and performance of the EMS is periodically evaluated using the *Procedure for Audit, Review, and Continual Improvement.*

    6.3.2 Nonconformance and Corrective and Preventive Action

    The Company uses and maintains procedure for *Audit, Review, and Continual Improvement.* This procedure and the *Procedure for Prevention of Pollution* are used to prevent, identify, control, and correct EMS nonconformances. It is our policy to correct all nonconformances in an expeditious manner.

    6.3.3 Environmental Management System Audit

    The Company uses and maintains *Procedure for Audit, Review, and Continual Improvement* to ensure that thorough and timely audits are performed. It is our policy to conduct internal audits at appropriate intervals to evaluate the performance of the EMS and to identify and correct nonconformances. It is also our policy to conduct a third-party certification audit every 3 years.

7.0 Relationship Between the Procedures Manual and ISO 14001

The following list shows how the Environmental Management Policy Manual and Procedures Manual are related to the elements of ISO 14001.

    7.1 *Environmental Policy*

    Environmental Management Policy Manual

    7.2.1 *Environmental Aspects*

    Procedures 1, 5, 6, 7, 8, 11, 12

    7.2.2 *Legal and Other Requirements*

    Procedures 2, 4, 5, 6, 7, 10, 11, 12

    7.2.3 *Environmental Management Program*

    Environmental Management Policy Manual and All Procedures

    7.3.1 *Structure and Responsibility*

    Environmental Management Policy Manual and Procedures 4, 8, 9, 10, 12, 13, 18, 22

    7.3.2 *Training, Awareness, and Competence*

    Procedure 5 and All Procedures

    7.3.3 *Communication*

    Procedure 6 and All Procedures

    7.3.4 *Environmental Management System Documentation*

    Environmental Management Policy Manual and All Procedures

ABC Food Company

7.1.8   *Document Control*
         Procedure 7
7.1.9   *Operational Control*
         Procedures 1, 2, 3, 4, 5, 6, 8, 12, 13, 14, 15, 16, 17, 21, 25
7.1.10   *Emergency Preparedness and Response*
         Procedures 5, 9, 10, 11, 12, 13
7.1.11   *Monitoring and Measurement*
         Procedures 5, 7, 9, 11, 12, 13
7.1.12   *Nonconformance and Corrective and Preventive Action*
         Procedures 1, 2, 3, 4, 5, 6, 7, 8, 11, 12, 13, 14, 16, 20
7.1.13   *Environmental Management System Audit*
         Procedure 12 and All Procedures

APPENDIX C

EXAMPLES OF PROCEDURES

The procedures that follow are intended to provide examples of how procedures might be structured around the requirements of the standard. They are not intended to be used in their present form but can serve as a template for customizing procedures. Each organization will need to develop its own procedures based on how it operates. In addition to the procedures required by the standard, each organization will have a unique set of operational procedures for the different areas of its operations (for example, sanitation, air pollution control equipment, wastewater treatment facility, and so forth). The examples of related procedures, documentation, and records that are included in each procedure are only for illustrative purposes. These documents are not included in the Appendix.

EXAMPLE PROCEDURES

Identifying Environmental Aspects and Impacts

Identifying Legal and Other Requirements

Developing and Achieving Environmental Targets and Objectives

Documenting Roles, Responsibilities and Authority

Environmental Training

Internal and External Communication

Document and Record Maintenance and Control

Operational Control

Emergency Response Planning

Spill and Disaster Management

Monitoring and Measurement

Pollution Prevention

Audit, Review, and Continual Improvement

ABC Company

Procedure No.: ___

Revision No.:   ___
Revision Date: ___

Title: Procedure for Identifying Environmental
Aspects and Impacts
Page ___ of ___

> This procedure is a controlled document. The contents of this procedure are
> authorized and approved by __Name and Title__. This procedure is issued to
> __Name__ on __Date__. It must not be copied in whole or part without
> written authorization. It must be returned to the ABC Company when
> requested. Other persons in possession of this document have an
> uncontrolled copy and should return it to __Department Name__.

PROCEDURE FOR IDENTIFYING ENVIRONMENTAL ASPECTS AND IMPACTS

1.0 Purpose

    1.1 To facilitate the identification of the environmental aspects and impacts of organizational operations.

    1.2 This procedure is used to evaluate all environmental aspects and impacts and to identify those that are significant so they can be assigned high priority for mitigation.

2.0 Scope

    2.1 This procedure applies to all operations performed at Company facilities.

    2.2 Suppliers, vendors, and contractors are encouraged to identify the environmental aspects and impacts of their operations.

3.0 Responsibilities

    3.1 The Director of Environmental Health and Safety is responsible for overseeing the process of identifying environmental aspects and impacts.

    3.2 Supervisory and management personnel are responsible for working with the Environmental Department to identify the environmental aspects and impacts in the areas for which they have responsibility.

    3.3 All employees are responsible for reporting actual potential environmental aspects and impacts to management.

    3.4 The Purchasing Department is responsible for encouraging suppliers, vendors, and contractors to identify environmental aspects and impacts for the services solicited by the Company.

4.0 Procedure

    4.1 The process for identifying environmental aspects and impacts begins with a review of existing documentation by the Environmental Department. This review includes monitoring reports; documents prepared for regulatory agencies; existing programs, procedures, and work instructions; audit reports; and records that are kept to document operations.

ABC Company

Procedure No.: ___   Title: Procedure for Identifying Environmental
Aspects and Impacts

Revision No.:   ___   Page ___ of ___

Revision Date: ___

4.2   The results of the document review are evaluated to identify those operations for which there is inadequate information to evaluate environmental aspects and impacts.

4.3   When information needs are identified, measures are taken to evaluate operations in detail. These measures may include interviews with workers, process analysis, flowcharting, monitoring, or other methods that provide the needed information.

4.4   The identification process considers normal operating conditions, abnormal operating conditions, failure of control mechanisms, and emergency conditions.

4.5   The identification process evaluates emissions to air, discharges to water, waste management, contamination of soils, the impact of releases on the community, use of raw materials and natural resources, local issues, and other issues determined to be relevant.

4.6   When all environmental aspects and impacts are identified, they are evaluated to determine which are significant environmental aspects.

4.7   Environmental aspects and impacts are ranked by their potential to cause harm to human health or the environment. Allocation of environmental resources gives priority to impact that is significant.

5.0   Related Procedures
     5.1   Internal and External Communications
     5.2   Legal and Other Requirements
     5.3   Process Analysis

6.0   Documentation
     6.1   Computerized Regulations Databases
     6.2   Work Instructions Manual
     6.3   Monitoring Reports
     6.4   Audit Reports
     6.5   Process Flowcharts

7.0   Records
The priority listing of aspects and impacts is filed by the Environmental Department and retained until revised, then archived for 3 years.

ABC Company

Procedure No.: ____

Revision No.: ____
Revision Date: ____

Title: Procedure for Identifying Legal
and Other Requirements
Page ____ of ____

> This procedure is a controlled document. The contents of this procedure are authorized and approved by __ Name and Title __. This procedure is issued to __Name__ on __ Date __. It must not be copied in whole or part without written authorization. It must be returned to the ABC Company when requested. Other persons in possession of this document have an uncontrolled copy and should return it to __ Department Name __.

PROCEDURE FOR IDENTIFYING LEGAL AND OTHER REQUIREMENTS

1.0 Purpose
    1.1   To identify legal and other requirements affecting the organization in order to include these requirements in the provisions of the EMS.

2.0 Scope
    2.1   This procedure identifies which statutes and regulations apply to the operations of the organization.
    2.2   This procedure identifies other requirements such as operating permits, consent decrees, agreed orders, industry codes of practice, and applicable voluntary guidelines and standards that apply to the operations of the organization.

3.0 Responsibilities
    3.1   The Director of Environmental Health and Safety has overall responsibility for the operation of this procedure.

4.0 Procedure
    4.1   The Director of Environmental Health and Safety supervises the review of records to identify any requirements that are included in operating permits, agreements with government agencies, court settlements, or other requirements subscribed to by the organization.
    4.2   The organization subscribes to computerized databases that include all pertinent laws, regulations, policies, and other appropriate documents. The databases are updated at least quarterly.
    4.3   The Director of Environmental Health and Safety ensures that an initial review and periodic reviews of the database are performed to identify all applicable requirements.
    4.4   The requirements identified by Sections 4.1 and 4.3 of this procedure are communicated to all appropriate parties in such a way that action can be taken to achieve compliance with these requirements.

5.0 Related Procedures
    5.1   Identification of Environmental Aspects and Impacts

ABC Company

Procedure No.: ___   Title: Procedure for Identifying Legal
and Other Requirements

Revision No.:   ___   Page ___ of ___
Revision Date: ___

    5.2   Internal and External Communication
    5.3   Developing and Achieving Environmental Objectives and Targets
    5.4   Environmental Training
    5.5   Emergency Response Planning
    5.6   Hazardous Waste Management
    5.7   All Operational Procedures

6.0   Documentation
    6.1   Regulations Manual
    6.2   Database Work Instructions
    6.3   Voluntary Guidelines Manual
    6.4   All Records

7.0   Records
    7.1   The listing of regulations and voluntary guidelines is filed by the Environmental Department and retained until revised, then archived for 5 years.

ABC Company

Procedure No.: ___

Revision No.: ___
Revision Date: ___

Title: Procedure for Developing and Achieving
Environmental Targets and Objectives
Page ___ of ___

This procedure is a controlled document. The contents of this procedure are authorized and approved by __Name and Title__. This procedure is issued to __Name__ on __Date__. It must not be copied in whole or part without written authorization. It must be returned to the ABC Company when requested. Other persons in possession of this document have an uncontrolled copy and should return it to __Department Name__.

PROCEDURE FOR DEVELOPING AND ACHIEVING ENVIRONMENTAL TARGETS AND OBJECTIVES

1.0 Purpose
    1.1 To identify a framework by which environmental objectives and targets are identified and achieved to accomplish the goals of the Environmental Policy.

2.0 Scope
    2.1 This procedure applies to any operation that has an environmental aspect associated with it.
    2.2 This procedure applies to vendors, suppliers, and contractors.

3.0 Responsibilities
    3.1 The Director of Environmental Health and Safety is responsible for developing environmental objectives and targets for each environmental aspect, impact, or potential impact identified during process analysis.
    3.2 All management staff are responsible for providing assistance to the Environmental Department in the development of environmental objectives and targets. Managers are also responsible for achievement of environmental objectives and targets within their area of responsibility.
    3.3 All employees are responsible for reviewing environmental objectives and targets related to their work areas and for making suggestions to help in the achievement of those objectives and targets.
    3.4 The Environmental Department is responsible for the development of an implementation plan to achieve environmental objectives and targets and to track progress toward achieving the objectives and targets.
    3.5 The Board of Directors and the Chief Executive Officer are responsible for the approval of environmental objectives and targets, and the implementation plan.

ABC Company

Procedure No.: ___

Title: Procedure for Developing and Achieving Environmental Targets and Objectives

Revision No.:   ___

Page ___ of ___

Revision Date: ___

4.0   Procedure
4.1   All environmental objectives and targets are consistent with the environmental policy and in the spirit of the corporate vision, mission, and values.
4.2   Input is sought from all levels within the organization during the development process.
4.3   The views of external interested parties are taken into account when developing objectives and targets.
4.4   Objectives and targets consider the legal requirements applicable to environmental aspects of operations, and objectives and targets are established for achieving regulatory compliance.
4.5   When developing the implementation plan for achieving objectives and targets, total-cost accounting is used to determine the benefits and costs of implementation. The implementation plan utilizes the most cost-effective methods that will lead to achievement of objectives and targets.
4.6   Risk-based prioritization of implementation activities is used to maximize the effectiveness of available resources.
4.7   The implementation plan for achievement of objectives and targets uses pollution prevention and waste minimization techniques wherever possible.
4.8   Targets established are specific and measurable.
4.9   Environmental objectives are developed for the design process so that environmental impact is minimized in production, use, and disposal of products and processes.
4.10 Environmental indicators are established that can be used to measure the effectiveness of objectives and targets in accomplishing the goals of the environmental policy.
4.11 Objectives and targets are reviewed and modified whenever audits or other EMS reviews indicate that they are not adequate to achieve the goals of the environmental policy.
4.12 Environmental objectives and targets are communicated to all employees and interested parties.
5.0   Related Procedures
5.1   Identification of Environmental Aspects and Impacts
5.2   Internal and External Communications
5.3   Audit, Review, and Continual Improvement
5.4   Pollution Prevention
6.0   Documentation
6.1   Implementation Plan for Objectives and Targets

ABC Company

Procedure No.: ___

Title: Procedure for Developing and Achieving
Environmental Targets and Objectives

Revision No.:   ___

Page ___ of ___

Revision Date: ___

    6.2   Regulations Manual
    6.3   Total-Cost Accounting Manual
    6.4   Audit and Review Records
    6.5   Monitoring Reports
7.0   Records
    7.1   The listing of objectives and targets is filed by the Environmental Department and retained until revised, then archived for 3 years.

ABC Company

Procedure No.: ___

Revision No.:   ___
Revision Date: ___

Title: Procedure for Documenting Roles,
Responsibilities, and Authority
Page ___ of ___

> This procedure is a controlled document. The contents of this procedure are
> authorized and approved by _Name and Title_. This procedure is issued to
> _Name_ on __Date__. It must not be copied in whole or part without
> written authorization. It must be returned to the ABC Company when
> requested. Other persons in possession of this document have an
> uncontrolled copy and should return it to __Department Name__.

PROCEDURE FOR DOCUMENTING ROLES, RESPONSIBILITIES, AND AUTHORITY

1.0   Purpose
   1.1   To document roles, responsibilities, and authority for the implementation of the EMS.

2.0   Scope
   2.1   This procedure applies to all operations within the organization.

3.0   Responsibilities
   3.1   The Vice President for Operations has overall responsibility for the development of the appropriate documentation.
   3.2   Management staff provides assistance in the identification of environmental roles and responsibilities.
   3.3   The Director of Environmental Health and Safety ensures that appropriate documentation of roles, responsibilities, and authority is included in the EMS.

4.0   Procedure
   4.1   Organizational charts are developed for each department in the organization.
   4.2   Responsibilities associated with each environmental objective and target are developed and documented.
   4.3   Those responsibilities that are not directly environmental but that contribute to the overall environmental management effort (such as performance appraisal or resource allocation) are identified and documented.
   4.4   The authority delegated with each of the environmental responsibilities is identified and documented. These authorities must be adequate to carry out the assigned responsibilities.
   4.5   The role of each job classification in implementing the EMS is identified and documented.

ABC Company

Procedure No.: ___   Title: Procedure for Documenting Roles,
                                            Responsibilities, and Authority
Revision No.:   ___   Page ___ of ___
Revision Date: ___

    4.6   The environmental responsibilities and authorities (such as the authority to stop production) of employees are clearly explained during orientation and training programs.

5.0   Related Procedures
    5.1   All procedures

6.0   Documentation
    6.1   Job Description Manual
    6.2   Organizational Chart

7.0   Records
    7.1   Records of roles and responsibilities are filed by the Environmental Department and retained until revised, then archived for 10 years.
    7.2   Organizational charts are filed by the Environmental Department and retained until revised, then archived for 3 years.

ABC Company

Procedure No.: ___   Title: Procedure for Environmental Training
Revision No.:   ___   Page ___ of ___
Revision Date: ___

This procedure is a controlled document. The contents of this procedure are authorized and approved by __Name and Title__. This procedure is issued to __Name__ on __Date__. It must not be copied in whole or part without written authorization. It must be returned to the ABC Company when requested. Other persons in possession of this document have an uncontrolled copy and should return it to __Department Name__.

ENVIRONMENTAL TRAINING PROCEDURE

1.0 Purpose

  1.1 To provide a mechanism to identify the training needs of employees, provide training, track training, and provide appropriate record keeping.

  1.2 The training program focus is meeting the applicable regulatory training requirements and providing training that will contribute to meeting the objectives and targets of the EMS.

2.0 Scope

  2.1 The training covered by this procedure includes all training that has an environmental component.

  2.2 Whenever appropriate and possible, the training covered by this procedure is incorporated into the orientation and job training provided to all employees.

3.0 Responsibilities

  3.1 The Director of Environmental Health and Safety is responsible for designing a system to identify training needs throughout the organization.

  3.2 All supervisory personnel are responsible for working with the Environmental Department to identify these needs and to assure that they are met.

  3.3 The Environmental Department conducts training that relates to Environmental Department matters of environmental health and safety, or they arrange for outside providers to conduct the training.

  3.4 When environmental training is integrated into other corporate training programs (preferred if possible), the Environmental Department will work with the Training Department to develop the appropriate training materials.

4.0 Procedure

  4.1 A training needs assessment is conducted and periodically reviewed to ensure that training programs are meeting the needs of the company.

ABC Company

Procedure No.: ___

Revision No.: ___

Revision Date: ___

Title: Procedure for Environmental Training

Page ___ of ___

4.2 Regulations are reviewed on an ongoing basis to ensure that the training being provided meets all requirements.

4.3 When an EMS nonconformance is documented in an audit finding, the nonconformance is evaluated to determine whether additional or different training can help correct the nonconformance.

4.4 Each employee's training record is evaluated prior to job placement to determine the adequacy of the training for the job to which the employee is assigned. Training deficiencies are noted, and arrangements made for the needed training.

4.5 Trainers must demonstrate by education and experience that they are qualified to teach the material in the courses for which they are responsible. Trainers must also demonstrate proficiency in instructional methods appropriate to the material being taught. Trainers must be certified by the appropriate certification body if such certification is required to teach a particular course.

4.6 Each record includes the following information: employee name, employee number, department of employment, name and number of course, date(s) of training, number of hours of training, location of the training, name of instructor, instructor's affiliation, exam results, results of other learning evaluations, and instructor comments.

4.7 All training is tracked. Employees and their supervisors are given notice at least 30 days before refresher training is due. Tracking includes identifying which training is required for each employee and whether or not the employee receives the training. Notices of training deficiencies are given to each employee who is not trained and to that employee's supervisor.

5.0 Related Procedures

5.1 All Procedures

6.0 Documentation

6.1 Job Descriptions Manual

6.2 Employee Training Records

6.3 Work Instructions Manual

6.4 Instructor Manuals

6.5 Instructor Qualifications and Certifications

6.6 Deficiency Notices

6.7 Refresher Notices

6.8 Course Evaluations

6.9 Needs Assessment

7.0 Records

7.1 Training Records are filed by the Training Department and retained for a period of 10 years or as required by regulations, whichever is longer.

ABC Company

Procedure No.: ___   Title: Procedure for Internal and External Communication
Revision No.: ___   Page ___ of ___
Revision Date: ___

> This procedure is a controlled document. The contents of this procedure are authorized and approved by <u>Name and Title</u>. This procedure is issued to <u>Name</u> on <u>Date</u>. It must not be copied in whole or part without written authorization. It must be returned to the ABC Company when requested. Other persons in possession of this document have an uncontrolled copy and should return it to <u>Department Name</u>.

PROCEDURE FOR INTERNAL AND EXTERNAL COMMUNICATION

1.0   Purpose
  1.1   To foster understanding of the EMS and to ensure that employees are aware of the purpose and value of achieving the Company's environmental objectives and targets.
  1.2   To help external interested parties and the community understand the EMS and its role in community improvement.

2.0   Scope
  2.1   This procedure applies to all communications related to the EMS.

3.0   Responsibilities
  3.1   The Director of Environmental Health and Safety is responsible for the development and implementation of the Environmental Communication Plan. The Public Information Office provides assistance in this development.
  3.2   The Public Information Office is responsible for answering inquiries and for supplying press releases to the media.
  3.3   Employees and supervisors are responsible for communicating environmental problems and concerns to the Environmental Department.

4.0   Procedure
  4.1   The plan for environmental communication with employees includes communicating the environmental policy, environmental aspects and impacts, environmental objectives and targets, legal requirements, training programs, and other pertinent information. This plan ensures that two-way communications are maintained.
  4.2   The communications plan includes a confidential mechanism for employees to provide information to the Environmental Department.
  4.3   The company newsletter is used to communicate environmental progress, changes in the EMS, difficulties being encountered in implementation, and the significant contributions of individuals.
  4.4   All communications that are received by the Environmental Department are evaluated, and responses are sent in a timely fashion.

ABC Company

Procedure No.: ___ Title: Procedure for Internal and External Communication
Revision No.: ___
Revision Date: ___ Page ___ of ___

4.5 The annual report lists progress toward environmental objectives and targets.

4.6 The Environmental Department develops environmental training materials for use in orientation and training classes. These materials include exercises, group projects, and videos.

4.7 With the exception of confidential and proprietary information, as well as information protected by attorney/client privilege, the Environmental Department provides, for a reasonable copying fee, information upon request to interest groups or the public.

4.8 The Environmental Department maintains an effective communication program with regulatory agencies.

4.9 The Public Information Office answers inquiries from the media. Inquiries are answered in a timely manner. Information requests are recorded and tracked. For issues where information is not immediately available to the Public Information Office, additional information is requested from the Environmental Department. If gathering information delays the reply, the requestor is notified of the delay and the reason for it.

4.10 All responses to information requests, both internal and external, are documented.

5.0 Related Procedures
   5.1 All Procedures

6.0 Documentation
   6.1 Newsletters
   6.2 Training Manual
   6.3 Information Request Log

7.0 Records
   7.1 Privileged communications are filed by the Legal Department and retained for a period of 10 years or until closure of legal actions, whichever is longer.
   7.2 Nonsensitive communications are filed by the Public Information Office and retained for a period of 3 years.

ABC Company

Procedure No.: ___

Title: Procedure for Document and Record
Maintenance and Control

Page ___ of ___

Revision No.: ___
Revision Date: ___

This procedure is a controlled document. The contents of this procedure are authorized and approved by __Name and Title__. This procedure is issued to __Name__ on __Date__. It must not be copied in whole or part without written authorization. It must be returned to the ABC Company when requested. Other persons in possession of this document have an uncontrolled copy and should return it to __Department Name__.

PROCEDURE FOR DOCUMENT AND RECORD MAINTENANCE AND CONTROL

1.0 Purpose

    1.1   To determine which documents and records are needed, where they should be maintained, and how long they should be retained.

2.0 Scope

    2.1   This procedure applies to all documentation of the EMS, in whatever form, paper or electronic.

3.0 Responsibilities

    3.1   It is the responsibility of the Environmental Department to determine which records exist and to collate them into the correct format.

    3.2   It is the responsibility of the Legal Department to determine what records are required by law or regulation.

    3.3   It is the responsibility of all employees to prepare required documentation in the proper format.

    3.4   It is the responsibility of the Environmental Department, in cooperation with the Training Department, to inform employees of proper record-keeping procedures.

4.0 Procedure

    4.1   The Legal Department determines which records are required by law or regulation and the appropriate retention times.

    4.2   The Environmental Department determines which other records are required to document the EMS properly. The Environmental Department develops a retention schedule for these records.

    4.3   Control and security measures are in place to ensure that all confidential records, such as medical records, remain confidential.

    4.4   Documents that are part of the EMS are organized into an EMS manual that includes policies, procedures, and other documents that are required to implement the EMS.

ABC Company

Procedure No.: ____

Revision No.: ____

Revision Date: ____

Title: Procedure for Document and Record Maintenance and Control

Page ____ of ____

    4.5   Records are indexed and stored in appropriate locations. Electronic copies of all documents and records are entered into the electronic file system.

    4.6   A work instruction manual is maintained. The manual does not contain all work instructions, but it does identify all instructions and state their location.

    4.7   All documents are controlled with revision numbers, dates, titles, and other items needed to track and control the document.

    4.8   Distribution records of all documents are maintained and updated as needed.

    4.9   The Environmental Department distributes and recalls all copies of documents except those protected by attorney/client privilege.

    4.10 Documents protected by attorney/client privilege are distributed by the Legal Department.

5.0   Related Procedures

    5.1   All Procedures

6.0   Documentation

    6.1   All Records

7.0   Records

    7.1   All records are filed by the appropriate department and retained according to the retention schedule, which is filed by the Environmental Department, then archived for 3 years.

ABC Company

Procedure No.: ___   Title: Procedure for Operational Control
Revision No.:   ___   Page ___ of ___
Revision Date: ___

This procedure is a controlled document. The contents of this procedure are authorized and approved by __Name and Title__. This procedure is issued to __Name__ on __Date__. It must not be copied in whole or part without written authorization. It must be returned to the ABC Company when requested. Other persons in possession of this document have an uncontrolled copy and should return it to ___Department Name___.

PROCEDURE FOR OPERATIONAL CONTROL

1.0 Purpose
    1.1 To ensure that the facility is operated in a way that protects the environment and facilitates the achievement of environmental objectives and targets.

2.0 Scope
    2.1 This procedure applies to all operations, including design of new equipment and processes.

3.0 Responsibilities
    3.1 The Director of Environmental Health and Safety evaluates operational procedures and recommends changes as needed.
    3.2 The Director of Engineering works with the Environmental Department when new equipment or processes are designed to ensure the inclusion of environmental considerations in their design.
    3.3 The Vice President for Operations evaluates and implements the recommendations of the Director of Environmental Health and Safety. If the Vice President for Operations concludes that a recommendation cannot be implemented, the Vice President for Operations works with the Director of Environmental Health and Safety to create an alternative that is equally protective of the environment.
    3.4 The Procurement Department works with the Environmental Department to develop procedures that minimize the impacts of purchased goods and services.

4.0 Procedure
    4.1 The Engineering Department involves the appropriate qualified people from the Environmental Health and Safety Department in the research and development phases of new projects. This applies to development of new equipment, processes, procedures, or other areas where it is necessary.
    4.2 The Purchasing Department develops procedures that are appropriate for each type of acquisition to ensure that environmental impacts from purchased goods and services are minimized.

ABC Company

Procedure No.: ___   Title: Procedure for Operational Control
Revision No.:   ___   Page ___ of ___
Revision Date: ___

4.3   Raw materials are handled and stored in ways that minimize or prevent environmental impacts.

4.4   Raw materials are selected that will have the lowest toxicity among those products suitable for the job.

4.5   Processes are designed so that they use the smallest amount of raw materials possible.

4.6   All production lines are operated in a way that minimizes environmental impacts. If a problem arises, production is stopped until normal operation can be restored.

4.7   A thorough maintenance plan is developed and updated as needed.

    4.7.1   Appropriate maintenance schedules are followed.

    4.7.2   Preventive maintenance is the keystone of the maintenance plan. Manufacturer's recommendations for preventive maintenance are followed.

    4.7.3   Appropriate supply and spare parts inventories are maintained.

    4.7.4   The maintenance plan includes procedures for emergency maintenance.

    4.7.5   Preventive and corrective maintenance procedures are reviewed for cost-effectiveness.

    4.7.6   Maintenance staff is trained on all current procedures and receives additional training as needed.

    4.7.7   Management staff from the Maintenance Department is involved in the budget process to ensure that adequate funding for maintenance activities is available.

4.8   The laboratory has a Laboratory Chemical Hygiene Plan and a Laboratory Safety Plan.

    4.8.1   All hazardous materials are properly labeled, stored, transported, and disposed of.

    4.8.2   All employees follow the health and safety procedures.

4.9   Finished products are stored in a manner that eliminates environmental or health and safety impacts.

4.10   All raw materials and finished products are transported in accordance with regulatory requirements.

4.11   The Marketing Department uses the performance of the EMS as a marketing tool.

4.12   To help satisfy our customers, information about the EMS is available to them. Customer service is handled in a way that does not create undue environmental impacts.

4.13   Before property is acquired, a thorough investigation of its history is made to determine whether there are potential environmental problems associated with the property. When problems are

ABC Company

Procedure No.: ___
Revision No.: ___
Revision Date: ___

Title: Procedure for Operational Control
Page ___ of ___

discovered, the Company works with governmental agencies to create a "Brownfield" redevelopment project.

4.14 Proper cleanups are made of any environmental contamination before property divestiture.

4.15 All property owned by the company is managed so as to prevent or mitigate environmental contamination.

4.16 The Director of Environmental Health and Safety reviews procedures to ensure that they reflect the most current policies and regulations.

4.17 Correct operational controls are communicated to all affected parties.

5.0 Related Procedures
   5.1 Internal and External Communications
   5.2 Audit, Review, and Continual Improvement
   5.3 All Operational Procedures

6.0 Documentation
   6.1 Manufacturers' Operation Manuals
   6.2 Maintenance Plan
   6.3 Design specifications
   6.4 Work Instructions Manual
   6.5 All records for maintenance, operations, training, and communications

7.0 Records
   7.1 Maintenance records are filed by the Maintenance Department and retained for the life of the equipment.
   7.2 Records of new product and process development projects are filed by the Engineering Department and retained until project implementation, then archived for 3 years.
   7.3 Records of orders and receiving are filed by the Purchasing Department and retained for 10 years.
   7.4 Shipping records are filed by the Shipping Department and retained for 10 years.
   7.5 Property records are filed by the Legal Department and retained for 30 years.

ABC Company

Procedure No.: ___

Title: Procedure for Emergency Response Planning

Revision No.: ___

Page ___ of ___

Revision Date: ___

This procedure is a controlled document. The contents of this procedure are authorized and approved by __Name and Title__. This procedure is issued to __Name__ on __Date__. It must not be copied in whole or part without written authorization. It must be returned to the ABC Company when requested. Other persons in possession of this document have an uncontrolled copy and should return it to __Department Name__.

PROCEDURE FOR EMERGENCY RESPONSE PLANNING

1.0 Purpose

To minimize the impacts of unplanned releases and other emergency events on the environment and on the health and safety of employees and the community.

2.0 Scope

This procedure covers emergency planning for unplanned environmental releases.

3.0 Responsibilities

   3.1 At each facility operated by ABC Company, the Director of Occupational Safety and Health is responsible for ensuring that an emergency response plan is developed and implemented to meet the requirements of all applicable rules and that it is adequate to meet any additional needs that arise because of operations at that facility.

   3.2 Each supervisor is responsible for implementation of the plan in areas that they supervise. This includes all training requirements listed in Section 6.0.

   3.3 Each employee is responsible for understanding the plan and for following the provisions contained therein.

4.0 Emergency Response Plan Procedure

   4.1 Each plan must contain five basic elements:

      4.1.1 General considerations related to the facility

      4.1.2 Planning for the response

      4.1.3 Training of emergency response staff

      4.1.4 Medical surveillance

      4.1.5 Postresponse operations

   4.2 The Occupational Safety and Health Administration (OSHA) publishes minimum requirements for these operations in 29 CFR Sections 1910.38 and 1910.120. All persons responsible for emergency response are familiar with these documents.

ABC Company

Procedure No.: ___   Title: Procedure for Emergency Response Planning
Revision No.:   ___   Page ___ of ___
Revision Date: ___

    4.3   An Emergency Response Plan, as outlined here, is required by 29 CFR Section 1910.120(q) unless the facility evacuates all employees from the workplace when an emergency occurs and does not permit any of its employees to assist in the handling of the emergency. Facilities that evacuate employees are required to provide an Emergency Action Plan as required by 29 CFR Section 1910.38.

    4.4   The Emergency Response Plan must be written, and available for inspection and release to employees, their representatives, and OSHA personnel. The emergency response plan developed by each facility must, as a minimum, address twelve areas in detail. These areas are:

        4.4.1   Pre-emergency planning and coordination with outside parties, such as police, fire department, or consultants

        4.4.2   Personnel roles, lines of authority, training, and communication

        4.4.3   Emergency recognition and prevention

        4.4.4   Safe distances and places of refuge

        4.4.5   Site security and control

        4.4.6   Evacuation routes and procedures

        4.4.7   Decontamination

        4.4.8   Emergency medical treatment and first aid

        4.4.9   Emergency alerting and response procedures

        4.4.10 Critique of response and follow-up

        4.4.11 Personal protective equipment (PPE) and emergency equipment

        4.4.12 Portions of other emergency response plans, such as local or state emergency response plans, that are being substituted into the plan

  5.0   The Incident Command System

    5.1   If a response is completely contained within the facility, causes no off-site discharges, and causes no danger to the health and welfare of people not on the facility, the response is directed by the senior facility response official who is present. If there is potential off-site impact, the Incident Command System (ICS) is followed. The ICS requires the control of response activities and communications through the senior emergency response official responding to the emergency. The individual in charge of the ICS is assisted by the senior facility response official.

    5.2   The person in charge of the response is responsible for controlling all hazards, including, but not limited to:

        5.2.1   Identification of all hazardous substances or conditions present

ABC Company

Procedure No.: ___

Revision No.: ___

Revision Date: ___

Title: Procedure for Emergency Response Planning

Page ___ of ___

5.2.2 A site analysis that includes use of engineering controls, maximum exposure limits, hazardous substance handling procedures, and use of appropriate technologies

5.2.3 Implementation of appropriate emergency operations

5.2.4 Assuring use of appropriate PPE, including fire-fighting equipment covered under 29 CFR Section 1910.156(e)

5.2.5 Air monitoring to assure that appropriate respiratory protection is used

5.2.6 Minimizing the number of personnel at the emergency site while maintaining use of the buddy system

5.2.7 Assuring that backup personnel are available for assistance or rescue and that first aid personnel, medical equipment, and medical transportation are available

5.2.8 Designation of a safety official with adequate knowledge of the operation to evaluate hazards and to provide direction with respect to the safety of the operation

5.2.9 Delegation of responsibility to the safety official for determination of conditions that are immediately dangerous to life or health, and with that responsibility, the authority to terminate activities as necessary

5.2.10 Implementation of decontamination procedures

5.3 It may be necessary at the site of the response to utilize individuals who are skilled in the use of certain types of equipment, such as earth-moving or digging equipment, or crane and hoisting equipment. These individuals are not required to have the training required for other emergency responders, but they are given an initial briefing prior to performing any site work. The site briefing includes instruction in the use of the appropriate PPE, what chemical hazards are involved, and what duties are to be performed. All appropriate safety and health precautions provided to the facility's own employees are used to assure the safety and health of these employees.

6.0 Training

6.1 There are several classifications of emergency response workers. Each of these classifications requires different levels of training and experience. Supervisors are responsible for ensuring that each employee has the appropriate training.

6.2 There may be types of training or categories of employees requiring training that are not listed below. Supervisors are responsible for identifying and implementing these special requirements. The Company requires more training than the minimum requirements of the OSHA

ABC Company

Procedure No.: ___
Revision No.: ___
Revision Date: ___

Title: Procedure for Emergency Response Planning
Page ___ of ___

regulations. The basic categories of responder and minimum training are:

6.2.1 **First responder awareness level**—This is a designation for staff members who witness or discover a hazardous substance release and who initiate an emergency response. They take no actions to contain or control the release.
Training requirements are—Employees should be able to:

6.2.1.a Understand what hazardous materials are involved and their risks

6.2.1.b Understand the potential outcomes of an emergency

6.2.1.c Recognize the presence of hazardous materials in an emergency

6.2.1.d Identify the hazardous materials, if possible

6.2.1.e Understand their roles in a response

6.2.1.f Realize the need for additional resources and the need to make appropriate notifications

6.2.1.g Minimum training: 8 hours of initial training and 8 hours of annual refresher training

6.2.2 **First responder operations level**—This is the designation for staff who are part of the initial response and are responsible for protecting nearby persons, property, or the environment from the effects of the release. Their function is to contain the release from a safe distance, keep it from spreading, and prevent exposures. They do not try to stop the release.
Training requirements are—Employees at this level should have knowledge of the awareness level and knowledge that includes:

6.2.2.a Knowledge of basic hazard and risk assessment techniques

6.2.2.b Selection and use of PPE

6.2.2.c An understanding of basic hazardous materials terms

6.2.2.d Basic control, containment, and/or confinement operations within the scope of their job

6.2.2.e Basic decontamination procedures

6.2.2.f An understanding of basic standard operation procedures and termination procedures. Required: 8 hours of initial training.

6.2.2.g Minimum training: 24 hours of initial training and 8 hours of annual refresher training

6.2.3 **Hazardous materials technician**—These staff members respond to releases or potential releases for the purpose of

ABC Company

Procedure No.: ___
Title: Procedure for Emergency Response Planning
Revision No.: ___
Page ___ of ___
Revision Date: ___

stopping the release. They approach the point of the release in order to plug, patch, or otherwise stop the release of a hazardous substance. Training requirements are—At this level an employee should, in addition to the knowledge required for awareness and operations level:

6.2.3.a Know how to implement the emergency response plan

6.2.3.b Know the classification, identification, and verification of known and unknown materials by using field survey equipment

6.2.3.c Be able to function within an assigned role in the ICS

6.2.3.d Know how to select and use specialized PPE

6.2.3.e Understand hazard risk assessment techniques

6.2.3.f Be able to perform advance control, containment, and/or confinement operations

6.2.3.g Understand and implement decontamination procedures

6.2.3.h Understand termination procedures

6.2.3.i Understand basic chemical and toxicological terminology and behavior

6.2.3.j Minimum training: 40 hours of initial training and 16–24 hours of annual refresher training

6.2.4 **Hazardous materials specialist**—This classification responds with and provides support to hazardous materials technicians. Though their duties parallel those of a hazardous materials technician, they are required to have a more specific knowledge of the various substances that they are required to contain. The hazardous materials specialist also functions as liaison with other groups involved with the response. Training requirements are—In addition to the training required for hazardous materials technicians, hazardous materials specialists must be knowledgeable and certified by the employer in the following areas:

6.2.4.a Implementation of the local emergency response plan

6.2.4.b Classification, identification, and verification of known and unknown materials by using advanced survey instruments and equipment

6.2.4.c The state emergency response plan

6.2.4.d Selection and use of PPE

6.2.4.e In-depth risk and hazard assessment techniques

ABC Company

Procedure No.: ___
Revision No.:   ___
Revision Date: ___

Title: Procedure for Emergency Response Planning

Page ___ of ___

6.2.4.f   Performance of specialized control, containment, and/or confinement operations

6.2.4.g   Determination and implementation of decontamination procedures

6.2.4.h   Development of site safety and control plans

6.2.4.I   Chemical, radiological, and toxicological terminology and behavior

6.2.4.j   Minimum training: 40 hours of initial training, experience as a hazardous materials technician, and 24 hours of annual refresher training

6.2.5   **On-scene incident commander**—This is the person who assumes control of the incident scene. The person at each facility with overall responsibility for emergency response is trained as an On-Scene Incident Commander (IC).
Training requirements are—The IC should have knowledge and certification in the following areas:

6.2.5.a   Implementation of the facility ICS

6.2.5.b   Implementation of the facility's emergency response plan

6.2.5.c   Understanding of the risks associated with working in chemical protective clothing

6.2.5.d   Implementation of the local emergency response plan

6.2.5.e   The state emergency response plan and working with the federal regional response team

6.2.5.f   Knowledge and understanding of decontamination procedures

6.2.5.6   Minimum training: 40 hours of initial training, experience with the operation and associated hazards, 24 hours of annual refresher training

7.0   Medical Surveillance

7.1   All members of the hazardous materials response team receive a baseline physical examination, annual physical, and additional medical consultation if there is a suspected exposure to hazardous substances.

7.2   Minimum requirements for these examinations are contained in the NIOSH/OSHA/USCG/EPA manual, *Occupational Safety and Health Guidance Manual for Hazardous Waste Site Activities*. In addition, testing is done for specific compounds to which the employee has the potential for exposure; for emergency responders who engage in strenuous activities, a cardiac stress (treadmill) test is required.

ABC Company

Procedure No.: ___   Title: Procedure for Emergency Response Planning
Revision No.:   ___   Page ___ of ___
Revision Date: ___

7.3   Medical records are maintained for 30 years after the employee leaves the company. These records are available to the employee or to the employee's representative for inspection or copying.

8.0   Postemergency Response

8.1   When the emergency situation is stabilized, hazardous substances or contaminated materials are removed as needed. These removal actions must:

8.1.1   meet all of the requirement of Sections (b) through (o) of 29 CFR Section 1910.120, unless

8.1.2   the cleanup is done entirely on plant property using only plant or workplace employees, and

8.1.3   is documented in a corrective action plan for mitigating all environmental damage.

8.2   In cases meeting the criteria of 8.1.2, the involved employees are trained in the requirements of 29 CFR Sections 1910.38(a), 1910.134, 1910.1200, and any other appropriate areas based on the work to be performed.

9.0   Related Documentation

(A list of documents, such as respirator fit testing records, medical records, state and local emergency response plans, spill control plan, evacuation maps, training records, respirator fit testing records, and so forth, goes here)

10.0   Records

10.1   Emergency Response Plans are maintained by the Environmental Department and are retained for 10 years after revision.

10.2   The emergency contact list is maintained by the Environmental Department until revised.

10.3   Agreements with other environmental response groups are maintained by the Environmental Department and retained for 10 years after expiration.

10.4   Spill reports are filed in the Environmental Department and retained for 30 years.

10.5   All government-required documents are maintained in the Environmental Department for the legally required times or for 10 years, whichever is longer. Privileged documents are maintained in the Legal Department for a period of 30 years or until closure of legal cases.

ABC Company

Procedure No.: ___   Title: Procedure for Spill and Disaster Management

Revision No.: ___   Page ___ of ___

Revision Date: ___

This procedure is a controlled document. The contents of this procedure are authorized and approved by __Name and Title__. This procedure is issued to __Name__ on __Date__. It must not be copied in whole or part without written authorization. It must be returned to the ABC Company when requested. Other persons in possession of this document have an uncontrolled copy and should return it to __Department Name__.

PROCEDURE FOR SPILL AND DISASTER MANAGEMENT

1.0 Purpose

    1.1   To ensure that supplies, systems, and management procedures are in place to allow the rapid cleanup of chemical spills and mitigation of the effects of the spills.

    1.2   To develop procedures for management of natural and human-created disasters.

2.0 Scope

    2.1   This procedure is a supplement to the Emergency Response Plan.

    2.2   This procedure applies to any spill or disaster, including: chemical spills, fires, earthquakes, weather emergencies, and other circumstances with the potential for severe impact to human health or to the environment.

3.0 Responsibilities

    3.1   The Director of Environmental Health and Safety is responsible for forming emergency response and disaster response teams and for developing logistical systems to support them.

    3.2   The Director of Security is responsible for developing a site security plan to control access to areas presenting potential health or safety threats during emergency or disaster conditions.

    3.3   The Emergency Response Team is responsible for verifying the proper cleanup of large and small spills.

    3.4   Properly trained staff working in the area of a small spill is responsible for containment of the spill and for following proper procedures to clean up the spill and dispose of the waste.

    3.5   When there is a spill that is larger than the reportable quantity, workers in the area are responsible for immediately reporting the spill to their supervisor and to the Emergency Response Team.

    3.6   In the event of a disaster that causes damage to company property and may present a threat to employees or the public, the Director of Security is responsible for coordinating company activities with local emergency personnel.

ABC Company

Procedure No.: ___
Revision No.: ___
Revision Date: ___

Title: Procedure for Spill and Disaster Management
Page ___ of ___

3.7 The Public Information Office is responsible for coordinating with media representatives to distribute appropriate information in the event of a spill or disaster.

4.0 Procedure

4.1 Based on the chemical inventory and the locations of chemicals in the facility; including delivery, storage, and transport locations, proper supplies in adequate amounts are stocked to contain and clean up spills.

4.2 Storage areas have secondary containment to prevent any spills from escaping the immediate area.

4.3 When there is a spill outside of a contained area, priority actions, after ensuring protection of the workers, focus on containing the spill to prevent it from spreading or spilling over into sewers.

4.4 All workers in areas handling chemicals receive training on the properties of the chemicals they handle, including toxicity, symptoms of exposure, behavior in the environment, flammability, and other hazardous properties, and personal protective measures.

4.5 A disaster plan is available that includes provisions for site control, notification of proper authorities, evacuations, fire control, chemical control, and other activities specific to operational departments.

4.6 Documentation for the disaster plan and records of chemical inventories, maps of the site, contact names and phone numbers, personnel records, and other records deemed essential in a disaster are maintained in duplicate on site by the Environmental Department and Administration Department. A third set is maintained off site by the Local Emergency Planning Committee to ensure that a copy that is available if the normal storage location is damaged during a disaster.

4.7 Cellular phones and hand-held radios are available and in operating condition in the event that telephones are not operational during disaster circumstances.

4.8 In disaster conditions, the following groups are notified and briefed: senior management, police, fire department, utility companies, medical services, local emergency management agencies, the company insurance provider, and other groups appropriate to the disaster.

4.9 When the disaster conditions are stabilized, staff from the environmental department, in cooperation with other appropriate personnel, performs a reconnaissance of the area to determine whether any off-site environmental impact exists.

ABC Company

Procedure No.: ___   Title: Procedure for Spill and Disaster Management
Revision No.:   ___   Page ___ of ___
Revision Date: ___

    4.10 The Department of Environmental Health and Safety is responsible for preparing all notifications, interim and final reports, to governmental agencies as required by law.

    4.11 When the disaster conditions are controlled, the disaster recovery team evaluates damage to the facility and estimates the costs to return the facility to normal operation. They are responsible for developing a recovery plan, and upon senior management approval, they hire contractors to perform the required work.

    4.12 If environmental cleanups are necessary when the emergency response team completes its work, the Environmental Department evaluates the extent of the contamination and develops a cleanup plan. The plan is reviewed by the appropriate regulatory agencies as legally required. After the plan is approved by the appropriate agency (agencies), it is submitted to senior management for financial approval. When the plan is approved, the Environmental Department hires contractors to perform the work. All contractors must perform needed services, including the disposal of waste in compliance with any and all regulatory requirements.

    4.13 A postemergency or postdisaster report is prepared and submitted to senior management. The Director of Environmental Health and Safety prepares the report if the event is an environmental, safety, or industrial hygiene problem. The Director of Security prepares the report if the event involves a natural disaster or fire. If the event involves both areas of responsibility, the report is prepared jointly.

    4.14 Management reviews the final report.

    4.15 The final report is used by all departments for continual improvement.

    4.16 The final report is used as a training tool for the emergency response and disaster response teams.

5.0 Related Procedures
    5.1 Emergency Response Plan
    5.2 EMS Roles, Responsibilities, and Authority
    5.3 Internal and External Communications
    5.4 Training
    5.5 Audit, Review, and Continual Improvement
    5.6 Documentation and Document control

6.0 Documentation
    6.1 Spill reports
    6.2 Disaster reports
    6.3 Environmental Sampling and Analysis Work Instructions and Records
    6.4 Cleanup Plans

ABC Company

Procedure No.: ___

Revision No.:   ___

Revision Date: ___

Title: Procedure for Spill and Disaster Management

Page ___ of ___

    6.5   Injury and Illness Reports

    6.6   Contracts

    6.7   Communications Records

    6.8   Identification, Maintenance, and Disposition of Environmental Records

    6.9   Inventory of Supplies Used

    6.10 Invoices

7.0   Records

    7.1   Spill reports, disaster reports, and related data are filed by the Environmental Department and retained for a period of 10 years.

    7.2   Injury and illness reports are filed by the Environmental Department and retained until the employee leaves the company, then archived for 30 years.

    7.3   Cleanup and repair plans are filed by the Environmental Department and retained for 10 years.

    7.4   Contracts are filed by the Purchasing Department and retained for 10 years after expiration.

    7.5   Invoices are filed by the Purchasing Department and retained for 3 years.

    7.6   Privileged communication records are filed by the Legal Department and retained for 10 years. Nonsensitive communication records are filed by the Public Information Office and retained for 3 years.

    7.7   Inventories of supplies used are filed by the Environmental Department and retained for 3 years.

ABC Company

Procedure No.: ___
Revision No.:   ___
Revision Date: ___

Title: Procedure for Monitoring and Measurement

Page ___ of ___

> This procedure is a controlled document. The contents of this procedure are authorized and approved by _Name and Title_. This procedure is issued to _Name_ on _Date_. It must not be copied in whole or part without written authorization. It must be returned to the ABC Company when requested. Other persons in possession of this document have an uncontrolled copy and should return it to _Department Name_.

PROCEDURE FOR MONITORING AND MEASUREMENT

1.0   Purpose
      1.1   To track performance of the EMS and key characteristics of operations that might affect performance of the EMS.

2.0   Scope
      2.1   This procedure covers all environmental aspects and impacts identified in the EMS.

3.0   Responsibilities
      3.1   The Director of Environmental Health and Safety is responsible for implementation of the monitoring and measurement program.
      3.2   Supervisors are responsible for monitoring operations in their areas of responsibility and for reporting any conditions that might affect the performance of the EMS.

4.0   Procedure
      4.1   All information related to performance of the EMS is recorded. This includes flow data, raw material use, maintenance records, calibration data, and other information.
      4.2   Work instructions are developed and followed for each piece of monitoring equipment. This includes setup, calibration, operation, and maintenance.
      4.3   A Quality Assurance/Quality Control Plan is available that includes all monitoring and measurement activities.
      4.4   Appropriate quality control/quality assurance procedures are followed to ensure the reliability of the environmental data that are generated.
      4.5   Performance in relation to legislative and regulatory requirements is monitored and recorded.
      4.6   Environmental performance is compared to environmental objectives and targets to evaluate the need for corrective action or changes in the EMS.
      4.6   The appropriateness of established environmental performance indicators is evaluated on an ongoing basis. New indicators are developed as needed.

ABC Company

Procedure No.: ___

Revision No.: ___

Revision Date: ___

Title: Procedure for Monitoring and Measurement

Page ___ of ___

    4.7   Deficiencies identified by monitoring and measurement are corrected.

5.0   Related Procedures

    5.1   Audit, Review, and Continual Improvement

    5.2   Developing and Achieving Environmental Objectives and Targets

    5.3   Identification of Environmental Aspects and Impacts

    5.4   Identification of Legal and Other Requirements

    5.5   Document and Record Maintenance and Control

6.0   Documentation

    6.1   Monitoring and measurement work instructions

    6.2   Corrective action plans

    6.3   Environmental indicator lists

    6.4   Performance reports

7.0   Records

    7.1   Calibration forms are filed by the Environmental Department and retained with equipment for the life of the equipment.

    7.2   Monitor audit forms are filed by the Environmental Department and retained with equipment for the life of the equipment.

    7.3   Monitoring data are filed by the Environmental Department and retained for at least 10 years.

ABC Company

Procedure No.: ___   Title: Procedure for Pollution Prevention
Revision No.:   ___   Page ___ of ___
Revision Date: ___

This procedure is a controlled document. The contents of this procedure are authorized and approved by __Name and Title__. This procedure is issued to __Name__ on __Date__. It must not be copied in whole or part without written authorization. It must be returned to the ABC Company when requested. Other persons in possession of this document have an uncontrolled copy and should return it to __Department Name__.

PROCEDURE FOR POLLUTION PREVENTION

1.0 Purpose
    1.1 To minimize the use of natural resources, reduce costs, and prevent environmental impacts.

2.0 Scope
    2.1 This procedure applies to all operations of the company.

3.0 Responsibilities
    3.1 The Director of Environmental Health and Safety designs and implements pollution prevention and waste minimization programs.
    3.2 Each supervisor and manager evaluates the possibilities for pollution prevention and the reduction, reuse, or recycling of waste in their area of responsibility.
    3.3 The Engineering Department works with the appropriate Department of Environmental Health and Safety on the design equipment and processes to ensure that waste production is minimized.
    3.4 The Purchasing Department ensures that environmental impacts from raw material purchases are minimized.

4.0 Procedure
    4.1 Whenever possible, materials of low toxicity are substituted for more hazardous substances.
    4.2 All manufacturing processes are designed to minimize the uses of raw materials, release of pollution, and production of waste.
    4.3 A thorough preventive maintenance program is established and maintained.
    4.4 All departments implement procedural, administrative, or other practices that minimize production of waste. These practices include use minimization, loss prevention, storage, and inventory control.
    4.5 Hazardous and nonhazardous wastes are never mixed.
    4.6 Products are designed to minimize potential environmental impacts and to facilitate recycling; packaging material is minimized.

ABC Company

Procedure No.: ___
Revision No.: ___
Revision Date: ___

Title: Procedure for Pollution Prevention
Page ___ of ___

4.7 Whenever possible, raw material wastes and other wastes are reused.

4.8 A comprehensive recycling program is in place. This program includes process waste, construction waste, shipping and receiving waste, cafeteria waste, and office waste.

4.9 A waste minimization team is in place with members representing all departments. This team identifies waste sources, evaluates and classifies waste, develops waste minimization options for management consideration, and oversees the implementation of approved options.

4.10 Waste disposal costs are tracked by department and are a line item in each department's annual budget.

5.0 Related Procedures

5.1 Operational Control

5.2 Developing and Achieving Environmental Objectives and Targets

5.3 Audit, Review, and Continual Improvement

5.4 Identification of Environmental Aspects and Impacts

5.5 Identifying Legal and Other Requirements

5.6 All Operational Procedures

6.0 Documentation

6.1 Pollution Prevention and Waste Minimization Manual

6.2 Waste Manifests

6.3 Shipping and receiving records

7.0 Records

7.1 Waste manifests are filed by the Environmental Department and retained for 3 years, then archived permanently.

7.2 Shipping and receiving records are filed by the Environmental Department and retained for 3 years, then archived for 10 years.

7.3 Recycling records are filed by the Environmental Department and retained for 10 years.

ABC Company

Procedure No.: ___   Title: Procedure for Audit, Review, and Continual
                                                                    Improvement
Revision No.:   ___   Page ___ of ___
Revision Date: ___

> This procedure is a controlled document. The contents of this procedure are
> authorized and approved by __Name and Title__ . This procedure is issued to
> __Name__ on __Date__ . It must not be copied in whole or part without
> written authorization. It must be returned to the ABC Company when
> requested. Other persons in possession of this document have an
> uncontrolled copy and should return it to ___Department Name___ .

PROCEDURE FOR AUDIT, REVIEW, AND CONTINUAL IMPROVEMENT

1.0   Purpose
    1.1   To establish a process by which the EMS can be continually improved
           to reduce costs and environmental impact.

2.0   Scope
    2.1   This procedure applies to all operations performed by permanent or
           temporary employees of the company.
    2.2   This procedure applies to all suppliers, vendors, and contractors.

3.0   Responsibilities
    3.1   The Director of Environmental Health and Safety is responsible for
           assembling an audit team for routine performance audits. This team
           may include the appropriate company employees or third-party
           contractors.
    3.2   The Director of Environmental Health and Safety is responsible for
           contracting with a third-party registrar every 3 years for ISO 14001
           certification audits.
    3.3   The Purchasing Department, in cooperation with the Department of
           Environmental Health and Safety, is responsible for obtaining audit in-
           formation from suppliers if the total purchase cost for goods and
           services exceeds $25,000. For procurement actions of less than
           $25,000, suppliers and vendors are encouraged to undergo an audit;
           having an audited EMS is a positive factor in these purchasing
           decisions.
    3.4   The Vice President of Support Services whose oversight includes
           environmental affairs is responsible for reviewing audit reports and for
           initiating corrective actions to eliminate any nonconformances that
           are noted in the audit findings.
    3.5   The Director of Environmental Health and Safety is responsible for
           scheduling internal audits at appropriate intervals. These intervals

ABC Company

Procedure No.: ___

Revision No.: ___

Revision Date: ___

Title: Procedure for Audit, Review, and Continual Improvement

Page ___ of ___

include an audit approximately 6 months prior to any certification audit to allow time for correction of any nonconformances.

4.0 Procedure

    4.1 Information about the purposes of the audit are communicated to all employees.

    4.2 The lead auditor is responsible for organizing the audit and for making audit team assignments. The audit is structured so as to minimize disruptions of operations.

    4.3 The Director of Environmental Health and Safety works with the audit team to ensure that they have access to any documentation, staff, and resources needed for the audit.

    4.4 The lead auditor convenes a preaudit conference that includes the audit team and appropriate management staff before auditing each major operation. At these conferences, the lead auditor explains the process to management staff and provides opportunities for questioning. The audit team is responsible for answering any questions presented to them during the course of the audit. The lead auditor is responsible for resolving any conflicts prior to the conduct of the audit.

    4.5 The audit team prepares audit checklists for each segment of the audit. These checklists cover all environmental aspects and provide adequate information to evaluate progress toward objectives and targets.

    4.6 Preliminary findings that raise questions about potential liability are turned over to the Legal Department for evaluation. This does not prevent or delay any cleanup or mitigation efforts. Any findings of events or activities that are required to be reported to a regulatory agency are reported within the appropriate time frames.

    4.7 At the conclusion of the audit, the preliminary findings are discussed with management staff from each of the audited areas.

    4.8 The audit team works with staff from audited areas to develop a corrective action plan for any nonconformances that are identified during the audit.

    4.9 A final audit report that discusses the status of the EMS is prepared by the lead auditor for the Director of Environmental Health and Safety. The topics in the final report are: progress toward objectives and targets, improvements in performance since the last audit, nonconformances noted during the audit (some of which may be protected by attorney/client privilege), corrective action plans, and any other items that are beneficial to understanding the status of the EMS.

ABC Company

Procedure No.: ___

Revision No.: ___
Revision Date: ___

Title: Procedure for Audit, Review, and Continual
Improvement

Page ___ of ___

4.10 The Director of Environmental Health and Safety prepares a summary report for senior management that provides information on EMS status, resources needed for the EMS continual improvement program, recommendations for modification of the EMS, and any other information that is necessary for a thorough management review.

4.11 When the audit is completed, the audit team evaluates the checklists and audit procedures that are used to conduct the audit so that they may determine ways to improve the audit system prior to the next audit.

4.12 The Director of Environmental Health and Safety works with management and staff in areas where nonconformances have been noted to implement and evaluate the effectiveness of corrective action plans.

4.13 All audit findings are thoroughly documented. This documentation includes audit checklists and forms, reports, corrective action plans, sampling data, and other records generated during the audit. All documentation is controlled.

4.14 With the exception of confidential and proprietary information and information protected by attorney/client privilege, the Public Information Office makes audit findings available upon request for a reasonable copying fee.

5.0 Related Procedures
  5.1 Internal and External Communications
  5.2 Operational Control
  5.3 Pollution Prevention
  5.4 Training
  5.5 Monitoring and Measurement
  5.6 Developing and Achieving Environmental Objectives and Targets
  5.7 Documentation
  5.8. All Operational Procedures

6.0 Documentation
  6.1 Audit Plan
  6.2 Corrective Action Plans
  6.3 Audit checklists and forms
  6.4 Audit reports

7.0 Records
  7.1 Audit reports are filed by the Environmental Department and are retained for 10 years.
  7.2 Records of corrective actions are filed by the Environmental Department and retained for 10 years.

APPENDIX D

POLLUTION PREVENTION WORKSHEETS

Firm _____ Site _____ Date _____	**Pollution Prevention Assessment Worksheets** Proj. No. _____	**Prepared By** _____ **Checked By** _____ **Sheet** ___ **of** ___ **Page** ___**of** ___
	SITE DESCRIPTION WORKSHEET	

Firm:
Plant:
Department:
Area:
Street Address:
City:
State/Zip Code:
Telephone: (   )
Major Products:
SIC Codes:
EPA Generator Number:
Major Unit:
Product or Service:
Operations:
Facilities/Equipment Age:

Source: Adapted from U.S. Environmental Protection Agency, Hazardous Waste Engineering Research Laboratory, *Waste Minimization Opportunity Assessment Manual,* EPA/625/7-88/003 (Cincinnati, OH: U.S. Environmental Protection Agency, 1988).

Firm _____	**Pollution Prevention**	**Prepared By** _____
Site: _____	**Assessment Worksheets**	**Checked By** _____
Date _____	**Proj. No.** _____	**Sheet** ____ **of** ___ **Page** ___**of** ___

| | **PROCESS INFORMATION WORKSHEET** | |

Process Unit/Operation: _____
Operation Type: ❏ Continuous   ❏ Discrete
                        ❏ Batch or Semibatch   ❏ Other _____

| | **STATUS** | | | | | |
Document	**Complete? (Y/N)**	**Current? (Y/N)**	**Last Revision**	**Used in this Report (Y/N)**	**Document Number**	**Location**
Process Flow Diagram						
Material/Energy Balance						
Design						
Operating						
Flow/Amount Measurements						
Stream						
Analyses/Assays						
Stream						

(Continued)

Process Description						
Operating Manuals						
Equipment List						
Equipment Specifications						
Piping and Instrument Diagrams						
Plot and Elevation Plan(s)						
Work Flow Diagrams						
Hazardous Waste Manifests						
Emission Inventories						
Annual/Biennial Reports						
Environmental Audit Reports						
Permit/Permit Applications						
Batch/Sheet(s)						
Materials Application Diagrams						
Product Composition Sheets						
Material Safety Data Sheets						
Inventory Records						
Operator Logs						
Production Schedules						

Source: Adapted from U.S. Environmental Protection Agency, Hazardous Waste Engineering Research Laboratory, *Waste Minimization Opportunity Assessment Manual,* EPA/625/7-88/003 (Cincinnati, OH: U.S. Environmental Protection Agency, 1988).

Firm _____	**Pollution Prevention**	**Prepared By** _____
Site _____	**Assessment Worksheets**	**Checked By** _____
Date _____	**Proj. No.** _____	**Sheet** ____ of ___ **Page** ___of ___

| | **INPUT MATERIALS SUMMARY WORKSHEET** | |

	Description		
Attribute	**Stream No.** ___	**Stream No.** ___	**Stream No.** ___
Name/ID			
Source/Supplier			
Component/Attribute of Concern			
Annual Consumption Rate			
Overall			
Component(s) of Concern			
Purchase Price, $ per ___			
Overall Annual Cost			
Delivery Mode[1]			
Shipping Container Size & Type[2]			

(Continued)

Storage Mode[3]			
Transfer Mode[4]			
Empty Container Disposal Management[5]			
Shelf Life			
Supplier would			
• accept expired material? (Y/N)			
• accept shipping containers? (Y/N)			
• accept expiration date? (Y/N)			
Acceptable Substitute(s), if any			
Alternate Supplier(s)			

Notes: [1]e.g., pipeline, tank car, 100-bb tank truck, truck, etc.
[2]e.g., 55-gal drum, 100-lb paper bag, tank etc.
[3]e.g., outdoor, warehouse, underground, aboveground, etc.
[4]e.g., pump, forklift, pneumatic transport, conveyor, etc.
[5]e.g., crush and landfill, clean and recycle, return to supplier, etc.

Source: Adapted from U.S. Environmental Protection Agency, Hazardous Waste Engineering Research Laboratory, *Waste Minimization Opportunity Assessment Manual,* EPA/625/7-88/003 (Cincinnati, OH: U.S. Environmental Protection Agency, 1988).

Firm _____	Pollution Prevention	Prepared By _____
Site _____	Assessment Worksheets	Checked By _____
Date _____	Proj. No. _____	Sheet ____ of ___ Page ___ of __
	PRODUCTS SUMMARY WORKSHEET	

Attribute	Description		
	Stream No. ___	Stream No. ___	Stream No. ___
Name/ID			
Component/Attribute of Concern			
Annual Consumption Rate			
Overall			
Component(s) of Concern			
Annual Revenues, $ ___			
Shipping Mode			
Shipping Container Size & Type			
On-site Storage Mode			
Containers Returnable (Y/N)			

(*Continued*)

Shelf Life			
Rework Possible (Y/N)			
Customer would			
• relax specification? (Y/N)			
• accept larger containers? (Y/N)			

Source: Adapted from U.S. Environmental Protection Agency, Hazardous Waste Engineering Research Laboratory, *Waste Minimization Opportunity Assessment Manual,* EPA/625/7-88/003 (Cincinnati, OH: U.S. Environmental Protection Agency, 1988).

Firm _____	**Pollution Prevention**	**Prepared By** _____
Site _____	**Assessment Worksheets**	**Checked By** _____
Date _____	**Proj. No.** _____	**Sheet** ____ **of** ___ **Page** ___**of** ___

| | **WASTE STREAM SUMMARY WORKSHEET** | |

	Description		
Attribute	**Stream No.** ___	**Stream No.** ___	**Stream No.** ___
Waste ID/Name:			
Source/Origin			
Component or Property of Concern			
Annual Generation Rate (units ___)			
Overall			
Components of Concern			
Cost of Disposal			
Unit Cost ($ per: ___)			
Overall (per year)			
Method of Management[1]			

(Continued)

Priorirty Rating Criteria[2]	Relative Wt. (W)	Rating (R)	R x W	Rating (R)	R x W	Rating (R)	R x W
Regulatory Compliance							
Treatment/Disposal Cost							
Potential Liability							
Waste Quantity Generated							
Waste Hazard							
Safety Hazard							
Minimization Potential							
Potential to Remove Bottleneck							
Potential By-product Recovery							
Sum of Priority Rating Scores		Σ(R x W)		Σ(R x W)		Σ(R x W)	
Priority Rank							

Notes: [1]For Example, sanitary landfill, hazardous waste landfill, on-site recycle, incineration, combustion with heat recovery, distillation, dewatering, etc.
[2]Rate each stream in each category on a scale from 0 (none) to 10 (high).

Source: Adapted from U.S. Environmental Protection Agency, Hazardous Waste Engineering Research Laboratory, *Waste Minimization Opportunity Assessment Manual,* EPA/625/7-88/003 (Cincinnati, OH: U.S. Environmental Protection Agency, 1988).

Firm _____	**Pollution Prevention**	Prepared By _____
Site _____	**Assessment Worksheets**	Checked By _____
Date _____	Proj. No. _____	Sheet _____ of ___ Page ___of ___
	OPTION GENERATION WORKSHEET	

Meeting format (e.g., brainstorming, nominal group technique)

Meeting Coordinator

Meeting Participants

List Suggestion Option	**Rationale/Remarks on Option**

(Continued)

Source: Adapted from U.S. Environmental Protection Agency, Hazardous Waste Engineering Research Laboratory, *Waste Minimization Opportunity Assessment Manual,* EPA/625/7-88/003 (Cincinnati, OH: U.S. Environmental Protection Agency, 1988).

Firm _____ Site _____ Date _____	**Pollution Prevention** **Assessment Worksheets** **Proj. No.** _____	**Prepared By** _____ **Checked By** _____ **Sheet** ____ **of** ___ **Page** ___**of** ___
	OPTION DESCRIPTION WORKSHEET	

Option Name
_____

Briefly Describe the Option
_____
_____
_____
_____
_____
_____

Waste Stream(s) Affected
_____
_____

Input Material(s) Affected
_____
_____

Product(s) Affected
_____
_____

Indicate Type: ❏Source Reduction

         ___ Equipment-Related Change

(*Continued*)

_____ Personnel/Procedure-Related Change

_____ Materials-Related Change

❐ Recycling/Reuse

_____ On-site   _____ Material reused for original purpose

_____ Off-site   _____ Material used for a lower-quality purpose

                       _____ Material sold

                       _____ Material burned for heat recovery

Originally Proposed by _____   Date _____

Reviewed by _____   Date _____

Approved for Study?   yes _____   no _____   By _____

Reason for Acceptance or Rejection _____

Source: Adapted from U.S. Environmental Protection Agency, Hazardous Waste Engineering Research Laboratory, *Waste Minimization Opportunity Assessment Manual,* EPA/625/7-88/003 (Cincinnati, OH: U.S. Environmental Protection Agency, 1988).

Firm _____	**Pollution Prevention**	**Prepared By** _____
Site _____	**Assessment Worksheets**	**Checked By** _____
Date _____	**Proj. No.** _____	**Sheet** ____ of ___ **Page** ___of __

PROFITABILITY WORKSHEET

Capital Costs

Purchased Equipment

_____

Materials

_____

Installation

_____

Utility Connections

_____

Engineering

_____

Startup and Training

_____

Other Capital Costs

_____

   Total Capital Costs

_____

Incremental Annual Operating Costs

  Change in Disposal Costs

_____

  Change in Raw Material Costs

_____

  Change in Other Costs

_____

   Annual Net Operating Cost Savings

_____

$$\text{Payback Period (years)} = \frac{\text{Total Capital Costs}}{\text{Annual Net Operating Cost Savings}} = \underline{\hspace{4cm}}$$

Source: Adapted from U.S. Environmental Protection Agency, Hazardous Waste Engineering Research Laboratory, *Waste Minimization Opportunity Assessment Manual,* EPA/625/7-88/003 (Cincinnati, OH: U.S. Environmental Protection Agency, 1988).

APPENDIX E

AIR POLLUTION LISTS

Initial List of Section 112 Hazardous Air Pollutants (HAPs)

Chemical Name (listed alphabetically)	CAS No.
acetaldehyde	75-07-0
acetamide	60-35-5
acetonitrile	75-05-8
acetophenone	98-86-2
2-acetylaminofluorene	53-96-3
acrolein	107-02-8
acrylamide	79-06-1
acrylic acid	79-10-7
acrylonitrile	107-13-1
allyl chloride	107-05-1
4-aminobiphenyl	92-67-1
aniline	62-53-3
o-anisidine	90-04-0
asbestos	1332-21-4
benzene (including benzene from gasoline)	71-42-2
benzidine	92-87-5
benzotrichloride	98-07-7
benzyl chloride	100-44-7
biphenyl	92-52-4
bis (2-ethylhexyl) phthalate (DEHP)	117-81-7
bis (chloromethyl) ether	542-88-1
bromoform	75-25-2
1,3-butadiene	106-99-0
calcium cyanamide	156-62-7
caprolactam	105-60-2
captan	133-06-2
carbaryl	63-25-2
carbon disulfide	75-15-0
carbon tetrachloride	56-23-5
carbonyl sulfide	463-58-1
catechol (1,2-dihydroxylbenzene)	120-80-9

chloramben	133-90-4
chlordane	57-74-9
chlorine	7782-50-5
chloroacetic acid	79-11-8
2-chloroacetophenone	532-27-4
chlorobenzene	108-90-7
chlorobenzilate	510-15-6
chloroform	67-66-3
chloromethyl methyl ether	107-30-2
chloroprene	126-99-8
cresols/cresylic acid (isomers and mixtures)	1319-77-3
o-cresol	95-48-7
m-cresol	108-39-4
p-cresol	106-44-5
cumene	98-82-8
2,4-D (2,4-dichlorophenoxyacetic acid, including salts and esters)	94-75-7
DDE (1,1-dichloro-2,2-bis(p-chlorophenyl) ethylene)	72-55-9
diazomethane	334-88-3
dibenzofurans	132-64-9
1,2-dibromo-3-chloropropane	96-12-8
dibutylphthalate	84-74-2
1,4-dichlorobenzene	106-46-7
3,3'-dichlorobenzidene	91-94-1
dichloroethyl ether [bis(2-chloroethyl) ether]	111-44-4
1,3-dichloropropene	542-75-6
dichlorvos (DDVP)	62-73-7
diethanolamine (2,2'-iminodiethanol)	111-42-2
N,N-dimethylaniline	121-69-7
diethyl sulfate	64-67-5
3,3'-dimethoxybenzidine	119-90-4
dimethyl aminoazobenzene	60-11-7
3,3'-dimethylbenzidene	119-93-7
dimethylcarbamoyl chloride	79-44-7
dimethylformamide (N,N-)	68-12-2
1,1-dimethylhydrazine	57-14-7
dimethyl phthalate	131-11-3
dimethyl sulfate	77-78-1
4,6-dinitro-o-cresol, and salts	
2,4-dinitrophenol	51-28-5

2,4-dinitrotoluene	121-14-2
1,4-dioxane (1,4-diethyleneoxide)	123-91-1
1,2-diphenylhydrazine	122-66-7
epichlorohydrin (1-chloro-2,3-epoxypropane)	106-89-8
1,2-epoxybutane	106-88-7
ethyl acrylate	140-88-5
ethylbenzene	100-41-4
ethyl carbamate (urethane)	51-79-6
ethyl chloride (chloroethane)	75-00-3
ethylene dibromide (dibromoethane)	106-93-4
ethylene dichloride (1,2-dichloroethane)	107-06-2
ethylene glycol	107-21-1
ethyleneimine (aziridine)	151-56-4
ethylene oxide	75-21-8
ethylene thiourea	96-45-7
ethylidene dichloride (1,1-dichloroethane)	75-34-3
formaldehyde	50-00-0
heptachlor	76-44-8
hexachlorobenzene	118-74-1
hexachlorobutadiene	87-68-3
hexachlorocyclopentadiene	77-47-4
hexachloroethane	67-72-1
hexamethylene-1,6-diisocyanate	822-06-0
hexamethylphosphoramide	680-31-9
hexane	110-54-3
hydrazine	302-01-2
hydrochloric acid	7647-01-0
hydrogen fluoride (hydrofluoric acid)	7664-39-3
hydrogen sulfide	7783-06-4
hydroquinone	123-31-9
isophorone	78-59-1
lindane (all isomers)	58-89-9
maleic anhydride	108-31-6
methanol	67-56-1
methoxychlor	72-43-5
methyl bromide (bromomethane)	74-83-9
methyl chloride (chloromethane)	74-87-3
methyl chloroform (1,1,1-trichloroethane)	71-55-6
methyl ethyl ketone (2-butanone)	78-93-3
methylhydrazine	60-34-4

methyl iodide (iodomethane)	74-88-4
methyl isobutyl ketone (hexone)	108-10-1
methyl isocyanate	624-83-9
methyl methacrylate	80-62-6
methyl tert-butyl ether	1634-04-4
4,4'-methylene bis (2-chloroaniline)	101-14-4
methylene chloride (dichloromethane)	75-09-2
4,4'-methylene diphenyl diisocyanate (MDI)	101-68-8
4,4-methylenedianiline	101-77-9
naphthalene	91-20-3
nitrobenzene	98-95-3
4-nitrobiphenyl	92-93-3
4-nitrophenol	100-02-7
2-nitropropane	79-46-9
N-nitroso-N-methylurea	684-93-5
N-nitrosodimethylamine	62-75-9
N-nitrosomorpholine	59-89-2
parathion	56-38-2
pentachloronitrobenzene (Quintobenzene)	82-68-8
pentachlorophenol	87-86-5
phenol	108-95-2
p-phenylenediamine	106-50-3
phosgene	75-44-5
phosphine	7803-51-2
phosphorus	7723-14-0
phthalic anhydride	85-44-9
polychlorinated biphenyls (arochlors)	1336-36-3
1,3-propane sultone	1120-71-4
beta-propiolactone	57-57-8
propionaldehyde	123-38-6
propoxur (baygon)	114-26-1
propylene dichloride (1,2-dichloropropane)	78-87-5
propylene oxide	75-56-9
1,2-propyleneimine (2-methyl aziridine)	75-55-8
quinoline	91-22-5
quinone	106-51-4
styrene	100-42-5
styrene oxide	96-09-3
2,3,7,8-tetrachlorodibenzo-p-dioxin	1746-01-6
1,1,2,2-tetrachloroethane	79-34-5

tetrachloroethylene (perchloroethylene)	127-18-4
titanium tetrachloride	7550-45-0
toluene	108-88-3
2,4-toluene diamine	95-80-7
2,4-toluene diisocyanate	584-84-9
o-toluidine	95-53-4
toxaphene (chlorinated camphene)	8001-35-2
1,2,4-trichlorobenzene	120-82-1
1,1,2-trichloroethane	79-00-5
trichloroethylene	79-01-6
2,4,5-trichlorophenol	95-95-4
2,4,6-trichlorophenol	88-06-2
triethylamine	121-44-8
trifluraline	1582-09-8
2,2,4-trimethylpentane	540-84-1
vinyl acetate	108-05-4
vinyl bromide	593-60-2
vinyl chloride	75-01-4
vinylidene chloride (1,1-dichloroethylene)	75-35-4
xylenes (isomers and mixtures)	1330-20-7
o-xylene	95-47-6
m-xylene	108-38-3
p-xylene	106-42-3

NOTE: For all listings below that contain the word *compounds* and for glycol ethers, unless otherwise specified, these listings are defined as including any unique chemical substance that contains the named chemical (for example, antimony, arsenic, beryllium, and so forth) as part of that chemical's infrastructure.

antimony compounds
arsenic compounds (inorganic, including arsine)
beryllium compounds
cadmium compounds
chromium compounds
cobalt compounds
coke oven emissions
cyanide compounds[1]
glycol ethers[2]
lead compounds
manganese compounds
mercury compounds

fine mineral fibers[3]
nickel compounds
polycyclic organic matter[4]
radionuclides (including radon)[5]
selenium compounds

[1] X′CN where X = H′ or any other group where a formal dissociation may occur. For example KCN or Ca(CN)$_2$.

[2] Includes mono- and di-ethers of ethylene glycol, diethylene glycol, and triethylene glycol R-(OCH$_2$CH$_2$)$_n$-OR′ where: n = 1, 2, or 3; R = alkyl or aryl groups; and R′ = R, H, or groups which, when removed, yield glycol ethers with the structure R-(OCH$_2$CH$_2$)$_n$-OH. Polymers are excluded from the glycol category.

[3] Includes mineral fiber emissions from facilities manufacturing or processing glass, rock, or slag fibers (or other mineral-derived fibers) of average diameter 1 micrometer or less.

[4] Limited to or referring to products from incomplete combustion of organic compounds (or material) and pyrolysis processes having more than one benzene ring and a boiling point ≥100°C.

[5] A type of atom that spontaneously undergoes radioactive decay.

Appendix E.2
MACT Sources and Scheduled Promulgation Dates

Categories of Major Sources (listed alphabetically)	Scheduled Promulgation Date
acetal resin production	11/15/97
acrylic fiber/modacrylic fiber production	11/15/97
acrylonitrile-butadiene-styrene production	11/15/94
aerosol can-filling facilities	11/15/2000
aerospace industries surface coating	11/15/94
alkyd resin production	11/15/2000
alumina processing	11/15/2000
amino resin production	1/15/97
ammonium sulfate production—caprolactam byproduct plant	11/15/2000
antimony oxides manufacturing	11/15/2000
asphalt concrete manufacturing	11/15/2000
asphalt processing	11/15/2000
asphaltic roofing manufacturing	11/15/2000
automobile and light-duty truck surface coating	11/15/2000
baker's yeast manufacturing	11/15/2000
benzyltrimethylammonium chloride production	11/15/2000
boat manufacturing	11/15/2000
butadiene dimer production	11/15/97
butadiene-furfural cotrimer (R-11) production	11/15/2000
butyl rubber production	11/15/94
captafol production	11/15/97
captan production	11/15/97
carboxymethylcellulose production	11/15/2000
carbonyl sulfide production	11/15/2000
cellophane production	11/15/2000
cellulose ether production	11/15/2000
cellulose food casing manufacturing	11/15/2000
chelating agent production	11/15/2000
chlorinated paraffin production	11/15/2000
chlorine production	11/15/97
4-chloro-2-methylphenoxyacetic acid production	11/15/97
chloroneb production	11/15/97
chlorothalonil	11/15/97
chromic acid anodizing	11/15/94
chromium chemical manufacturing	11/15/97
chromium refractories production	11/15/2000
clay product manufacturing	11/15/2000

coke byproduct plants	11/15/2000
coke ovens: charging top side and door leaks	12/31/92
coke ovens: pushing, quenching, and battery stacks	11/15/2000
commercial dry cleaning (perc) transfer machines	11/15/92
commercial sterilization facilities	11/15/94
cyanuric chloride production	11/15/97
2,4-D salt and ester production	11/15/97
dacthal production	11/15/97
decorative chromium electroplating	11/15/94
4,6-dinitro-o-cresol production	11/15/97
dodecanedioic acid production	11/15/2000
dry cleaning (petroleum solvent)	11/15/2000
engine test facilities	11/15/2000
epichlorohydrin elastomer production	11/15/94
epoxy resin production	11/15/94
ethylene-propylene rubber production	11/15/94
ethylidene norborene production	11/15/2000
explosives production	11/15/2000
ferroalloy production	11/15/97
flat wood paneling surface coating	11/15/2000
flexible polyurethane foam production	11/15/97
fume silica production	11/15/2000
gasoline distribution	11/15/94
halogenated solvent cleaners	11/15/94
hard chromium electroplating	11/15/94
hazardous waste incineration	11/15/2000
hydrazine production	11/15/2000
hydrochloric acid production	11/15/2000
hydrogen cyanide production	11/15/97
hydrogen fluoride production	11/15/2000
hypalon production	11/15/2000
industrial boilers	11/15/2000
industrial dry cleaning (perc) dry-to-dry machines	11/15/92
industrial dry cleaning (perc) transfer machines	11/15/92
industrial process cooling towers	11/15/94
institutional/commercial boilers	11/15/2000
integrated iron and steel manufacturing	11/15/2000
iron foundries	11/15/2000
large appliance surface coating	11/15/2000
lead acid battery manufacturing	11/15/2000

lime manufacturing	11/15/2000
magnetic tape surface coating	11/15/94
maleic anhydride copolymer production	11/15/2000
manufacture of paints, coating, and adhesives	11/15/2000
metal can surface coating	11/15/2000
metal coil surface coating	11/15/2000
metal furniture surface coating	11/15/2000
methylcellulose production	11/15/2000
methyl methacrylate-acrylonitrile-butadiene-styrene production	11/15/94
methyl methacrylate-butadiene-styrene terpolymer production	11/15/94
mineral wool production	11/15/97
miscellaneous metal parts and products (surface coating)	11/15/2000
municipal landfills	11/15/2000
neoprene production	11/15/94
nitrile butadiene rubber production	11/15/94
nonnylon polyamide production	11/15/94
nonstainless steel manufacturing electric arc furnace (EAF) operation	11/15/97
nylon 6 production	11/15/97
oil and natural gas production	11/15/97
organic liquids distribution (nongasoline)	11/15/2000
oxybisphenoxarsine (OBPA)/1,3-diisocyanate production	11/15/2000
paper and other webs surface coating	11/15/2000
paint stripper users	11/15/2000
petroleum refineries—catalytic cracking (fluid and other) units, catalytic reforming units and sulfur plant units	11/15/97
petroleum refineries—other sources not distinctly listed	11/15/94
pharmaceutical production	11/15/97
phenolic resin production	11/15/97
phosphate fertilizers	11/15/2000
phosphoric acid manufacturing	11/15/2000
photographic chemical production	11/15/2000
phthalate plasticizer production	11/15/2000
plastic parts and products surface coating	11/15/2000
plywood/particle board manufacturing	11/15/2000
polyether polyol production	11/15/97
polybutadiene rubber production	11/15/94
polycarbonates production	11/15/97
polyester resin production	11/15/2000
polyether polyol production	11/15/97
polyethylene terephthalate production	11/15/94

polymerized vinylidene chloride production	11/15/2000
polymethyl methacrylate resin production	11/15/2000
polystyrene production	11/15/94
polysulfide rubber production	11/15/94
polyvinyl acetate emulsion production	11/15/2000
polyvinyl alcohol production	11/15/2000
polyvinyl butyral production	11/15/2000
polyvinyl chloride and copolymer production	11/15/2000
portland cement manufacturing	11/15/97
primary aluminum production	11/15/2000
primary copper smelting	11/15/97
primary lead smelting	11/15/97
primary magnesium refining	11/15/2000
printing, coating, and dyeing of fabrics	11/15/2000
printing/publishing surface coating	11/15/94
process heaters	11/15/2000
publicly owned treatment works	11/15/95
pulp and paper production	11/15/97
quaternary ammonium compound production	11/15/2000
rayon production	11/15/2000
reinforced plastic composite production	11/15/97
rocket engine test firing	11/15/2000
rubber chemical manufacturing	11/15/2000
secondary aluminum production	11/15/97
secondary lead smelting	11/15/97
semiconductor manufacturing	11/15/2000
sewage sludge incineration	11/15/2000
ship-building and ship repair surface coating	11/15/94
site remediation	11/15/2000
sodium cyanide production	11/15/97
sodium pentachlorophenate production	11/15/97
solid waste treatment, storage, and disposal facilities	11/15/94
spandex production	11/15/2000
stationary internal combustion engines	11/15/97
stationary turbines	11/15/2000
stainless steel manufacturing electric arc furnace operation	11/15/97
steel foundries	1/15/2000
steel pickling—HCl process	11/15/97
styrene-acrylonitrile production	11/15/94
styrene-butadiene rubber and latex production	11/15/94

symmetrical tetrachloropyridine production	11/15/2000
synthetic organic chemical manufacturing	11/15/92
taconite iron ore processing	11/15/2000
tire production	11/15/2000
tordon acid production	11/15/97
uranium hexafluoride production	11/15/2000
vegetable oil production	11/15/2000
wood furniture surface coating	11/15/94
wood treatment	11/15/97
wool fiberglass manufacturing	11/15/97

Appendix E.3
Stratospheric Ozone-Depleting Chemicals

A. **Class I**   **Controlled Substances**   **CAS No.**
   Group I:

$CFCl_3$—trichlorofluoromethane (CFC-11)	75-69-4
CF_2Cl_2—dichlorodifluormethane (CFC-12)	75-71-8
$C_2F_3Cl_3$—trichlorotrifluorethane (CFC-113)	76-13-1
$C_2F_4Cl_2$—dichlorotetrafluorethane (CFC-114)	76-14-2
C_2F_5Cl—monochloropentafluorethane (CFC-115)	76-15-3
all isomers of the above chemicals	

   Group II:

CF_2ClBr—bromochlorodifluoromethane (Halon-1211)	421-01-2
CF_3Br—bromotrifluoromethane (Halon-1301)	75-63-8
$C_2F_4Br_2$—dibromotetrafluoroethane (Halon-2402)	124-73-2
all isomers of the above chemicals	

   Group III:

CF_3Cl—chlorotrifluoromethane (CFC-13)	75-72-9
C_2FCl_5—pentachlorofluoroethane (CFC-111)	954-56-3
$C_2F_2Cl_4$—tetrachlorodifluoroethane (CFC-112)	76-12-0
C_3FCl_7—heptachlorofluoropropane (CFC-211)	422-78-6
$C_3F_2Cl_6$—hexachlorodifluoropropane (CFC-212)	3182-26-1
$C_3F_3Cl_5$—pentachlorotrifluoropropane (CFC-213)	2354-06-5
$C_3F_4Cl_4$—tetrachlorotetrafluoropropane (CFC-214)	29255-31-0
$C_3F_5Cl_3$—trichloropentafluoropropane (CFC-215)	4259-42-2
$C_3F_6Cl_2$—dichlorohexafluoropropane (CFC-216)	661-97-2
C_3F_7Cl—trichloroheptafluoropropane (CFC-217)	422-86-6
all isomers of the above chemicals	

   Group IV:

CCl_4—carbon tetrachloride	56-23-5

   Group V:

$C_2H_3Cl_3$—1,1,1 trichloroethane (methyl chloroform)	71-55-6
all isomers of the above chemical except	
1,1,2—trichloroethane	79-00-5

B. **Class II**   **Controlled Substances**   **CAS No.**

$CHFCl_2$—dichlorofluoromethane (HCFC-21)	75-43-4
CHF_2Cl—chlorodifluoromethane (HCFC-22)	75-45-6
$CHFCl$—chlorofluoromethane (HCFC-31)	593-70-4
C_2HFCl_4—tetrachlorofluoroethane (HCFC-121)	*
$C_2HF_2Cl_3$—trichlorodifluoroethane (HCFC-122)	*
$C_2HF_3Cl_2$—dichlorotrifluoroethane (HCFC-123)	*

C_2HF_4Cl—chlorotetrafluoroethane (HCFC-124)   *

$C_2H_2FCl_3$—trichlorofluoroethane (HCFC-131)   *

$C_2H_2F_2Cl_2$—dichlorodifluoroethane (HCFC-132b)   1649-08-7

$C_2H_2F_3Cl$—chlorotrifluoroethane (HCFC-133a)   75-88-7

$C_2H_3FCl_2$—dichlorofluoroethane (HCFC-141b)   1717-00-6

$C_2H_3F_2Cl$—chlorodifluoroethane (HCFC-142b)   75-68-3

C_3HFCl_6—hexachlorofluoropropane (HCFC-221)   *

$C_3HF_2Cl_5$—pentachlorodifluoropropane (HCFC-222)   *

$C_3HF_3Cl_4$—tetrachlorotrifluoropropane (HCFC-223)   *

$C_3HF_4Cl_3$—trichlorotetrafluoropropane (HCFC-224)   *

$C_3HF_5Cl_2$—dichloropentafluoropropane (HCFC-225ca)   *
    (HCFC-225cb)

C_3HF_6Cl—chlorohexafluoropropane (HCFC-226)   *

$C_3H_2FCl_5$—pentachlorofluoropropane (HCFC-231)   *

$C_3H_2F_2Cl_4$—tetrachlorodifluoropropane (HCFC-232)   *

$C_3H_2F_3Cl_3$—trichlorotrifluoropropane (HCFC-233)   *

$C_3H_2F_4Cl_2$—dichlorotetrafluoropropane (HCFC-234)   *

$C_3H_2F_5Cl$—chloropentafluoropropane (HCFC-235)   *

$C_3H_3FCl_4$—tetrachlorofluoropropane (HCFC-241)   *

$C_3H_3F_2Cl_3$—trichlorodifluoropropane (HCFC-242)   *

$C_3H_3F_3Cl_2$—dichlorotrifluoropropane (HCFC-243)   *

$C_3H_3F_4Cl$—chlorotetrafluoropropane (HCFC-244)   *

$C_3H_4FCl_3$—trichlorofluoropropane (HCFC-251)   *

$C_3H_4F_2Cl_2$—dichlorodifluoropropane (HCFC-252)   *

$C_3H_4F_3Cl$—chlorotrifluoropropane (HCFC-253)   *

$C_3H_5FCl_2$—dichlorofluoropropane (HCFC-261)   *

$C_3H_5F_2Cl$—chlorodifluoropropane (HCFC-262)   *

C_3H_6FCl—chlorofluoropropane (HCFC-271)   *

all isomers of the above chemicals

*multiple CAS numbers

Appendix E.4
National Emission Standards for Hazardous Air Pollutants (NESHAPS)

Pollutant	Facility or Emission Unit Type	40 CFR 61 Subpart
Radon	underground uranium mines; Department of Energy facilities; phosphorus fertilizer plants; and facilities processing or disposing of uranium ore and tailings	B, Q, R, T, W
Beryllium	beryllium extraction plants; ceramic plants, foundries, incinerators, propellant plants, and machine shops that process beryllium-containing material; and rocket motor firing test sites	C, D
Mercury	mercury ore processing; manufacturing processes using mercury chloralkali cells; and sludge incinerators	E
Vinyl Chloride	ethylene dichloride manufacturing via oxygen and HCl with ethylene; vinyl chloride manufacturing; and polyvinyl chloride manufacturing	F
Radionuclides	Department of Energy; Nuclear Regulatory Commission licensed facilities; other federal facilities; and elemental phosphorus plants	H, I, K
Benzene	fugitive process, storage, and transfer equipment leaks; coke by-product recovery plants; benzene storage vessels; benzene transfer operations; and benzene waste operations	J, L, Y, BB, FF
Asbestos*	asbestos mills; roadway surfacing with asbestos tailings; manufacture of products containing asbestos; demolition; renovation; and spraying and disposal of asbestos waste	M
Inorganic Arsenic	glass manufacturing; primary copper smelter; arsenic trioxide; and metallic arsenic production facilities	N, O, P
Volatile Hazardous Air Pollutants (VHAP)	pumps, compressors, pressure relief devices, connections, valves, lines, flanges, product accumulator vessels, and so forth in VHAP service	V

*A Part 70 permit is not needed if the facility is subject only to 40 CFR Section 61.145 (Subpart M standard for demolition and renovation) because it is engaged in asbestos abatement.

Source: 40 CFR Part 61 National Emission Standards for Hazardous Air Pollutants, 1996

Appendix E.5
New Source Performance Standards (NSPS) Source Categories

Source Categories	40 CFR 60 Subpart	Effective Date Constructed, Modified, or Reconstructed
Fossil-fuel fired steam generators >250 MMBtu	D	After: 08/17/71
Electric utility steam generators >250 MMBtu	Da	After: 09/18/78
Industrial-commercial-institutional steam generators >1,000 MMBtu	Db	After: 06/19/84
Small industrial- commercial-institutional steam generators ≥10 MMBtu but <100 MMBtu	Dc	After: 06/09/89
Incinerators	E	After: 08/17/71
Municipal waste combustors	Ea,Ca	After: 12/20/89 and on or before 09/20/94
Municipal waste combustors	Eb	After: 09/20/94
Portland cement plants	F	After: 08/17/71
Nitric acid plants	G	After: 08/17/71
Sulfuric acid plants	H,Cb	After: 08/17/71
Hot mix asphalt facilities	I	After: 06/11/73
Petroleum refineries	J	After: 06/11/73
Storage vessels for petroleum liquids	K,Ka	After: 06/11/73 and prior to 5/19/78
Volatile organic liquid storage vessels (including petroleum liquids)	Kb	After: 07/23/84
Secondary lead smelters	L	After: 06/11/73
Secondary brass and bronze production plants	M	After: 06/11/73
Oxygen process furnaces	N	After: 06/11/73
Oxygen process steel-making facilities	Na	After: 01/20/83
Sewage treatment plants	O	After: 06/11/73
Primary copper smelters	P	After: 10/16/74
Primary zinc smelters	Q	After: 10/16/74
Primary lead smelters	R	After: 10/16/74
Primary aluminum reduction plants	S	After: 10/23/74
Phosphate fertilizer industry	T,U,V,W,X	After: 10/22/74
Coal preparation plants	Y	After: 10/24/74
Ferroalloy production facilities	Z	After: 10/21/74
Steel plants	AA,AAa	After: 10/21/74
Kraft pulp mills	BB	After: 09/24/76
Glass manufacturing plants	CC	After: 06/15/79

Grain elevators	DD	After: 08/03/78
Surface coating of metal furniture	EE	After: 11/28/80
Stationary gas turbines	GG	After: 10/03/77
Lime manufacturing plants	HH	After: 05/03/77
Lead-acid battery manufacturing plants	KK	After: 01/14/80
Metallic mineral processing plants	LL	After: 08/24/82
Automobile and light-duty truck surface coating operations	MM	After: 10/05/79
Phosphate rock plants	NN	After: 09/21/79
Ammonium sulfate manufacture	PP	After: 02/04/80
Graphics arts industry: publication rotogravure printing	QQ	After: 08/28/80
Pressure-sensitive tape and label surface coating operations	RR	After: 12/30/80
Industrial surface coating: large appliances	SS	After: 12/24/80
Metal coil surface coating	TT	After: 01/05/81
Asphalt processing and asphalt roofing manufacture	UU	After: 11/18/80
Equipment leaks of VOC in the synthetic organic chemicals manufacturing industry	VV	After: 01/05/81
Beverage can surface coating industry	WW	After: 11/26/80
Bulk gasoline terminals	XX	After: 12/17/80
New residential wood heaters	AAA	After: 07/01/88
Rubber tire manufacturing industry	BBB	After: 01/20/83
VOC emissions from the polymer manufacturing industry	DDD	After: 09/30/87
Flexible vinyl and urethane coating and printing	FFF	After: 01/18/83
Equipment leaks of VOC in petroleum refineries	GGG	After: 01/04/83
Synthetic fiber production facilities	HHH	After: 11/23/82
VOC emissions from the synthetic organic chemical manufacturing industry (SOCMI) air oxidation unit processes	III	After: 10/21/83
Petroleum dry cleaners	JJJ	After: 12/14/82
On-shore natural gas processing plants: VOC equipment leak and SO_2 emissions	KKK,LLL	After: 01/20/84
VOC emissions from synthetic organic chemical manufacturing industry (SOCMI) distillation operations	NNN	After: 12/30/83
Nonmetallic mineral processing plants (including sand and gravel processing)	OOO	After: 08/31/93
Wool fiberglass insulation manufacturing plants	PPP	After: 02/07/84

VOC emissions from petroleum refinery wastewater systems	QQQ	After: 05/04/87
VOC emissions from the synthetic organic chemical manufacturing industry (SOCMI) reactor processes	RRR	After: 07/29/90
Magnetic tape coating facilities	SSS	After: 01/22/86
Industrial surface coating: surface coating of plastic parts for business machines	TTT	After: 01/08/86
Calciners and dryers in mineral industries	UUU	After: 04/23/86
Polymeric coating of supporting substrates facilities	VVV	After: 04/30/87

Source: 40 CFR Part 60. Standards of Performance for New Stationary Sources, 1996.

Appendix E.6
Section 112(r) Regulated Toxic and Flammable Substances with Threshold Quantities for Accidental Release Prevention

Chemical Name	CAS No.	Threshold Quantity (lbs)
acetaldehyde	75-07-0	10,000
acetylene [ethyne]	74-86-2	10,000
acrolein [2-propenal]	107-02-8	5,000
acrylonitrile [2-propenenitrile]	107-13-1	20,000
acrylyl chloride [2-propenoyl chloride]	814-68-6	5,000
allyl alcohol [2-propen-l-ol]	107-18-61	15,000
allylamine [2-propen-l-amine]	107-11-9	10,000
ammonia [anhydrous]	7664-41-7	10,000
ammonia (conc 20% or greater)	7664-41-7	20,000
arsenous trichloride	7784-34-1	15,000
arsine	7784-42-1	1,000
boron trichloride [borane, trichloro-]	10294-34-5	5,000
boron trifluoride [borane, trifluoro-]	7637-07-2	5,000
boron trifluoride compound with methyl ether (1:1) [boron, trifluoro(oxybis[metane])-,T-4]	353-42-4	15,000
bromine	7726-95-6	10,000
bromotrifluorethylene [ethene, bromotrifluoro-]	598-73-2	10,000
1,3-butadiene	106-99-0	10,000
butane	106-97-8	10,000
1-butene	106-98-9	10,000
2-butene	107-01-7	10,000
butene	25167-67-3	10,000
2-butene-cis	590-18-1	10,000
2-butene-trans [2-butene,(E)]	624-64-6	10,000
carbon disulfide	75-15-0	20,000
carbon oxysulfide [carbon oxide sulfide(COS)]	463-58-1	10,000
chlorine	7782-50-5	2,500
chlorine dioxide [chlorine oxide (ClO_2)]	10049-04-4	1,000
chlorine monoxide [chlorine oxide]	7791-21-1	10,000
chloroform [methane, trichloro-]	67-66-3	20,000
chloromethyl ether {methane, oxybis[chloro-]}	542-88-1	1,000
chloromethyl methyl ether [methane, chloromethoxy-]	107-30-2	5,000
2-chloropropylene [1-propene, 2-chloro-]	557-98-2	10,000
1-chloropropylene [1-propene, 1-chloro-]	590-21-6	10,000

crotonaldehyde [2-butenal]	4170-30-3	20,000
crotonaldehyde, (E)- [2-butenal, (E)-]	123-73-9	20,000
cyanogen [ethanedinitrile]	460-19-5	10,000
cyanogen chloride	506-77-4	10,000
cyclohexylamine [cyclohexanamine]	108-91-8	15,000
cyclopropane	75-19-4	10,000
diborane	19287-45-7	2,500
dichlorosilane [silane, dichloro-]	4109-96-0	10,000
difluoroethane [ethane, 1,1-difluoro-]	75-37-6	10,000
dimethyldichlorosilane [silane, dichlorodimethyl]	75-78-5	5,000
1,1-dimethylhydrazine [hydrazine, 1,1-dimethyl-]	57-14-7	15,000
dimethylamine [methanamine, N-methyl-]	124-40-3	10,000
2,2-dimethylpropane [propane, 2,2-dimethyl-]	463-82-1	10,000
epichlorohydrin [oxirane,(chloromethyl)-]	106-89-8	20,000
ethane	74-84-0	10,000
ethyl acetylene [1-butyne]	107-00-6	10,000
ethylamine [ethanamine]	75-04-7	10,000
ethyl chloride [ethane, chloro-]	75-00-3	10,000
ethylene [ethene]	74-85-1	10,000
ethylenediamine [1,2-ethanediamine]	107-15-3	20,000
ethyleneimine [aziridine]	151-56-4	10,000
ethylene oxide [oxirane]	75-21-8	10,000
ethyl ether [ethane, 1,1'-oxybis-]	60-29-7	10,000
ethyl mercaptan [ethanethiol]	75-08-1	10,000
ethyl nitrite [nitrous acid, ethyl ester]	109-95-5	10,000
fluorine	7782-41-4	1,000
formaldehyde (solution)	50-00-0	15,000
furan	110-00-9	5,000
hydrazine	302-01-2	15,000
hydrochloric acid (conc 30% or greater)	7647-01-0	15,000
hydrocyanic acid	74-90-8	2,500
hydrogen	1333-74-0	10,000
hydrogen chloride (anhydrous) [hydrochloric acid]	7647-01-0	5,000
hydrogen fluoride/hydrofluoric acid (conc 50% or greater) [hydrofluoric acid]	7664-39-3	1,000
hydrogen selenide	7783-07-5	500
hydrogen sulfide	7783-06-4	10,000
iron, pentacarbonyl-[iron carbonyl $(Fe(CO)_5,(TB-5-11)-]$	13463-40-6	2,500
isobutane [propane, 2-methyl]	75-28-5	10,000
isobutyronitrile [propanenitrile, 2-methyl-]	78-82-0	20,000

isopentane [butane, 2-methyl-]	78-78-4	10,000
isoprene [1,3-butadinene, 2-methyl-]	78-79-5	10,000
isopropylamine [2-propanamine]	75-31-0	10,000
isopropyl chloride [propane, 2-chloro-]	75-29-6	10,000
isopropyl chloroformate [carbonochloridic acid, 1-methylethyl ester]	108-23-6	15,000
methacrylonitrile [2-propenenitrile, 2-methyl-]	126-98-7	10,000
methane	74-82-8	10,000
methylamine [methanamine]	74-89-5	10,000
3-methyl-1-butene	563-45-1	10,000
2-methyl-1-butene	563-46-2	10,000
methyl chloride [methane, chloro-]	74-87-3	10,000
methyl chloroformate [carbonochloridic acid, methylester]	79-22-1	5,000
methyl ether [methane, oxybis-]	115-10-6	10,000
methyl formate [formic acid, methyl ester]	107-31-3	10,000
methyl hydrazine [hydrazine, methyl-]	60-34-4	15,000
methyl isocyanate [methane, isocyanato-]	624-83-9	10,000
methyl mercaptan [methanethiol]	74-93-1	10,000
2-methylpropene [1-propene, 2-methyl-]	115-11-7	10,000
methyl thiocyanate [thiocyanic acid, methyl ester]	556-64-9	20,000
methyltrichlorosilane [silane, trichloromethyl-]	75-79-6	5,000
nickel carbonyl	13463-39-3	1,000
nitric acid (conc 80% or greater)	7697-37-2	15,000
nitric oxide [nitrogen oxide (NO)]	10102-43-9	10,000
oleum (fuming sulfuric acid) [sulfuric acid, mixture with sulfur trioxide]	8014-95-7	10,000
1,3-Pentadienne	504-60-9	10,000
pentane	109-66-0	10,000
1-pentene	109-67-1	10,000
2-pentene, (E)-	646-04-8	10,000
3-pentene, (Z)-	627-20-3	10,000
peracetic acid [ethaneperoxoic acid]	79-21-0	10,000
perchloromethylmercaptan [methanesulfenyl chloride, trichloro-]	594-42-3	10,000
phosgene [carbonic dichloride]	75-44-5	500
phosphine	7803-51-2	5,000
phosphorous oxychloride [phosphoryl chloride]	10025-87-3	5,000
phosphorous trichloride	7719-12-2	15,000
piperidine	110-89-4	15,000
propadiene [1,2-propadiene]	463-49-0	10,000

propane	74-98-6	10,000
propionitrile [propanenitrile]	107-12-0	10,000
propyl chloroformate [carbonochloridic acid, propylester]	109-61-5	15,000
propylene [1-propene]	115-07-1	10,000
propyleneimine [aziridine, 2-methyl-]	75-55-8	10,000
propylene oxide [oxirane, methyl-]	75-56-9	10,000
propyne [1-propyne]	74-99-7	10,000
silane	7803-62-5	10,000
sulfur dioxide (anhydrous)	7446-09-5	5,000
sulfur tetrafluoride [sulfur fluoride (SF_4),(T-4)-]	7783-60-0	2,500
sulfur trioxide	7446-11-9	10,000
tetrafluoroethylene [ethene, tetrafluoro-]	116-14-3	10,000
tetramethyllead [plumbane, tetramethyl-]	75-74-1	10,000
tetramethylsilane [silane, tetramethyl-]	75-76-3	10,000
tetranitromethane [methane, tetranitro-]	509-14-8	10,000
titanium tetrachloride [titanium chloride ($TiCl_4$) (T-4)-]	7550-45-0	2,500
toluene 2,4-diisocyanate [benzene, 2,4-diisocyanato-1-methyl-]	584-84-9	10,000
toluene, 2,6 diisocyanate [benzene, 1,3-diisocyanato-2-methyl-]	91-08-7	10,000
toluene diisocyanate (unspecified isomer) [benzene, 1,3-diisocyanatomethyl-]	26471-62-5	10,000
trichlorosilane [silane, trichloro-]	10025-78-2	10,000
trifluorochloroethylene [ethene, chlorotrifluoro-]	79-38-9	10,000
trimethylamine [methanamine, N, N-dimethyl-]	75-50-3	10,000
trimethylchlorosilane [silane, chlorotrimethyl-]	75-77-4	10,000
vinyl acetate monomer [acetic acid ethenyl ester]	108-05-4	15,000
vinyl acetylene [1-buten-3-yne]	689-97-4	10,000
vinyl chloride [ethene, chloro-]	75-01-4	10,000
vinyl ethyl ether [ethene, ethoxy-]	109-92-2	10,000
vinyl fluoride [ethene, fluoro-]	75-02-5	10,000
vinylidene chloride [ethene, 1,1-dichloro-]	75-35-4	10,000
vinylidene fluoride [ethene, 1,1-difluoro-]	75-38-7	10,000
vinyl methyl ether [ethene, methoxy-]	107-25-5	10,000

Source: 40 CFR Part 68. Chemical Accident Prevention Provisions, 1996.

APPENDIX F

RESOURCES

There are many resources (libraries, consulting and engineering firms, universities, government offices, industry associations) that can provide helpful guidance for the development and improvement of an EMS and an OHSMS. Many different types of information can be obtained free of charge or at low cost through the Internet or from organizational resources, including government agencies and professional/industrial associations. These two important sources of information are highlighted in the following sections.

THE INTERNET

A BRIEF OVERVIEW OF THE INTERNET

The Internet is a huge system of interconnected computer networks covering most of the world. It is not a single network, but a network or networks that have agreed on communication standards. The exact size of the Internet is difficult to determine—it includes thousands of business, research, and education networks containing millions of computers in more than 100 countries, and it continues to grow! There are a multitude of companies whose sole purpose is to provide connectivity to the Internet. To say that growth in the past few years has been phenomenal is an understatement. The Internet truly provides access to a world of information.

The Internet began as a project in the U.S. Department of Defense. In 1969, the Advanced Research Projects Agency began the creation of ARPANET. The objective was to allow researchers to share information and to create a computerized command and control system that would remain functional in the event that a nuclear attack destroyed part of the network. ARPANET connected the military, defense contractors, and universities through specially developed communications software. The communications software had two components: the Transmission Control Protocol (TCP) and the Internet Protocol (IP). This software is generally referred to as TCP/IP, and a personal computer must utilize software that understands and interprets TCP/IP to use all of the Internet's features.

When the National Science Foundation wanted to connect its six supercomputers, TCP/IP communications was a logical way to do it. This supercomputer network was expanded to include others, and the "backbone" (high-speed communication line) was improved to a T3 line, which transmits data at 45 million bits per second. This speed is equivalent to about 14,000 single-spaced, typed pages per second! These expansions and improvements put the Internet on a course to its current level of development.

The Internet performs three basic functions:

Electronic Mail: Electronic mail, or e-mail, is a service that allows users to send messages to individuals or groups anywhere in the world, to send and receive text files, and more.

Telnet Sessions: Telnet allows users to log in to a remote computer and work interactively. Users can search library and university systems or participate in on-line discussions. Some Telnet functions are being replaced by similar capabilities on the World Wide Web (WWW).

File Transfer: The file transfer protocol (ftp) allows users to download files from many sources. There are so many sources of files to download that a search tool is necessary unless the user knows the network address of the source file. The WWW also facilitates file transfers.

WHO RUNS THE INTERNET?

An often asked question is, Who is in charge of the Internet? The short answer is that no one is; there is no centralized management. However, cooperation between networks requires the development of operational standards, and the primary organization for standards development is the Internet Society, a private, not-for-profit corporation. The Internet Society supports the Internet Activities Board (IAB), which handles architecture and protocol issues. The IAB has assigned specific responsibilities to the Internet Engineering Task Force, Internet Research Task Force, Internet Assigned Numbers Authority, and Internet Registry to ensure smooth operation of the Internet.

The World Wide Web Consortium (W3 Consortium) develops standards for the WWW. The W3 Consortium is an industry consortium led by the Laboratory for Computer Science at the Massachusetts Institute of Technology. The WWW is the fastest-growing part of the Internet, and it is displacing some older services and protocols.

Each of the thousands of individual networks has its own management structure, but regional networks are often formed to enhance resource sharing. Such a structure requires a high level of cooperation. The backbones that make the Internet possible are paid for by governments, corporations, and groups such as the National Science Foundation in the U.S. Without these high-speed communication lines, traffic on the Internet would be slow and impractical.

WHAT DO INTERNET ADDRESSES MEAN?

To send and receive messages, users must have an address that is derived from the communications software. The IP addresses are four numbers separated by periods, such as 163.45.122.64. This type of address is difficult for people to remember, so the Domain Name System was created to allow the use of words. Name Servers translate between the domain address and the IP address.

A typical e-mail address might be: tpeters@inca.epa.gov. The part to the left of the @ (at) represents the *login name* of the user (receiver of the message), in this case, T. Peters. The coding to the right of the @ gives the user's address. At the far right is the name of the largest domain of the address which, in this example, is the federal government (*gov*). Working back from the largest domain, the *epa* represents a smaller domain, the user's organization, which is the Environmental Protection Agency. To the left of the organization information is the host name (*inca*). The host name (or server) tells which computer within the domain should receive the message. The host then looks at the user name to deliver the message.

Some standard organization domain names are:

com   Commercial domains in the U.S. The *.com* domain represents a corporate or private sector site. For example, *www.yahoo.com* is a private search engine for the WWW.

edu   Colleges and universities in the U.S. For example, *fsu.edu* is Florida State University.

gov   Government sites for the federal government in the US. For example, *epa.gov* is the Environmental Protection Agency.

mil   Military sites in the U.S.

net   A network organization.

org   Organizations that do not fit into other categories. For example, *aiha.org* is the American Industrial Hygiene Association.

A few examples of country domain names are:

au	Australia
br	Brazil
ca	Canada
ch	Switzerland
cr	Costa Rica
de	Germany
fr	France
gh	Ghana
it	Italy
jp	Japan
kp	Korea
my	Malaysia
ru	Russian Federation
ug	Uganda
uk	United Kingdom
us	United States
zw	Zimbabwe

THE WORLD WIDE WEB

The WWW is by far the fastest-growing part of the Internet, primarily because it is user-friendly. It provides access to a wealth of information through user-friendly *browsers*. A browser is a search engine that simplifies the process of finding information. A number of search engines are available at low cost or at no charge. As connection prices have decreased, growth of the WWW has increased exponentially. The explosion of information on the WWW defies the imagination—a topic search will often yield thousands of "hits." WWW sites generally have a home page—a starting point from which to navigate. It is often possible to find many more links (literally hundreds) to other sites from a home page.

UNIVERSAL RESOURCE LOCATORS

Universal Resource Locators (URLs), sometimes called *Uniform Resource Locators*, were developed as part of the specification for the WWW. The URL is a site-naming convention or, in other words, an address with a specific format. A typical URL is:

http://www.epa.gov/epahome/text.htm

Reading from left to right this URL gives the:

- access method, in this case, hypertext transport protocol (http),
- site IP address, in this case, the U.S. Environmental Protection Agency site on the WWW (*www.epa.gov*),
- directory tree to the file to be accessed, in this case, the Environmental Protection Agency home page (*epahome*), and
- file name, in this case, *text.htm*.

CONNECTING TO THE INTERNET

There are many ways to access the Internet. The traditional way is through a UNIX shell account, a character-based and less user-friendly method, but easier methods are becoming popular. There are two types of connections that allow a personal computer to use TCP/IP and graphical client programs: Serial Line Internet Protocol (SLIP) and Point-to-Point Protocol (PPP). As software improvements are made, SLIP/PPP connections are becoming the best way to negotiate cyberspace. This is the type of connection used by Internet service companies.

Most users connect to the Internet through their organizational server, an Internet service company, or an on-line service. On-line services originally connected the user only to the service's mainframe computer and allowed them to access the files that it contained. Most now provide software to access ftp, Telnet, e-mail, and the WWW.

INTERNET NAVIGATION TIPS

As you work your way through the sites on the Internet, some of them will ask for a *login name*. Many will accept *anonymous* or *guest* and will let you visit the site. Some home pages give you the opportunity to register if you want to be on their mailing list.

Because many sites have links to other sites, you can often find a useful site far from the point of origin. During the course of navigating on the Internet, it is important to write down the site address or the series of links that got you to a particular site or to add the site to the list of readily accessible sites in your browser. The trail to a site can be complicated; for example:

- The site *www.aiha.org* is the home page of the American Industrial Hygiene Association.
- This home page contains a link to the British Occupational Hygiene Society (BOHS).
- BOHS has a link to the Directory of Sites in Occupational and Environmental Health, which has about 250 links to other sites.
- That directory allows the user to link to another directory containing Occupational and Environmental Health Directories (also British).
- These directories provide a link to the OSHWEB (Finland).
- The OSHWEB provides links to Other Resource Lists.
- Other Resource Lists allow the user to link to Environmental Health and Safety Links at the University of Wisconsin-Milwaukee.
- The University has even more links to other sites.

This example clearly illustrates that because of the complexity and length of a chain of connections, it is much simpler to note useful site addresses than to try and retrace the chain of links.

If you plan to download large files, it is best to do it during off-peak hours *at the site from which you are downloading*. This is approximately 6:00 P.M. until 7:00 A.M. *at the site from which you are downloading*. So, if you are downloading from a site in London, you need to know what time it is in London.

USEFUL DEFINITIONS

Backbone	High-speed connections that route long-distance traffic. Backbones connect to slower local lines.
BBS	Bulletin board system. A way for users to leave messages and files for others to access. Usually local systems.
Browser	A program used to navigate the WWW.
Cyberspace	The world of networked computers. When using the Internet, you are searching cyberspace.
Domain	A group of computers on the Internet. The Domain Name System gives structure to site addresses and provides a way to convert names to numeric IP addresses.

Downloading	Transfering a file from one computer to another.
e-mail	Electronic system of messaging. Networks generally have an e-mail program.
ftp	File transfer protocol. A tool for moving files.
Gopher	A tool developed by the University of Minnesota to aid in Internet searches. It is a menu-driven system.
Host	A computer connected directly to the Internet.
Hypertext	Data that provide links between key elements. Hypertext allows the WWW to be navigated.
HTML	Hypertext Markup Language. This is the language used to develop WWW home pages.
http	Hypertext transport protocol. This is the access method for moving from place to place on the WWW.
IP	Internet Protocol. One of the software packages that allows diverse computers to communicate.
PPP	Point-to-Point Protocol. One of two ways to access the Internet from a personal computer. PPP provides data compression.
protocol	An agreed-upon communication system for computers. A protocol defines how computers will work together to communicate.
SLIP	Serial Line Internet Protocol. One of two ways to access the Internet from a personal computer. Even though SLIP is widely used, it is not an official standard.
SMTP	Simple Mail Transfer Protocol. This is the Internet's standard way of handling e-mail.
TCP	Transmission Control Protocol. This protocol controls transmission of data to ensure that it arrives in its original form.
Telnet	An Internet protocol that allows a user to log onto a remote computer.
URL	Universal (Uniform) Resource Locator. Part of the WWW specification. A standard way of identifying protocol and address.
World Wide Web	A system working through hypertext links to allow users to explore the internet.

SOME USEFUL SITES

There are many Internet sites related to environment and occupational health and safety. The sites listed below provide some starting points to the resources that are available. Some interesting information on the ISO 14000 standards is available at *http://www.stoller.com.*

U.S. Environmental Protection Agency Sites Most of the EPA sites can be accessed through *http://www.epa.gov.* Files and software can often be downloaded in multiple formats to suit individual needs. This site also provides access to departments, people, addresses, and phone numbers. Other useful sites at the EPA include:

- gopher://gopher.epa.gov—the U.S. EPA gopher
- ftp://ftp.epa.gov—the U.S. EPA ftp server
- http://es.inel.gov/dfe/dfe.html—Design For Environment provides information on how to build environmental friendliness into industrial designs.
- http://www.epa.gov/docs/oppe/eaed/eedhmpg.htm—Economy and environment
- http://www.epa.gov/energystar.html—The Energy Star Program provides information for saving energy resources.
- http://www.epa.gov/docs/gpo.html—U.S. Government Printing Office
- http://www.epa.gov/epahome/programs.html—U.S. EPA Programs and Initiatives provides information about the agency's policies, programs, and future agenda.
- http://www.epa.gov/epahome/finding.html—National Center for Environmental Publications and Information
- http://www.epa.gov/epahome/data.html—EPA—Data Systems and Software
- http://www.epa.gov/epahome/business.htm—Business and Industry has a link to the U.S. Business Advisor. The goal of this site is to provide one-stop access to federal government information, services, and transactions.
- http://www.epa.gov/partners—Partners for the Environment provides information on the partnership programs sponsored by the U.S. EPA. Some of these programs include toxics reduction, agriculture, global climate change, energy efficiency and conservation, state and local initiatives, solid waste, and compliance. The EPA *WasteWi$e* and *Enviro$en$e* programs are two additional examples of partnership programs.
- http://es.inel.gov/index.html—*Enviroene* home page
- http://es.inel.gov/partners.wise/wise.html—*WasteWi$e* Program
- http://www.epa.gov/ow—Office of Water
- http://www.epa.gov/docs/ord—Office of Research and Development
- http://www.epa.gov/oar—Office of Air and Radiation
- http://www.epa.gov/docs/GCDOAR.OAR-APPD.html—Information and software to prevent air pollution
- http://earth1.epa.gov/chemfact/chemical—Chemical substance fact sheets
- http://www.epa.gov/regionXX—Regional U.S. EPA offices; substitute the appropriate region number for the *XX*

Energy and Transportation Related Sites

- http://solstice.crest.org/online/aeguide—Information on alternative energy
- http://wwwofe.er.doe.gov—U.S. Department of Energy
- http://www.epa.gov/energystar.html—U.S. EPA Energystar Program
- http://www.dot.gov—U.S. Department of Transportation

Manufacturing

- http://www.cygnus-group.com:9011—Waste reduction with a focus on source reduction and reuse
- http://euler.berkeley.edu/green/cgdm.html—Provides background information and studies on green manufacturing
- http://www.enviroindustry.com—Environmental business opportunities

Occupational Safety and Health

- http://www.osha.gov—Main link into OSHA databases
- http://www.osha.goc/oshapubs—Access to OSHA documents, most can be downloaded
- http://www.ed.ac.uk/~robin/bohs.html—British Occupational Hygiene Society
- http://turva.me.tut.fi/cis/home.html—International Occupational Safety and Health Information Center
- http://turva.me.tut.fi/~oshweb/—The OSHWEB provides many links to safety and health resources.
- http://www.ed.ac.uk/~rma/sites.html—The Directory of Sites in Occupational and Environmental Health, which has over 250 links to other sites.

On-Line Organizations

- http://www.clearlake.ibm.com/erc/—The Environmental Resource Center
- http://www.aiha.org—The American Industrial Hygiene Association
- http://www.awma.org—The Air and Waste Management Association
- http://www.acs.org—The American Chemical Society
- http://www.isea.rutgers.edu/isea/isea.htm—The International Society of Exposure Analysis
- http://kaos.erin.gov.au:80/erin.html—Australian Environmental Resources Information Network
- http://www.nofc.forestry.ca/—Canadian Forest Service
- http://www.und.ac.za/prg/prg.html—South African Pollution Research Group
- http://itre.ncsu.edu/itre/cte/cte.html—Center for Transportation and the Environment
- http://www.envirolink.org—A large environmental information resource that provides many links to other sites.

U.S. ENVIRONMENTAL PROTECTION AGENCY RESOURCES

- *ACCESS EPA* is a comprehensive directory to major information services, collections of the EPA, and other public sector organizations. It provides detailed descriptions of the resource with complete contact information. *ACCESS EPA* provides information on how to locate and use EPA's National Environmental Supercomputing Center, clearinghouses, hotlines, dockets, libraries, scientific models, documents, and EPA programs. The directory also includes directions for obtaining information from state environmental libraries, government document ordering services, and federal depository libraries. *ACCESS EPA* contains a selected acronym list and a detailed author/title/subject index. It can be ordered by contacting:

Government Printing Office
710 North Capitol Street, NW
Washington, DC 20401
Telephone: 202-783-3238; Fax: 202-512-2250

or

National Technical Information Service
5285 Port Royal Road
Springfield, VA 22161
Telephone: 703-487-4650; Fax: 703-321-8547

- *ACCESS Express*, a less detailed but quick reference guide to major EPA information contacts, can be obtained at no charge from the EPA's Public Information Center (PIC). The EPA PIC can provide referrals to Agency resources (some are listed below), and it distributes nontechnical information. Contact:

U.S. EPA
Public Information Center (PIC)
401 M Street SW, 3404
Washington, DC 20460
Telephone: 202-260-2080; Fax: 202-260-6257

To request information about specific topics, contact one of the following hot lines or clearinghouses—telephone numbers and addresses can be obtained through the EPA PIC (listed above, or through the U.S. EPA WWW site listed in the internet resources section).

AIR AND RADIATION

- Acid Rain Hotline
- Air Pollution Training Branch Information Line
- At Risk Information Support Center Hotline (AIR RISC)
- Clearinghouse for Inventories and Emisson Factors
- Control Technology Center (CTC)
- EPA Model Clearinghouse
- Green Lights Program
- Indoor Air Quality Information Clearinghouse (IAQ INFO)
- National Air Toxics Information Clearinghouse (NATICH)
- National Radon Hotline
- Office of Air Quality Planning and Standards Technology Transfer Network Bulletin Board System (OAQPS TTN BBS)

- Reasonably Available Control Technology, Best Available Control Technology and Lowest Achievable Emissions Rate (RACT/BACT/LAER)
- Stratospheric Ozone Information Hotline

HAZARDOUS AND SOLID WASTE

- Alternative Treatment Technology Information Center (ATTIC)
- Clean-Up Information Bulletin Board System (CLU-IN)
- Emergency Planning and Community Right-to-Know Act (EPCRA) Information Hotline
- Hazardous Waste Ombudsman Program
- Methods Information Communications Exchange (MICE)
- National Response Center (NRC)
- Resource Conservation and Recovery Act/SUPERFUND/Underground Storage Tank (RCRA/SF/OUST) Hotline
- Solid Waste Assistance Program
- Subsurface Remediation Information Center (SRIC)

INTERNATIONAL

- INFOTERRA

PESTICIDES AND TOXIC SUBSTANCES

- Asbestos Ombudsman Clearinghouse/Hotline
- National Lead Information Center
- National Pesticide Information Retrieval System (NPIRS)
- National Pesticide Telecommunications Network (NPTN)
- The 33/50 Program
- Toxic Release Inventory User Support (TRI-US)
- Toxic Substances Control Act (TSCA) Assistance Information Service

POLLUTION PREVENTION

- Pollution Prevention Information Clearinghouse (PPIC)
- Pollution Prevention Information Exchange System (PIES)

RESEARCH AND DEVELOPMENT

- Office of Research and Development Electronic Bulletin Board System (ORD BBS)

WATER

- Clean Lakes Clearinghouse

- National Small Flows Clearinghouse
- Nonpoint Source Electronic Bulletin Board System (NPS BBS)
- Office of Water Resource Center
- Safe Drinking Water Hotline
- Wastewater Treatment Information Exchange Bulletin Board System (WTIE-BBS)
- Wetlands Protection Hotline

CROSS-PROGRAM

- Environmental Financing Information Network (EFIN)
- EPA Institute
- Inspector General Hotline
- Office of Environmental Equity
- Office of Public Liaison, EPA Activities Update Hotline
- Small Business Ombudsman Clearinghouse/Hotline

REFERENCES

GENERAL

HAWKEN, PAUL. *The Ecology of Commerce*, New York, NY: HarperBusiness, 1993.

REPETTO, ROBERT. *Jobs, Competitiveness, and Environmental Regulation: What Are The Real Issues?* Washington, DC: World Resources Institute, 1995.

WORLD COMMISSION ON ENVIRONMENT AND DEVELOPMENT, *Our Common Future*, Oxford: Oxford University Press, 1987.

MANAGEMENT

ARNOLD, J.R. TONY. *Materials Management*. 2d Edition, Upper Saddle River, NJ: Prentice Hall, 1996.

COVEY, STEPHEN R. *The Seven Habits of Highly Effective People*, New York, NY: Fireside/Simon & Schuster, 1989.

DENTON, KEITH D. *Enviro-Management: How Smart Companies Turn Environmental Costs into Profits*, Englewood Cliffs, NJ: Prentice-Hall, 1994.

DEMING, W. EDWARDS. *Out of Crisis*, Cambridge, MA: Massachusetts Institute of Technology, Center for Advanced Engineering Study, 1986.

DRUCKER, PETER F. *Managing in a Time of Great Change*, New York, NY: Penguin Books USA Inc., Truman Talley Books/Dutton, 1995.

JABLONSKI, JOSEPH R. *Implementing TQM*, Albuquerque, NM: Technical Management Consortium, 1992.

JURAN, JOSEPH M. *Juran on Quality by Design: The New Steps for Planning Quality Into Goods and Services*, New York, NY: Maxwell Macmillan International, 1992.

LABOVITZ, GEORGE. *Making Quality Work*, New York, NY: Harper-Collins, 1993.

PETERS, THOMAS J. *The Tom Peters Seminar: Crazy Times Call for Crazy Organizations*, New York, NY: Vintage Books, 1994.

PETERS, THOMAS J. *The Pursuit of Wow: Every Person's Guide to Topsy-Turvy Times*, New York, NY: Vintage Books, 1994.

SENGE, PETER M. *The Fifth Discipline: The Art and Practice of the Learning Organization*, New York, NY: Doubleday/Currency, 1990.

456

STATISTICAL PROCESS CONTROL

BURR, JOHN T. *SPC Tools for Everyone*, Milwaukee, WI: ASQC Quality Press, 1993.

DOTY, LEONARD A. *Statistical Process Control*, New York, NY: Industrial Press, 1991.

FELLERS, GARY. *The Deming Vision: SPC/TQM for Administrators*, Milwaukee, WI: ASQC Quality Press, 1992.

ISHIKAWA, KAORU. *Guide to Quality Control*, Tokyo, Japan: Asian Productivity Organization, 1986.

ISHIKAWA, KAORU. *Introduction to Quality Control*, Tokyo, Japan: 3A Corporation, 1990.

ISO 14000

HEMENWAY, CAROLINE G, ed. *What is ISO 14000? Questions and Answers*, Fairfax, VA: CEEM Information Services and ASQC Quality Press, 1995.

KUHRE, W. LEE. *ISO 14001 Certification: Environmental Management Systems*, Upper Saddle River, NJ: Prentice Hall, 1995.

INTERNATIONAL STANDARDS ORGANIZATION. ISO/TC 207. *Section 1 Strategic Plan ISO/TC 207 on Environmental Management N47*, ISO/TC 207 N45, np: International Standards Organization, 1995.

ROTHERY, BRIAN. *BS 7750—Implementing the Environment Management Standard and the EC Eco-Management Scheme*, Brookfield, VT: Gower Press, 1993.

ISO 9000

PATTERSON, JAMES G. *ISO 9000: Worldwide Quality Standard*, Menlo Park, CA: Crisp Publications, 1995.

ARNOLD, KENNETH L. *The Manager's Guide to ISO 9000*. New York, NY: Macmillan, The Free Press.

OWEN, BRYAN; COTHRAN, TOM; MALKOVICH, PETER. *Achieving ISO 9000 Registration*, Knoxville, TN: SPC Press, 1994.

LAW

SULLIVAN, THOMAS F.P., editor, *Environmental Law Handbook*, 13th ed., Rockville, MD: Government Institutes, 1995.

RODGERS, WILLIAM H, JR. *Environmental Law*, 2d ed., St. Paul, MN: West Publishing, 1994. *Selected Environmental Law Statutes: 1994–95 Educational Edition*, St. Paul, MN: West Publishing, 1994.

AUDITING AND ENVIRONMENTAL TOPICS

BLAKESLEE, H.W.; GRABOWSKI, T.M. *A Practical Guide to Plan Environmental Audits*, New York, NY: Van Nostrand Reinhold, 1985.

CAHILL, LAWRENCE B.; KANE, RAYMOND W.; KARAS, JENNIFER L.; MAUCH, JAMES C.; PRICE, COURTNEY M.; RIEDEL, BRIAN P.; SCHOMER, DAWNE P.; VETRANO, THOMAS R. *Environmental Audits*, Rockville, MD: Government Institutes, 1996.

CSUROS, MARIA. *Environmental Sampling and Analysis for Technicians*, Boca Raton, FL: CRC Press, 1994.

KASE, DONALD W.; WIESE, KAY J. *Safety Auditing: A Management Tool*, New York, NY: Van Nostrand Reinhold, 1990.

KEITH, LAWRENCE H. *Environmental Sampling and Analysis: A Practical Guide*, Chelsea, MI: Lewis Publishers, 1991.

LEDGERWOOD, GRANT; STREET, ELIZABETH; THERIVEL, RIKI. *Implementing an Environmental Audit: How to Gain a Competitive Advantage Using Quality and Environmental Responsibility*, New York, NY: Irwin Professional Publishing, 1994.

NOLL, GREGORY G.; HILDEBRAND, MICHAEL S.; YVORRA, JAMES G. *Hazardous Materials*, Stillwater, OK: Oklahoma State University, Fire Protection Publications, 1995.

ODI, MARCIA J. *Air Operating Permits Guide*, Indianapolis, IN: Indiana Chamber of Commerce, 1995.

SHIELDS, J., ed., *Air Emissions, Baselines & Environmental Auditing*, New York, NY: Van Nostrand Reinhold, 1993.

U.S. ENVIRONMENTAL PROTECTION AGENCY, HAZARDOUS WASTE ENGINEERING LABORATORY. *Waste Minimization Opportunity Assessment Manual*, EPA/625/7-88003, Cincinnati, OH: U.S. Environmental Protection Agency, 1988.

U.S. ENVIRONMENTAL PROTECTION AGENCY, OFFICE OF ENFORCEMENT AND COMPLIANCE ASSURANCE. *Federal Facility Pollution Prevention Planning Guide*, Washington, DC: U.S. Environmental Protection Agency, 1994.

LOUVAR, JOSEPH; LOUVAR, B. DIANE, *Risk Analysis. Fundamentals with Applications*, Upper Saddle River, NJ: Prentice Hall, In Press.

U.S. ENVIRONMENTAL PROTECTION AGENCY, OFFICE OF ENFORCEMENT AND COMPLIANCE ASSURANCE. *Federal Facility Pollution Prevention Project Analysis: A Primer for Applying Life Cycle and Total Cost Assessment Concepts*, EPA 300-B-95-008, Washington, DC: U.S. Environmental Protection Agency, 1995.

U.S. ENVIRONMENTAL PROTECTION AGENCY, OFFICE OF WATER, OFFICE OF ENFORCEMENT AND PERMITS. *NPDES Compliance Inspection Manual*, Washington, DC: U.S. Environmental Protection Agency, 1988.

U.S. ENVIRONMENTAL PROTECTION AGENCY, SOLID WASTE AND EMERGENCY RESPONSE. *RCRA Inspection Manual 1993 Edition*, OSWER Directive 9938.02B, Washington, DC: U.S. Environmental Protection Agency, 1993.

U.S. ENVIRONMENTAL PROTECTION AGENCY, SOLID WASTE AND EMERGENCY RESPONSE. *Musts for USTs: A Summary of Federal Regulations for Underground Storage Systems*, EPA 510-K-95-002, Washington, DC: U.S. Environmental Protection Agency, 1995.

U.S. ENVIRONMENTAL PROTECTION AGENCY, SOLID WASTE AND EMERGENCY RESPONSE. *Detecting Leaks: Successful Methods Step-by-Step*, EPA/

530/UST-89/012, Washington, DC: U.S. Environmental Protection Agency, 1989.

WHITE, ALLEN L.; SAVAGE, DEBORAH; BECKER, MONICA. *Revised Executive Summary. Total Cost Assessment: Accelerating Industrial Pollution Prevention Through Innovative Project Financial Analysis with Applications to the Pulp and Paper Industry*, Washington, DC: U.S. Environmental Protection Agency, 1993.

THE INTERNET

GLISTER, PAUL. *The New Internet Navigator*, New York, NY: John Wiley and Sons, 1995.

GRALLA, PRESTON. *How the Internet Works*, Emeryville, CA: Ziff-Davis Press, an imprint of Macmillan Computer Publishing USA, 1996.

GLOSSARY OF ACRONYMS AND ABBREVIATIONS

ACGIH—American Conference of Governmental Industrial Hygienists; a professional organization in the United States that publishes recommended standards of exposure to hazardous materials in the workplace

AIHA—American Industrial Hygiene Association; a professional organization devoted to worker health and safety protection; AIHA has developed a guidance standard for a model OHSMS

ANSI—American National Standards Institute; a coordinating body composed of trade, technical, professional, and consumer groups that develops voluntary standards in the United States, the United State's representative to the ISO

AP-42—document that contains air pollution emission factors

ASTM—American Society for Testing and Materials; an organization composed of individuals, agencies, and industries concerned with materials that develops voluntary consensus standards for materials, products, systems, and services; ASTM is a resource for sampling and testing methods, health and safety, and safe performance guidelines for hazardous materials

BACT—best available control technology; in the U.S., an emission limitation based on the maximum degree of emission reduction, which (considering energy, environmental, and economic impacts and other costs) is achievable through the application of production processes and available methods, systems, and techniques; BACT does not permit emissions in excess of those allowed under any applicable CAA provisions; use of the BACT concept is allowable on a case-by-case basis for major new or modified emission sources in attainment areas and applies to each regulated pollutant

BAT—best available technology; under the CWA, the most stringent type of control for existing discharges; applies to toxic pollutants as well as conventional and some nonconventional pollutants; as used in the ISO 14001 standard BAT, refers to existing technology that can produce the lowest emissions

BCT—best conventional technology; under the U.S. CWA, available control technology for discharges of conventional pollutants; more stringent than BPT

BMP—best management practice

460

BOD—biochemical oxygen demand; a measure of the amount of oxygen consumed in the biological processes that break down organic matter in water or wastewater; the greater the BOD, the greater the degree of pollution

BPT—best practicable technology; under the CWA, minimum acceptable level of treatment technology for existing plants

BSI—British Standards Institute; the British standards setting organization

BTU—British thermal unit

CAA—Clean Air Act; U.S. law devoted to controlling air pollution

CAS—Chemical Abstract Service; CAS number is an identification number assigned by the CAS of the American Chemical Society; the CAS is used in various databases for identification purposes

CEO—chief executive officer; the highest-ranking individual working in a corporation

CERCLA—Comprehensive Environmental Response, Compensation and Liability Act; also known as the Superfund Act; U.S. law that establishes a fund for cleaning up the most contaminated sites in the U.S.

CERCLIS—Comprehensive, Environmental Response, Compensation and Liability System; the EPA list of contaminated sites that are subsequently ranked for cleanup according to severity under CERCLA (Superfund)

CESQG—Conditionally Exempt Small-Quantity Generator; under U.S. RCRA, the smallest of three generator categories; based on the amount of waste generated or accumulated during any one calendar month; CESQGs are exempt from certain RCRA requirements

CFM—cubic feet per minute

CFR—Code of Federal Regulations; collection of rules and regulations originally published in the *Federal Register* by various governmental departments and agencies in the U.S.; the CFR is updated once each year

COD—chemical oxygen demand; in water or wastewater, a measure of the amount of oxygen consumed by the breakdown of organic matter that is susceptible to oxidation by a strong chemical oxident

CSI—Common Sense Initiative; component of the U.S. EPA's Pollution Prevention Strategy that involves the regulated community, government officials, environmentalists, and the public to develop programs for preventing pollution in selected industries

CWA—Clean Water Act; U.S. law devoted to controlling water pollution

CZMA—Coastal Zone Management Act; U.S. law devoted to protecting coastal resources

DOT—Department of Transportation; U.S. government agency that regulates transportation sources; develops and enforces regulations governing the transport of hazardous materials

DFE—design for environment; a technique that considers potential environmental impact during the early stages of product development when the concept, need, and design are considered

DIS—draft international standard; name given to ISO standards before they are finalized

EC—European Commission; executive branch of the EU

EHS—extremely hazardous substance; any of over 400 chemicals identified by the EPA on the basis of toxicity and listed under SARA Title III; the list is subject to revision

EIS—Environmental Impact Statement; a document that describes the positive and negative effects of a proposed project and lists alternative actions; required of U.S. federal agencies by the NEPA for major projects or legislative proposals significantly affecting the environment

ELP—Environmental Leadership Program; component of EPA's Pollution Prevention Strategy that encourages the regulated community to develop and implement pollution prevention management practices and to establish environmental goals beyond those needed for compliance with regulations

EMAR—Eco-Management and Audit Regulation; environmental management regulation developed by the EC for companies within the EU

EMAS—Eco-Management and Audit Scheme; contained within the EMAR; establishes specifications for environmental management systems for companies with sites in the EU

EMS—environmental management system; in the ISO context, the part of the overall management system for developing, implementing, achieving, reviewing, and maintaining the environmental policy; encompasses organizational structure, planning activities, responsibilities, practices, procedures, processes, and resources

EPA—U.S. Environmental Protection Agency; established in 1970 to bring together parts of various government agencies involved with the control of pollution; regulates air, land, and water pollution, and toxic materials

EPCRA—Emergency Planning and Community Right-to-Know Act; U.S. law that establishes emergency planning, reporting, and notification requirements for industry to protect the public in the event of a release of hazardous substances

ERNS—Emergency Response Notification System; EPA's list of reported CERCLA hazardous substance releases or spills in quantities greater than the RQ; ERNS is maintained by the NRC

ERT—emergency response team; group of highly trained individuals who have the knowledge and skill to respond to emergency releases and spills of hazardous materials

ES—environmental surveillance

EU—European Union; a confederation of European member nations that have agreed to integrate their economies and eventually form a political union

FADE—Focus, Analyze, Develop, Execute; a problem-solving technique in which the information developed during brainstorming is analyzed

FESOP—federally enforceable state operating permit; under the U.S. CAA, a permit program that major sources can elect to comply with in lieu of a Part 70 operating permit

FFCA—Federal Facility Compliance Act; U.S. law ensuring that federal facilities comply with all federal, state, and local solid and hazardous waste laws

FIFRA—Federal Insecticide, Fungicide, and Rodenticide Act; U.S. law that regulates pesticides

FMECA—failure modes, effects, and criticality analysis; method of ranking failure modes according to criticality to determine which are the most likely to cause a serious accident

FR—*Federal Register;* a daily publication of the U.S. government that is the legal medium for recording and communicating the rules, regulations, and other business of the U.S.

FWPCA—Federal Water Pollution Control Act; U.S. law devoted to protecting water resources; also known as the CWA

GATT—General Agreement on Tariffs and Trade; international agreement for reducing barriers to trade

H—hydraulic head

HAP—hazardous air pollutant; air pollutants that are not covered by ambient air quality standards but which, as defined in the U.S. CAA, may reasonably be expected to cause or contribute to irreversible illness or death; includes pollutants such as asbestos, mercury, benzene, vinyl chloride, and so forth

HAZWOPER—Hazardous Waste Operations and Emergency Response regulation; under the U.S. OSH Act, a regulation that prescribes measures to protect worker health and safety at hazardous waste sites and for responses to emergencies

HEPA—high-efficiency particulate air; filter system that is capable of capturing 99.99% of 3-micron monodisperse particles; used in air purifying respirators and air filtration systems

HMTA—Hazardous Materials Transportation Act; U.S. law devoted to ensuring the safe transport of hazardous materials

HSWA—Hazardous and Solid Waste Amendments; U.S. law that amended RCRA to include a national ban on the land disposal of hazardous wastes, to promote waste minimization programs, to regulate small-quantity generators, and to regulate underground storage tanks

IEC—International Electrochemical Commission; develops voluntary consensus standards for the electrical and electronic engineering industries

IMS—Incident Management System; a highly structured approach for managing the response to an unplanned release

ISO—International Organization for Standardization; founded in 1946 to develop voluntary consensus standards for manufacturing, trade, and communication industries; developed ISO 14000 and 9000 series of standards

IWP permit—Industrial Waste Pretreatment permit; a permit issued by state environmental agencies to facilities that discharge to POTWs that do not operate approved pretreatment programs; part of the NPP under the U.S. CWA

JIT—just-in-time; a philosophy based on the principle of eliminating waste in production of goods; typically refers to a delivery system in which equipment and materials are ordered and received when needed to meet a production schedule

KG—kilogram

L—liter

LAER—lowest achievable emission rate; under the U.S. CAA, this is the rate of emissions that reflects a) the most stringent emission limitation contained in the implementation plan of any state for such source unless the owner or operator of the proposed source demonstrates that such limitations are not achievable; or

b) the most stringent emissions limitation achieved in practice, whichever is more stringent; LAER does not permit a proposed new or modified source to emit pollutants in excess of existing new source standards

LCA—life-cycle assessment; a systematic set of procedures for compiling and examining the inputs and outputs of materials and energy, and the associated environmental impact directly attributable to the functioning of a product or service system throughout its life cycle, from acquisition of raw materials through final disposal

LCL—lower control limit; designated limit on a control chart (typically three standard deviations below the mean), below which the system is out of control

LEPC—local emergency planning committee; a committee appointed by the state emergency response commission, as required by the U.S. SARA Title III, to formulate a comprehensive emergency plan for its jurisdiction

LFD—local fire department

LQG—Large-Quantity Generator; under RCRA, the largest category of waste generator subject to the most stringent requirements; based on the quantity of waste generated or stored in any one calendar month

LRWPAA—Low-Level Radioactive Waste Policy Act Amendments; U.S. law that amended the LRWPC

LRWPC—Low-Level Radioactive Waste Policy Act; U.S. law devoted to the safe disposal of commercial low-level radioactive waste

LUST—leaking underground storage tank

LWL—lower warning limit; designated limit on a control chart (typically two standard deviations below the mean); data between the UWL and the LWL are considered to be in control and acceptable

MACT—maximum achievable control technology; under the U.S. CAA Amendments of 1990, emission standards for all major sources of HAPs; standards must be phased in by the year 2000

MBO—management by objectives; management system based on evaluation of progress toward clearly identified objectives

MG—milligram

ML—milliliter

MPRSA—Marine Protection, Research and Sanctuaries Act; U.S. law devoted to the protection of ocean waters from pollution

MSDS—material safety data sheet; a compilation of information required under the U.S. Hazard Communication Standard on the identity of hazardous chemicals, health and physical hazards, exposure limits and precautions; Section 311 of SARA requires facilities to submit MSDSs under certain circumstances

NAAQS—National Ambient Air Quality Standards; air quality standards established by the EPA under the CAA to regulate outdoor air pollution

NEPA—National Environmental Policy Act; U.S. law that requires federal agencies to give environmental factors the same consideration as other factors in decision making about its facilities

NESHAPS—National Emission Standards for Hazardous Air Pollutants; emissions standards set by EPA for an air pollutant not covered by the NAAQS that may cause an increase in deaths or serious, irreversible, or incapacitating illness

NOV—notice of violation

NPDES—National Pollutant Discharge Elimination System; a provision of the CWA that prohibits discharge of pollutants into waters of the U.S. unless a special permit is issued by the EPA, a state, or—where delegated—a tribal government on an Indian reservation

NPL—National Priorities List; in the U.S., EPA's list of the most serious uncontrolled or abandoned contaminated sites identified for possible long-term remedial action under CERCLA (Superfund)

NPP—National Pretreatment Program; under the CWA, a program to protect POTWs by limiting industrial discharges to the sanitary sewer system

NRC—National Response Center; the federal operations center that receives notifications of all releases of oil and hazardous substances into the environment; the Center, open 24 hours a day, is operated by the U.S. Coast Guard, which evaluates all reports and notifies the appropriate agency

NSPS—New Source Performance Standards; uniform national U.S. EPA air emission and water effluent standards that limit the amount of pollution allowed from new sources or from existing sources that have been modified

NSR—new source review; under the CAA, a program to review major and minor new sources, reconstructions, and modifications of sources

NWPA—Nuclear Waste Policy Act; U.S. law devoted to the safe disposal of high-level radioactive wastes from commercial facilities

OHSMS—Occupational Health and Safety Management System; a management system for developing, implementing, achieving, reviewing, and maintaining worker health and safety; encompasses organizational structure, planning activities, responsibilities, practices, procedures, processes, and resources

OPA—Oil Pollution Act; U.S. law devoted to preventing oil spills; establishes strict liability for damage caused by oil spills in navigable waters

OSHA—Occupational Health and Safety Administration; U.S. government agency that is responsible for enforcing the regulations related to safety and health in the workplace

OSH Act—Occupational Safety and Health Act; U.S. law devoted to the protection of worker health and safety

PCB—polychlorinated biphenyl; a group of toxic, persistent chemicals containing carbon, hydrogen, and chlorine; used in transformers and capacitors for insulating purposes and in gas pipeline systems; sale of new use was banned by law in the United States in 1979

PDCA—Plan, Do, Check, Act; a systematic approach to solving problems

PM—preventive maintenance; maintenance performed at regularly scheduled intervals

POTW—publicly owned treatment works; a waste treatment works owned by a state, unit of local government, or Indian tribe; usually designed to treat domestic wastewater

PPA—Pollution Prevention Act; U.S. law that established the prevention of pollution as a national objective

P/PC—production/production capacity

PPE—personal protective equipment; respirators, clothing, and other items to isolate the worker from hazardous materials

PPM—parts per million

PRP—potentially responsible party; any individual or company—including owners, operators, transporters, or generators—potentially responsible for or contributing to the contamination problems at a Superfund site

PSD—Prevention of Significant Deterioration; U.S. EPA program in which state and/or federal permits are required and are intended to restrict emissions for new or modified sources in places where air quality is already better than required to meet primary and secondary ambient air quality standards

PSES—pretreatment standards for existing sources; under the U.S. CWA, effluent standards that must be met by existing industrial users discharging to a POTW

PSI—pounds per square inch

PSNS—pretreatment standards for new sources; under the U.S. CWA, effluent standards that must be met by new industrial users discharging to a POTW

PTE—potential to emit

QA/QC—quality assurance/quality control; a system of procedures, checks, audits, and corrective actions to ensure that environmental sampling, monitoring, laboratory analysis, reporting, and technical activities produce the highest quality of data

RCRA—Resource Conservation and Recovery Act; U.S. law devoted to managing all aspects of solid and hazardous wastes

RFP—request for proposal

RQ—reportable quantity; the quantity of a hazardous substance that triggers reporting requirements under CERCLA in the U.S.; substances released in excess of the RQ must be reported to the National Response Center, the State Emergency Response Commission, and community emergency coordinators for areas likely to be affected

SAC—Supplier Audit Confirmation; a modified recertification scheme under ISO 9000 that places greater reliance on internal audits; proposed for certified/registered companies that have demonstrated the effectiveness of their quality systems

SAGE—Strategic Action Group on the Environment; formed in 1991 by the ISO to make recommendations about international environmental standards

SARA—Superfund Amendments and Reauthorization Act; U.S. law that establishes emergency planning, reporting, and notification requirements to protect the public in the event of a release of hazardous substances

SDWA—Safe Drinking Water Act; U.S. law that regulates public drinking water supplies

SEC—Securities and Exchange Commission; U.S. government agencies that oversees the trading of securities

SEP—Supplemental Environmental Project; projects, including environmental audits, that can mitigate the civil penalties assessed by the EPA in the United States

SERC—State Emergency Response Commission; a commission appointed by each state governor according to the requirements of SARA Title III; the SERCs designate emergency planning districts, appoint local emergency planning committees, and supervise and coordinate their activities

SIC—Standard Industrial Classification Code; numerical code for industries

SIP—State Implementation Plan; EPA-approved state plans for the establishment, regulation, and enforcement of air pollution standards

SMCRA—Surface Mining Control and Reclamation Act; U.S. law that establishes a national program for surface mining operations; includes provisions to reclaim abandoned mines and controls the environmental impact of surface mining

SPCC—Spill Prevention Control and Countermeasure; plans required under the CWA and OPA to prevent oil and hazardous materials from discharging from their containers into the environment

SQG—Small-Quantity Generator; under the U.S. RCRA, the middle of three categories of waste generators; based on the amount of waste generated or accumulated during any one calendar month

SSOA—state source-specific operating agreement; under the U.S. CAA, a permit program available to certain sources (surface coating and printing operations, woodworking operations, abrasive cleaning operations, grain elevators) in lieu of a Part 70 permit or an FESOP

SWOT—Strengths, Weaknesses, Opportunities; Threats; a form of problem solving that involves identifying and creating lists of strengths, weaknesses, opportunities, and threats

TC—Technical Committee; the ISO work groups charged with developing standards

TCLP—Toxicity Characteristic Leaching Procedure; under the U.S. RCRA, the procedure used to identify wastes that have the characteristic of toxicity

TOC—total organic carbon; a measure of the total organic content of water or wastewater

TPQ—threshold planning quantity; a quantity designated for each chemical on the list of extremely hazardous substances that triggers notification by facilities to the SERC that such facilities are subject to emergency planning under the U.S. SARA Title III

TRI—Toxic Release Inventory; in the U.S., an inventory of toxic pollutants released to the air, land, and water; based on reporting requirements in EPCRA

TSCA—Toxic Substances Control Act; U.S. law that regulates the manufacture, distribution, processing, use, and disposal of certain designated chemical substances; includes programs for asbestos in schools and public buildings, and for radon indoors

TSD and TSDF—treatment, storage, and disposal facility; site where a hazardous substance is treated, stored, or disposed; TSD facilities are regulated by the U.S. EPA and states under RCRA

TSS—total suspended solids; a portion of the total solids in a sample of water or wastewater that is retained by a filter

UCL—upper control limit; designated limit on a control chart (typically three standard deviations above the mean), above which the system is out of control

UMTRCA—Uranium Mill Tailings Radiation Control Act; U.S. law that regulates the remedial actions for uranium mill tailings

µG—microgram

UN—United Nations

U.S.—United States of America

USGS—U.S. Geological Survey

UST—underground storage tank; a tank located all or partially under the ground and designed to hold petroleum products or other hazardous products

UWL—upper warning limit; designated limit on a control chart (typically two standard deviations above the mean); data between the UWL and the LWL are considered to be in control and acceptable

VOC—volatile organic compound; any organic compound that participates in atmospheric photochemical reactions except for those designated by the U.S. EPA Administrator as having negligible photochemical reactivity

VPP—Voluntary Protection Program; a certification program offered by the U.S. OSHA to companies that can demonstrate that they have an effective occupational safety and health program; certification exempts the company from pre-programmed OSHA inspections

INDEX

 469

capsule description of, 11–16
environmental auditing standards, 12–13
environmental labeling, 13–14
environmental management system standards, 12
life cycle assessment, 14–15
product standard development, 15–16
defined, 4
development of, 4–5
implementation of:
cons, 10–11
pros, 7–10
ISO 14001:
capsule description, 12
certification, 6–7
correspondence between ISO 901 and, 21–22
documentation needed to conform to, 53
EMAS and BS 7750 compared to, 8
EMS audits required by, 175
and implementation, 72
relationship to other standards, 7
ISO 14004, capsule description, 12
ISO 14010, capsule description, 12–13
ISO 14011:
capsule description, 12–13
protocols in conduct of audits, 174
ISO 14012, capsule description, 12–13
ISO 14024, capsule description, 13–14
ISO 14040, capsule description, 14–15
ISO 14060, capsule description, 15–16
ISO standards:
advantages of, 487–88
definitions used in, 17

J

Jobs, Competitiveness, and Environmental Regulation, 3
Just-in-time delivery, 91–92

K

"K" list, 277

L

Large-quantity generators (LQGs), 281
responsibilities of, 282–85
Lead auditor, 175–76, 180, 181, 184
Leaking underground storage tanks (LUSTs), 236, 238–39
Life cycle assessment (LCA), 14–15, 76
phases of, 15
Life Cycle Assessment Worksheet, 78
Listed wastes, 277–79
Local Emergency Planning Committee (LEPC), 224
Low-Level Radioactive Waste Policy Act, 221

M

Macroscale flowcharts, 40, 41
Maintenance activities, 91–92
Maintenance controls, evaluation of, checklist for, 119–26

Maintenance cost control system, 92
Maintenance of documents/records, procedure for, 385–86
Maintenance service, AIHA OHSMS guidance document, 198
Management review, 156–57
Manual tank gauging, 243
Maps, 178
Marine Protection, Research and Sanctuaries Act, 221
Material Safety Data Sheet:
form, 104–5
form letter for obtaining, 106
Methods for Estimating the Chronic Toxicity of Effluents and Receiving Waters to Freshwater Organisms, 269
Methods for Measuring the Acute Toxicity of Effluents to Freshwater and Marine Organisms, 267–69
Microscale flowcharts, 40, 42
Mission statement, 30
example, 33
writing, 32–33
Motivation, 81

N

National Environmental Policy Act, 221
National Pollution Discharge Elimination System (NPDES), 212, 256
audit process, 259–70
National Priorities List (NPL), 207
National Reporter System, The, 47
National Response Center, 107, 246
National Sanitation Foundation (NSF) International, 26
Needs analysis, 89
New Source Review (NSR) program, 327–29
agency review, 329
major NSR, 327–28
minor NSR, 328–29
Noncontact cooling water, 257
Notices of violation (NOVs), 260
Nuclear Waste Policy Act, 221

O

Occupational health and safety management systems, 187–200
AIHA OHSMS guidance document, 189–99
communication systems, 193
compliance and conformance review, 192
control of nonconforming process or device, 195–96
control of OHS inspection/measuring/test equipment, 195
control of OHS records, 197
corrective and preventive action, 196
design control, 192
document and data control, 192–93
handling/storage/packaging of hazardous materials, 196–97
hazard identification and traceability, 193–94
inspection and evaluation, 194–95
inspection and evaluation status, 195

LICENSE AGREEMENT AND LIMITED WARRANTY

READ THE FOLLOWING TERMS AND CONDITIONS CAREFULLY BEFORE OPENING THIS DISK PACKAGE. THIS LEGAL DOCUMENT IS AN AGREEMENT BETWEEN YOU AND PRENTICE-HALL, INC. (THE "COMPANY"). BY OPENING THIS SEALED DISK PACKAGE, YOU ARE AGREEING TO BE BOUND BY THESE TERMS AND CONDITIONS IF YOU DO NOT AGREE WITH THESE TERMS AND CONDITIONS DO NOT OPEN THE DISK PACKAGE. PROMPTLY RETURN THE UNOPENED DISK PACK-AGE AND ALL ACCOMPANYING ITEMS TO THE PLACE YOU OBTAINED THEM FOR A FULL REFUND OF ANY SUMS YOU HAVE PAID.

1. **GRANT OF LICENSE:** In consideration of your purchase of this book, and your agreement to abide by the terms and conditions of this Agreement, the Company grants to you a nonexclusive right to use and display the copy of the enclosed software program (hereinafter the "SOFTWARE") on a single computer (i.e., with a single CPU) at a single location so long as you comply with the terms of this Agreement. The Company reserves all rights not expressly granted to you under this Agreement.

2. **OWNERSHIP OF SOFTWARE:** You own only the magnetic or physical media (the enclosed disk) on which the SOFTWARE is recorded or fixed, but the Company and the software are developers retain all the rights, title, and ownership to the SOFTWARE recorded on the original disk copy(ies) and all subsequent copies of the SOFTWARE, regardless of the form or media on which the original or other copies may exist. This license is not a sale of the original SOFTWARE or any copy to you.

3. **COPY RESTRICTIONS:** This SOFTWARE and the accompanying printed materials and user manual (the "Documentation") are the subject of copyright. You may not copy the Documentation or the SOFTWARE, except that you may make a single copy of the SOFTWARE for backup or archival purposes only. You may be held legally responsible for any copying or copyright infringement which is caused or encouraged by your failure to abide by the terms of this restriction.

4. **USE RESTRICTIONS:** You may not network the SOFTWARE or otherwise use it on more than one computer or computer terminal at the same time. You may physically transfer the SOFTWARE from one computer to another provided that the SOFTWARE is used on only one computer at a time. You may not distribute copies of the SOFTWARE or Documentation to others. You may not reverse engineer, disassemble, decompile, modify, adapt, translate, or create derivative works based on the SOFTWARE or the Documentation without the prior written consent of the Company.

5. **TRANSFER RESTRICTIONS:** The enclosed SOFTWARE is licensed only to you and may not be transferred to any one else without the prior written consent of the Company. Any unauthorized transfer of the SOFTWARE shall result in the immediate termination of this Agreement.

6. **TERMINATION:** This license is effective until terminated. This license will terminate automatically without notice from the Company and become null and void if you fail to comply with any provisions or limitations of this license. Upon termination, you shall destroy the Documentation and all copies of the SOFTWARE. All provisions of this Agreement as to warranties, limitation of liability, remedies or damages, and our ownership rights shall survive termination.

7. **MISCELLANEOUS:** This Agreement shall be construed in accordance with the laws of the United States of America and the State of New York and shall benefit the Company, its affiliates, and assignees.

8. **LIMITED WARRANTY AND DISCLAIMER OF WARRANTY :** The Company warrants that the SOFTWARE, when properly used in accordance with the Documentation, will operate in substantial conformity with the description of the SOFTWARE set forth in the Documentation. The Company does not warrant that the SOFTWARE will meet your requirements or that the operation of the SOFTWARE will be uninterrupted or error-free. The Company warrants that the media on which the SOFTWARE is delivered shall be free from defects in materials and workmanship under normal use for a period of thirty (30) days from the date of your purchase. Your only remedy and the

Company's only obligation under these limited warranties is, at the Company's option, return of the warranted item for a refund of any amounts paid by you or replacement of the item. Any replacement of SOFTWARE or media under the warranties shall not extend the original warranty period. The limited warranty set forth above shall not apply to any SOFTWARE which the Company determines in good faith has been subject to misuse, neglect, improper installation, repair, alteration, or damage by you. EXCEPT FOR THE EXPRESSED WARRANTIES SET FOR THABOVE, THE COMPANY DISCLAIMS ALL WARRANTIES, EXPRESS OR IMPLIED, INCLUDING WITHOUT LIMITATION, THE IMPLIED WARRANTIES OF MERCHANTABILITY AND FITNESS FOR A PARTICULAR PURPOSE. EXCEPT FOR THE EXPRESS WARRANTY SET FOR THABOVE, THE COMPANY DOES NOT WARRANT, GUARANTEE, OR MAKE ANY REPRESENTATION REGARDING THE USE OR THE RESULTS OF THE USE OF THE SOFTWARE IN TERMS OF ITS CORRECTNESS, ACCURACY, RELIABILITY, CURRENTNESS, OR OTHERWISE.

IN NO EVENT, SHALL THE COMPANY OR ITS EMPLOYEES, AGENTS, SUPPLIERS, OR CONTRACTORS BE LIABLE FOR ANY INCIDENTAL, INDIRECT, SPECIAL, OR CONSEQUENTIAL DAMAGES ARISING OUT OF OR IN CONNECTION WITH THE LICENSE GRANTED UNDER THIS AGREEMENT, OR FOR LOSS OF USE, LOSS OF DATA, LOSS OF INCOME OR PROFIT, OR OTHER LOSSES, SUSTAINED AS A RESULT OF INJURY TO ANY PERSON, OR LOSS OF OR DAMAGE TO PROPERTY, OR CLAIMS OF THIRD PARTIES, EVEN IF THE COMPANY OR AN AUTHORIZED REPRESENTATIVE OF THE COMPANY HAS BEEN ADVISED OF THE POSSIBILITY OF SUCH DAMAGES. IN NO EVENT SHALL LIABILITY OF THE COMPANY FOR DAMAGES WITH RESPECT TO THE SOFTWARE EXCEED THE AMOUNTS ACTUALLY PAID BY YOU, IF ANY, FOR THE SOFTWARE.

SOME JURISDICTIONS DO NOT ALLOW THE LIMITATION OF IMPLIED WARRANTIES OR LIABILITY FOR INCIDENTAL, INDIRECT, SPECIAL, OR CONSEQUENTIAL DAMAGES, SO THE ABOVE LIMITATIONS MAY NOT ALWAYS APPLY. THE WARRANTIES IN THIS AGREEMENT GIVE YOU SPECIFIC LEGAL RIGHTS AND YOU MAY ALSO HAVE OTHER RIGHTS WHICH VARY IN ACCORDANCE WITH LOCAL LAW.

ACKNOWLEDGMENT

YOU ACKNOWLEDGE THAT YOU HAVE READ THIS AGREEMENT, UNDERSTAND IT, AND AGREE TO BE BOUND BY ITS TERMS AND CONDITIONS YOU ALSO AGREE THAT THIS AGREEMENT IS THE COMPLETE AND EXCLUSIVE STATEMENT OF THE AGREEMENT BETWEEN YOU AND THE COMPANY AND SUPERSEDES ALL PROPOSALS OR PRIOR AGREEMENTS, ORAL, OR WRITTEN, AND ANY OTHER COMMUNICATIONS BETWEEN YOU AND THE COMPANY OR ANY REPRESENTATIVE OF THE COMPANY RELATING TO THE SUBJECT MATTER OF THIS AGREEMENT.

Should you have any questions concerning this Agreement or if you wish to contact the Company for any reason, please contact in writing at the address below.

Robin Short
Prentice Hall PTR
One Lake Street
Upper Saddle River, New Jersey 07458